THE MECHANICS OF SCOUR IN THE MARINE ENVIRONMENT

ADVANCED SERIES ON OCEAN ENGINEERING

Series Editor-in-Chief
Philip L-F Liu (*Cornell University*)

Vol. 1 The Applied Dynamics of Ocean Surface Waves
by *Chiang C Mei* (MIT)

Vol. 2 Water Wave Mechanics for Engineers and Scientists
by *Robert G Dean* (Univ. Florida) and *Robert A Dalrymple* (Univ. Delaware)

Vol. 3 Mechanics of Coastal Sediment Transport
by *Jørgen Fredsøe and Rolf Deigaard* (Tech. Univ. Denmark)

Vol. 4 Coastal Bottom Boundary Layers and Sediment Transport
by *Peter Nielsen* (Univ. Queensland)

Vol. 5 Numerical Modeling of Ocean Dynamics
by *Zygmunt Kowalik* (Univ. Alaska) and *T S Murty* (Inst. Ocean Science, BC)

Vol. 6 Kalman Filter Method in the Analysis of Vibrations Due to Water Waves
by *Piotr Wilde and Andrzej Kozakiewicz* (Inst. Hydroengineering, Polish Academy of Sciences)

Vol. 7 Physical Models and Laboratory Techniques in Coastal Engineering
by *Steven A. Hughes* (Coastal Engineering Research Center, USA)

Vol. 8 Ocean Disposal of Wastewater
by *Ian R Wood* (Univ. Canterbury), *Robert G Bell* (National Institute of Water & Atmospheric Research, New Zealand) and *David L Wilkinson* (Univ. New South Wales)

Vol. 9 Offshore Structure Modeling
by *Subrata K. Chakrabarti* (Chicago Bridge & Iron Technical Services Co., USA)

Vol. 10 Water Waves Generated by Underwater Explosion
by *Bernard Le Méhauté and Shen Wang* (Univ. Miami)

Vol. 11 Ocean Surface Waves; Their Physics and Prediction
by *Stanislaw R Massel* (Australian Inst. of Marine Sci)

Vol. 12 Hydrodynamics Around Cylindrical Structures
by *B Mutlu Sumer and Jørgen Fredsøe* (Tech. Univ. of Denmark)

Vol. 13 Water Wave Propagation Over Uneven Bottoms
Part I — Linear Wave Propagation
by *Maarten W Dingemans* (Delft Hydraulics)

Part II — Non-linear Wave Propagation
by *Maarten W Dingemans* (Delft Hydraulics)

Vol. 14 Coastal Stabilization
by *Richard Silvester and John R C Hsu* (The Univ. of Western Australia)

Vol. 15 Random Seas and Design of Maritime Structures (2nd Edition)
by *Yoshimi Goda* (Yokohama National University)

Vol. 16 Introduction to Coastal Engineering and Management
by *J William Kamphuis* (Queen's Univ.)

Vol. 17 The Mechanics of Scour in the Marine Environment
by *B Mutlu Sumer and Jørgen Fredsøe* (Tech. Univ. of Denmark)

Advanced Series on Ocean Engineering — Volume 17

THE MECHANICS OF SCOUR IN THE MARINE ENVIRONMENT

B. Mutlu Sumer
Jørgen Fredsøe

Technical University of Denmark
Denmark

World Scientific
New Jersey • Singapore • London • Hong Kong

Published by

World Scientific Publishing Co. Pte. Ltd.

5 Toh Tuck Link, Singapore 596224

USA office: 27 Warren Street, Suite 401-402, Hackensack, NJ 07601

UK office: 57 Shelton Street, Covent Garden, London WC2H 9HE

British Library Cataloguing-in-Publication Data
A catalogue record for this book is available from the British Library.

First published 2002
Reprinted 2005, 2012

THE MECHANICS OF SCOUR IN THE MARINE ENVIRONMENT

Copyright © 2002 by World Scientific Publishing Co. Pte. Ltd.

All rights reserved. This book, or parts thereof, may not be reproduced in any form or by any means, electronic or mechanical, including photocopying, recording or any information storage and retrieval system now known or to be invented, without written permission from the Publisher.

For photocopying of material in this volume, please pay a copying fee through the Copyright Clearance Center, Inc., 222 Rosewood Drive, Danvers, MA 01923, USA. In this case permission to photocopy is not required from the publisher.

ISBN 978-981-02-4930-4

Printed in Singapore.

Contents

1 **Introduction and basic concepts**   5
   1.1 Introduction 5
   1.2 Amplification factor 7
   1.3 Equilibrium scour depth and time scale of scour 8
   1.4 Clear-water scour and live-bed scour 9
   1.5 Local and global scour 11
   1.6 References 12

2 **Scour below pipelines**   15
   2.1 Onset of scour 16
       2.1.1 Mechanism of onset of scour. Seepage flow and piping underneath the pipe 16
       2.1.2 Criterion for the onset of scour 23
   2.2 Tunnel erosion 30
   2.3 Two-dimensional scour 32
       2.3.1 Lee-wake erosion 33
       2.3.2 Scour depth 38
       2.3.3 Width of scour hole 66
       2.3.4 Time scale 69
       2.3.5 Change in wave climate 75
   2.4 Three-dimensional scour 76
       2.4.1 Scour, backfilling, and self-burial in the free-span areas 77
       2.4.2 Scour, backfilling and self-burial at span shoulders . . . 96
       2.4.3 Stimulated self-burial of pipelines 108
   2.5 Scale effects 111
   2.6 Scour protection for pipelines 113
   2.7 Mathematical modelling 117
       2.7.1 Potential-flow models 117

         2.7.2   Advanced models 125
         2.7.3   Integrated models 137
  2.8 References 138

3 **Scour around a single slender pile**   **149**
  3.1 Flow around a slender pile 150
         3.1.1   Horseshoe vortex in steady currents 151
         3.1.2   Horseshoe vortex in waves 160
         3.1.3   Lee-wake vortex flow 169
         3.1.4   Contraction of streamlines 174
  3.2 Scour around a slender pile 174
         3.2.1   Scour around a slender pile in steady currents 174
         3.2.2   Scour around a slender pile in waves 188
         3.2.3   Scour around a cone-shaped object 205
  3.3 Time scale 206
  3.4 Scour protection 212
  3.5 Mathematical modelling 218
  3.6 References 228

4 **Scour around a group of slender piles**   **239**
  4.1 Pile group in steady currents 239
         4.1.1   Two-pile group 240
         4.1.2   Three-pile group 247
  4.2 Pile group of in waves 248
         4.2.1   Two-pile group 249
         4.2.2   Three-pile group 257
         4.2.3   Four-pile group 258
         4.2.4   Effect of the KC number 260
  4.3 Global and local scour at pile groups 264
  4.4 References 272

5 **Examples of more complex configurations**   **275**
  5.1 Scour at pile-supported offshore structures 276
  5.2 Scour characteristics 279
  5.3 References 285

CONTENTS

6 Scour around large piles — 287
- 6.1 Large pile. Diffraction regime 288
- 6.2 Phase-resolved flow around the pile 291
- 6.3 Steady streaming around the pile 294
- 6.4 Scour around the pile 301
  - 6.4.1 Mechanism of streaming-induced scour 302
  - 6.4.2 Influence of KC and D/L 305
  - 6.4.3 Influence of combined waves and current, and cross-sectional shape 316
  - 6.4.4 Scour around a cone-shaped object 324
- 6.5 References 325

7 Scour around breakwaters — 329
- 7.1 Scour at the trunk section of a breakwater 333
  - 7.1.1 Scour at the trunk section of a vertical-wall breakwater 333
  - 7.1.2 Scour at the trunk section of a rubble-mound breakwater 347
  - 7.1.3 Scour protection at the trunk section of a breakwater . 361
  - 7.1.4 Mathematical modelling of scour at the trunk section of a breakwater 365
- 7.2 Scour around the head of a breakwater 371
  - 7.2.1 Scour around the head of a vertical-wall breakwater . . 371
  - 7.2.2 Scour around the head of a rubble-mound breakwater . 377
  - 7.2.3 Scour protection at the head section 385
  - 7.2.4 Influence of finite length of breakwater 387
- 7.3 Scour at jetties 390
- 7.4 References 391

8 Scour at seawalls — 399
- 8.1 Scour by normally incident breaking waves 402
- 8.2 Scour by normally incident nonbreaking waves 415
- 8.3 Scour induced by wave overtopping 417
- 8.4 Scour protection 419
- 8.5 References 419

9 Ship-propeller scour — 423
- 9.1 Scour due to unconfined propeller wash 426
- 9.2 Scour due to confined propeller wash 431
- 9.3 Scour protection 436

9.4	References	442

10 Impact of liquefaction — 445
- 10.1 Physics of liquefaction 446
  - 10.1.1 Liquefaction induced by the buildup of pore pressure. Residual liquefaction 446
  - 10.1.2 Liquefaction induced by the upward-directed pressure gradient. Momentary liquefaction 448
- 10.2 Biot equations and their solutions 450
  - 10.2.1 Biot equations 450
  - 10.2.2 Stresses in soil under a progressive wave 455
- 10.3 Residual liquefaction 464
  - 10.3.1 Peacock and Seed's (1968) experiment 464
  - 10.3.2 Equation governing the buildup of pore pressure 469
  - 10.3.3 Solution to the equation of buildup of pore pressure 474
- 10.4 Momentary liquefaction 483
  - 10.4.1 General description 483
  - 10.4.2 The case of completely saturated soil 484
  - 10.4.3 The case of unsaturated soil 485
- 10.5 Sinking/floatation of pipelines 487
  - 10.5.1 The case of residual liquefaction 488
  - 10.5.2 The case of momentary liquefaction 499
- 10.6 Sinking of armour blocks in liquefied soil 500
  - 10.6.1 The case of residual liquefaction 500
  - 10.6.2 The case of momentary liquefaction 503
- 10.7 Appendix I. Relationships among soil properties 504
- 10.8 Appendix II. Ranges of soil properties 507
- 10.9 Appendix III. Hsu & Jeng coefficients 509
- 10.10 References 511

Appendix A Small amplitude, linear waves — 521

Author index — 525

Subject index — 533

List of symbols

The main symbols used in the book are listed below. In some cases, the same symbol is used for more than one quantity. This is to maintain generally accepted conventions in different fields. In most cases, however, their use is restricted to a single chapter, as indicated in the following list.

A	amplitude of vibrations
a	amplitude of the horizontal component of orbital motion of water particles
B	width of breakwater
C	clearance between the propeller tip and the seabed
C_D	drag coefficient
C_s	Smagorinski constant in Large Eddy Simulation (LES)
c	concentration
c	celerity at which scour propagates along pipeline (Chapter 2)
c'	fluctuating component of concentration
c_p	pressure coefficient
c_v	coefficient of consolidation
D	pipe/pile diameter
D_p	propeller diameter
D_r	relative density of soil
d	grain size
d	soil depth (Chapter 10)
d_{50}	grain size
E	modulus of elasticity (Young's modulus)

e	clearance between pipeline and the seabed; self-burial depth of pipeline; burial depth of pipeline
e	void ratio (Chapter 10)
e_x, e_y, e_z	soil deformation in $x-, y-$ and $z-$directions
F_0	densimetric Froude number
Fr	Froude number
f	frequency of vibrations
f	source term in the equation governing the buildup of pore pressure (Chapter 10)
f_n	natural frequency
f_w	wave boundary layer friction coefficient
G	shear modulus (Chapter 10)
G	gap between piles in a pile group
g	acceleration due to gravity
H	wave height
H_0	deep-water wave height
H_s	significant wave height
h	water depth
h	bed elevation (Chapter 2); pile height (Chapter 3)
h_w	water depth at seawall
I	inertia moment
i	imaginary unit
K	bulk modulus of elasticity of water
K'	apparent bulk modulus of elasticity of water
KC	Keulegan-Carpenter number
$K_d, K_I..$	multiplying factors in the design method of Melville-Sutherland for pile scour-depth prediction
k	wave number, $k = 2\pi/L$
k	kinetic energy (Chapter 2)
k	coefficient of permeability
k_0	coefficient of lateral earth pressure
k_s	surface roughness
L	wave length
L	pipeline span length (Chapter 2)
L_0	deep-water wave length
m	side slope, beach slope
N	number of piles in a pile group (Chapter 4)
N	number of waves

LIST OF SYMBOLS

N	number of cycles for cyclic shear stress (Chapter 10)
N_ℓ	number of cycles to cause liquefaction
n	porosity
p	pressure
p	pore-water pressure (Chapter 10)
p_0	hydrostatic pressure
p_0	absolute (not excess) pore-water pressure, which can be taken equal to the initial value of pressure
p_1	pressure induced by waves on the seabed, the bed pressure
p_b	maximum value of the bed pressure
$\bar{p}$	period-averaged pore-water pressure (Chapter 10)
q, q_b, q_B	bed-load sediment transport rate
Re	Reynolds number
$\text{Re}_D, \text{Re}_\delta$	pipe/pile Reynolds number, boundary-layer Reynolds number, respectively
RE	wave-boundary layer Reynolds number
S	equilibrium scour depth
S_c	equilibrium scour depth in current-alone case (Chapter 3)
S_r	degree of saturation (Chapter 10)
S_t	scour depth at any instant
s	specific gravity of sediment grains
s_p	specific gravity of pipe
T	time scale of scour, or time scale of self-burial of pipeline
T	wave period (Chapter 10)
T_h	time scale of span development in pipeline scour
T_p	peak period in irregular waves
T_v	time scale of two-dimensional scour in pipeline scour
T_w	wave period
T^*	normalized time scale of scour, that of self-burial of pipeline
t	time
U	undisturbed flow velocity at the top of pipeline in current
U	undisturbed orbital velocity at seabed
U_c	undisturbed flow velocity at the center of pipeline in current
U_{cr}	critical value of U (see above for U) corresponding to the onset of scour below pipeline (Chapter 2)
U_{cr}	critical flow velocity corresponding to the initiation of motion at bed, the incipient sediment transport
U_{cw}	velocity ratio, $U_c/(U_c + U_m)$
U_f	bed shear velocity

LIST OF SYMBOLS

U_{fm}	maximum value of bed shear velocity in waves
U_m	maximum value of undisturbed orbital velocity at the bed
U_{max}	deflection/sagging at the middle section of pipeline span
U_r, U_θ	plan-view components of period averaged-velocity (plan-view components of steady-streaming velocity) (Chapter 6)
u	streamwise component of velocity
u	x–component of soil displacement (Chapter 10)
u_i	ith component of velocity
u_r, u_θ	plan-view components of velocity
$\bar{u}_r, \bar{u}_\theta$	plan-view components of ensemble-averaged velocity
V	mean flow velocity (cross-sectional-/depth-averaged velocity)
V_0	efflux velocity for ship propeller scour
V_p	sagging velocity of pipeline
V_r	reduced velocity
V_x, V_y, V_z	$x-, y-$ and $z-$components of ground-water velocity
W	half-width of scour hole in pipeline scour in waves
W_1, W_2	upstream and downstream widths of scour hole, respectively, in pipeline scour in steady current
w	$z-$component of soil displacement (Chapter 10)
w, w_s	fall velocity of sediment grains
X_m	distance from the propeller of the maximum-scour point
x, y, z	Cartesian coordinates
x_s	distance from the pile axis to the separation point at the bed in front of pile, a characteristic length representing the size of a horseshoe vortex in streamwise direction
z	depth measured from the mudline/seabed downwards (Chapter 10)
z	scour depth in cohesive sediment
z_{max}	maximum (equilibrium) scour depth in cohesive sediment
α	amplification factor
α	angle of attack (Chapter 2)
γ	specific weight of water
γ'	submerged specific weight of soil
γ_s	specific weight of sediment grains
$\gamma_x, \gamma_y, \gamma_z$	shear (angular) deformations in the $x-, y-$ and $z-$directions
δ	undisturbed boundary layer thickness
δ^*	displacement thickness of undisturbed boundary layer

LIST OF SYMBOLS

ε	rate of dissipation of turbulent kinetic energy
ϵ	volume expansion (Chapter 10)
η	surface elevation
θ	Shields parameter
θ	angle of wave propagation (Chapter 7)
θ_{cr}	critical value of the Shields parameter, corresponding to the initiation of motion at the bed
θ_s	critical value of the Shields parameter, corresponding to the initiation of suspension from the bed
θ_{su}	critical value of the Shields parameter, corresponding to the suction removal of the fine sediment from between armour blocks
λ	wave number (Chapter 10), $k = 2\pi/L$
ν	kinematic viscosity
ν	Poisson's ratio (Chapter 10)
ν_T	turbulence viscosity
ν_t	turbulence viscosity associated with small-scale, unresolved turbulence in Large Eddy Simulation (LES)
ρ	water density
ρ_s	density of sediment grains
$\sigma_x, \sigma_y, \sigma_z$	$x-, y-$ and $z-$components of normal stress
$\sigma'_x, \sigma'_y, \sigma'_z$	$x-, y-$ and $z-$components of effective stress in the soil
σ'_0	initial effective stress, the overburden pressure value
τ	bed shear stress
τ	amplitude of the cyclic shear stress in soil (Chapter 10)
τ_{cr}	critical value of shear stress, corresponding to the initiation of motion at bed
τ_x, τ_y, τ_z	$x-, y-$ and $z-$components of shear stress in the soil
τ_∞	undisturbed bed shear stress in current
$\tau_{max,\infty}$	maximum value of undisturbed bed shear stress in waves
Max τ	maximum value of bed shear stress in waves
ϕ	potential function
Φ	potential function (Chapter 2)
Φ	normalized bed-load sediment transport
ψ	stream function
ω	vorticity (Chapter 2)
ω	angular frequency of waves, $\omega = 2\pi/T_w$, or $\omega = 2\pi/T$

Preface

Scour around structures exposed to a steady current has received large attention at least during the last fifty years. This is because one of the most important man-made structures, namely the Bridge, the structure constructed to allow the crossing of a river, may fail due to scouring, which is one of the major failure modes. Many excellent text books can be found on the subject, among which we would like to mention the book by H.N.C. Breusers and A.J. Raudkivi: Scouring. A.A. Balkema, Rotterdam, 1991; that by G.J.C.M. Hoffmans and H.J. Verheij: Scour Manual. A.A. Balkema, Rotterdam, 1997; and, very recently, the book by B.W. Melville and S.E. Coleman: Bridge Scour. Water Resources Publications, LLC, CO, USA, 2000.

Wave scour has not received the same kind of attention. This is partly because the use of structures in the ocean is much more recent than the bridge crossing of rivers, and partly because scour has, not to the same degree, been recognized as a failure mode. The first important contribution to the topic is by J.B. Herbich, who published two early monographs on the subject, J.B. Herbich: Scour around pipelines and other objects. In: Offshore Pipeline Design Elements. Marcell Dekker, Inc. New York, NY, 1981; and J.B. Herbich, R.E. Schiller, Jr., R.K. Watanabe and W.A. Dunlap: Seafloor Scour. Design Guidelines for Ocean-Founded Structures, Marcell Dekker, Inc., New York, NY, 1984. In these books attention was concentrated on especially pipelines and piles. It has later been recognized that the scour pattern around pipelines is much more complex and three-dimensional in structure. Pipeline scour is one of the main chapters in the present book, simply because a pipeline is a very important marine structure.

However, our aim is to describe the wave scour around other kinds of coastal and offshore structures as well, such as vertical piles, breakwaters and seawalls, just to mention the most important ones. Vertical piles of different shapes will be a very important marine structure in the future when offshore windmill farms will increase in number.

While we were working on the present book, another one, R. Whitehouse: Scour at Marine Structures. Thomas Telford, 1998, appeared on the market. Fortunately, our treatment is fairly different. In the present book the hydrodynamic description is very detailed, and very much based on laboratory tests in conjunction with mathematical/numerical modelling. The reason for this is that one structure is seldom alike to another, and therefore the understanding of the processes is a "must" in order to be able to predict the scour for a new structure. We hope that, with our book, we have satisfied this goal.

Acknowledgement

A considerable portion of our research on wave scour during the last 15 years has been supported by the Danish Technical Research Council (STVF) including two ongoing programs, namely "Computational Hydrodynamics" and "Coast and Tidal Inlets". Without the support of STVF and without the time and effort of the scientific staff funded by STVF, the present book would never have been possible.

We would also like to appreciate the support from the EU programs of which we would like to acknowledge particularly the support of the following two:

1. "Scour Around Coastal Structures (SCARCOST)", which ran during 1997-2000, Contract No. MAS3-CT97-0097 of the Commission of the European Communities, Directorate-General XII for Science, Research and Development (Program Marine Science and Technology, MAST III); and

2. "Liquefaction Around Marine Structures (LIMAS)", which is currently running (2001-2004), Contract No. EVK3-CT-2000-00038, of the same commission (FP5 specific program "Energy, Environment and Sustainable Development").

PREFACE

B. Mutlu Sumer has been the program leader of these two programs. On this occasion, BMS would like to extend his appreciation to Mr. Christos Fragakis, the Scientific Officer in Charge of the programs, for the kind cooperation.

Ms. Kirsten Djørup edited the language, and Mr. Hans Jørn Poulsen prepared the figures and helped with the word processing.

Chapter 1

Introduction and basic concepts

1.1 Introduction

When a structure is placed in a marine environment, the presence of the structure will change the flow pattern in its immediate neighbourhood, resulting in one or more of the following phenomena:

1. the contraction of flow;

2. the formation of a horseshoe vortex in front of the structure;

3. the formation of lee-wake vortices (with or without vortex shedding) behind the structure;

4. the generation of turbulence;

5. the occurrence of reflection and diffraction of waves;

6. the occurrence of wave breaking; and

7. the pressure differentials in the soil that may produce "quick" condition/liquefaction allowing material to be carried off by currents.

These changes usually cause an increase in the local sediment transport capacity and thus lead to scour.

(The term "scour" is used instead of the more general term "erosion" to distinguish the process caused by the presence of a structure, Coastal Engineering Manual, 2001).

The scour is a threat to the stability of the structure.

The type of structure where such local scour is involved can vary considerably: it may be a simple structure such as a plain pipeline or a pile or the trunk section of a vertical-wall breakwater, or it may be a complex structure such as a group of piles, a subsea template, a protection structure with horizontal and vertical members, or an offshore platform.

Such structures are usually exposed to currents, waves, and combined waves and currents. Clearly, scour processes in the marine environment (with waves being the dominating flow effect) are more complex than in steady-current flows such as in rivers. In river hydraulics, a long tradition exists for studying scour around hydraulic structures. Scour at a bridge pier, for example, has been studied most extensively (Breusers and Raudkivi, 1991; Melville and Coleman, 2000), simply because it has been realized that this is an important cause of bridge failure. The scour problems in coastal and offshore engineering have not received the same kind of attention. One of the first important contributions is that of Herbich, who published two early monographs on the subject, Herbich (1981) and Herbich et al. (1984). However, at the time of publication of these monographs, the knowledge of the hydrodynamic processes was quite sparse, and many of the design rules were based on only empirical information. Recent years, however, have witnessed a rapid development of the knowledge of flow and scour processes around marine structures, particularly those which have simple geometries such as pipelines, piles, etc. A substantial volume of knowledge has accumulated as a result of this intensive research activity. The book by Whitehouse (1998) has covered developments which took place until mid nineties.

The present book is an attempt to give a comprehensive account of scour at marine structures, and also taking into consideration all state-of-the-art knowledge. It is our aim to describe the *hydrodynamic processes causing scour* in details. With a hydrodynamic understanding, it is easier for the consulting engineer to predict expected scour in those many cases, where physical model tests are not available.

We shall start off with the basic concepts (the present chapter). These include the amplification factor in the bed shear stress in the vicinity of a structure; the equilibrium scour depth and the time scale of scour; the clear-water scour versus the live-bed scour; and the local scour versus the global scour.

Next, we shall concentrate on scour at pipelines (Chapter 2), which will be followed by a full account of scour around piles (Chapters 3-6); namely,

1.2. AMPLIFICATION FACTOR

scour around slender piles (Chapter 3), scour around a group of slender piles (Chapter 4), scour at "complex" structures comprising vertical/inclined and horizontal slender cylindrical elements (such as a piled steel platform, a subsea template, or a wind turbine foundation) (Chapter 5), and scour around large piles (Chapter 6). In Chapters 7 and 8, attention will be concentrated on scour at breakwaters and seawalls, respectively. Chapter 9 will study ship-propeller scour. Chapter 10 will address the question of the impact of liquefaction.

It may be noted that some marine-engineering projects may include structural elements or flow conditions that are typically associated with inland waters and estuaries. Breusers and Raudkivi (1991), Hoffmans and Verheij (1997) and Melville and Coleman (2000) review techniques for estimating maximum scour characteristics for cases that may be applicable to marine-engineering projects, such as scour downstream of sills and stone blankets; scour downstream of hard bottoms due to horizontal submerged jets; scour at control structures due to plunging jets; scour at 2-D and 3-D culverts; and scour at abutments and spur dikes.

The topics covered in these latter books and the material presented in the present book form a complementary source of information on scour.

1.2 Amplification factor

Consider a structure placed in a marine environment. The presence of the structure will cause the flow in its neighborhood to change. This local change in the flow will generally cause an increase in the bed shear stress and in the turbulence level. The sediment transport close to the structure is increased mainly because:

1. the average bed shear stress is increased close to the structure, and

2. the degree of turbulence is increased in the vicinity of the structure.

Both features will lead to an increase in the local sediment transport capacity. Today, however, much more knowledge is available about item (1) than about item (2).

Usually the increase in the bed shear stress is expressed in terms of the so-called **amplification factor** defined by

$$\alpha = \frac{\tau}{\tau_\infty} \tag{1.1}$$

in which τ = the bed shear stress and τ_∞ = the bed shear stress for the undisturbed flow. This is illustrated in Fig. 3.10 in Chapter 3 for a pile exposed to a steady current. (Only one half plane is shown in the figure for reasons of symmetry). As seen, the amplification factor can be very large near the structure (as large as $O(10)$).

Owing to the local increase in α (i.e., $\alpha > 1$), the sediment transport capacity will increase (since the rate of sediment transport as bed load $q_b \sim \tau^{3/2}$), and presumably the bed will be eroded, the *scour process*. Fig. 3.20 in Chapter 3 shows an illustration of the scour hole generated in the vicinity of a pile subjected to a steady current.

This process will continue until the scour reaches such levels that the bed shear stress around the structure becomes $\alpha = O(1)$. The stage where the scour process comes to an end is called the equilibrium stage. (It should be noted that, in the preceding discussion, the formula $q_b \sim \tau^{3/2}$ is assumed to be valid in the present context also, although it is anticipated that the presence of the "structure-generated turbulence" may further increase the sediment transport capacity, as discussed in the preceding paragraphs).

1.3 Equilibrium scour depth and time scale of scour

From the preceding considerations, the scour develops towards the equilibrium stage through a transitional period, as illustrated schematically in Fig. 1.1. The scour depth corresponding to the equilibrium stage, S in Fig. 1.1, is called the **equilibrium scour depth**.

It is also seen from Fig. 1.1 that, for a substantial amount of scour to develop, a certain amount of time must elapse. This time is called the **time scale** of the scour process. The time scale of the scour process may be defined in several ways. The following definition will be adopted in the present treatment:

$$S_t = S\left(1 - \exp(-\frac{t}{T})\right) \qquad (1.2)$$

in which T = the time scale of the scour process, and corresponds to the time period T indicated in Fig. 1.1 where the dashed line is tangent to the scour-depth-versus-time curve at $t = 0$.

1.4. CLEAR-WATER SCOUR AND LIVE-BED SCOUR

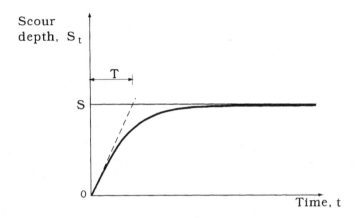

Figure 1.1: Time development of scour depth.

The aforementioned quantities, namely the equilibrium scour depth and the time scale, are two major parameters in scour studies.

The scour depth is important because, given the structure and the flow climate, it indicates the degree of scour potential. The assessment of scour depth is essential in the design of both (1) the foundation of the structure and (2) the scour protection work.

The time scale is also equally important. A scour hole produced after a storm may be backfilled. Normally, the question asked in practice is whether any substantial amount of scour would occur over the backfilled area during the next storm. Obviously, for a substantial amount of scour to occur, the storm should prevail over a space of time larger than the time scale of the scour process. Clearly, to answer the aforementioned question, the time scale of scour must be known.

1.4 Clear-water scour and live-bed scour

Scour may be classified in two categories: the clear-water scour and the live-bed scour.

In the case of the **clear-water scour**, no sediment motion takes place *far* from the structure ($\theta < \theta_{cr}$), while, in the case of the **live-bed scour**, the sediment transport prevails over the entire bed ($\theta > \theta_{cr}$). Here θ is the

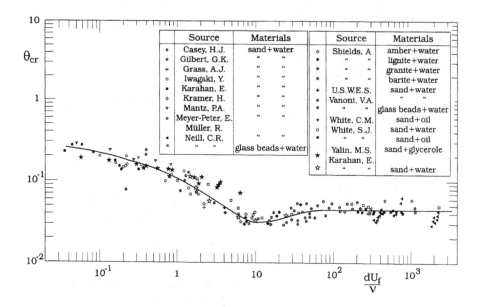

Figure 1.2: Initiation of motion at the bed. Data compiled by Yalin and Karahan (1979).

undisturbed Shields parameter defined by

$$\theta = \frac{U_f^2}{g(s-1)d} \quad (1.3)$$

in which $U_f = \sqrt{\tau_\infty/\rho}$, the undisturbed bed shear velocity (in the case of waves, τ_∞ should be replaced by $\tau_{max,\infty}$, the maximum value of the undisturbed bed shear stress), g = the acceleration due to gravity, s = the specific gravity of sediment grains and d = the grain size. θ_{cr} is the critical value of the Shields parameter corresponding to the initiation of sediment motion at the bed. θ_{cr} is a function of the grain Reynolds number, dU_f/ν, Fig. 1.2. (For basic concepts regarding the sediment transport, Chapter 7 in the book by Fredsøe and Deigaard (1992) may be consulted).

In the clear water case, the variation of the scour depth with θ is more pronounced (as illustrated in Fig. 1.3 for the case of scour below a pipeline): the scour depth increases from zero at very small values of θ up to θ_{cr} ($\simeq 0.05$

1.5. LOCAL AND GLOBAL SCOUR

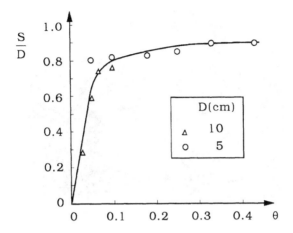

Figure 1.3: Variation of equilibrium scour depth (for a pipeline) versus the Shields parameter. The initial clearance from the bed is nil. Mao (1986).

for the experiments presented in Fig. 1.3, c.f. Fig. 1.2). At very low θ values, no scour will occur because, in this case, even the amplified local bed shear stress may still be too small to cause sediment transport. However, when the live-bed case is reached, and beyond ($\theta > \theta_{cr}$), a very small variation of the scour depth with θ is observed (see also, for example, Hjorth, 1975, for the current alone case, and Sumer and Fredsøe, 1990, for the waves alone case). This is because any change in θ results in corresponding changes in sediment transport, and these changes occur both inside and outside of the scour hole in equivalent amounts, eventually causing only small changes in the equilibrium scour hole.

1.5 Local and global scour

The local scour and the global scour will be described by reference to two examples: Scour at a piled steel platform and scour at a bridge pier.

Consider a piled steel platform comprising horizontal and vertical members (Fig. 1.4). When this structure is exposed to flow action, two kinds of scour will take place; the local scour around the individual structural elements such as that around the supporting piles, and the global scour beneath

12   CHAPTER 1. INTRODUCTION AND BASIC CONCEPTS

Figure 1.4: Scour around a piled steel platform. A conceptual picture. Angus and Moore (1982). By courtesy of the Offshore Technology Conference.

and around the structure in the form of a saucer-shaped depression, as illustrated in the conceptual picture in Fig. 1.4. The global scour here is due to the combined action of all the flow effects generated by the individual structural elements, namely the contraction of flow and the "turbulence" generated by the structural elements.

Likewise, scour at a bridge occurs as local scour and global scour (Fig. 1.5). Local scour occurs around the individual piers and at the abutments, while the global scour occurs as the general lowering of the river bed, as sketched in Fig. 1.5. The global scour in this example may, in addition to the contraction scour, occur due to hydrometeorological changes (e.g., prolonged high flows), geomorphological changes (e.g., lowering of channel base level due to catchment wide adjustment in geomorphology). human activities (e.g., dam construction), bank erosion (caused by channel widening, meander migration, a change in the river controls, or a sudden change in the river course, e.g., with the formation of a meander-loop cut-off) (Melville and Coleman, 2000).

1.6 References

1. Angus, N.M. and Moore, R.L. (1982): Scour repair methods in the

1.6. REFERENCES

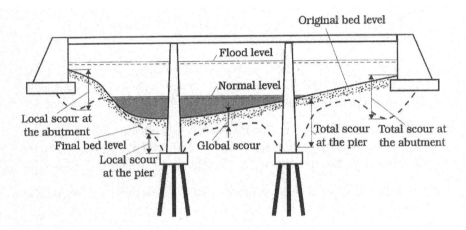

Figure 1.5: The types of scour that can occur at a bridge. Adapted from Melville and Coleman (2000).

Southern North Sea. Proc. 14th Annual Offshore Technology Conference, Houston, Texas, May 3-6, 1982, Paper No. 4410, 385-399.

2. Breusers, H.N.C. and Raudkivi, A.J. (1991): Scouring. A.A. Balkema, Rotterdam, viii + 143 p.

3. Coastal Engineering Manual (2001). Scour and Scour Protection. Chapter VI-5-6, Engineer Manual EM 1110-2-1100, Headquarters, U.S. Army Corps of Engineers, Washington, D.C.

4. Fredsøe, J. and Deigaard, R. (1992): Mechanics of Coastal Sediment Transport. Advanced Series on Ocean Engineering, vol. 8, World Scientific, xviii + 369 p.

5. Herbich, J.B. (1981): Scour around pipelines and other objects. In: Offshore Pipeline Design Elements. Marcell Dekker, Inc. New York, NY., xvi + 233 p.

6. Herbich, J.B., Schiller, R.E., Jr., Watanabe, R.K. and Dunlap, W.A. (1984): Seafloor Scour. Design Guidelines for Ocean-Founded Structures, Marcell Dekker, Inc., New York, NY, xiv + 320 p.

7. Hoffmans, G.J.C.M. and Verheij, H.J. (1997): Scour Manual. A.A. Balkema, Rotterdam, xv + 205 p.

8. Hjorth, P. (1975): Studies on the nature of local scour. Bull. Series A, No. 46, viii + 191 p., Department of Water Resources Engineering, Lund Institute of Technology/University of Lund, Lund, Sweden.

9. Mao, Y. (1986): The interaction between a pipeline and an erodible bed. Series Paper 39, Tech. Univ. of Denmark, ISVA, in partial fulfillment of the requirement for the degree of Doctor of Philosophy.

10. Melville, B.W. and Coleman, S.E. (2000): Bridge Scour. Water Resources Publications, LLC, CO, USA, xxii + 550 p.

11. Sumer, B.M. and Fredsøe, J. (1990): Scour below pipelines in waves. J. Waterway, Port, Coastal and Ocean Engineering, ASCE, vol. 116, No. 3, 307-323.

12. Whitehouse, R. (1998): Scour at Marine Structures. Thomas Telford. xix + 198.

13. Yalin, M.S. and Karahan, E. (1979): Inception of sediment transport. J. Hydraulic Division, ASCE, vol. 105, HY 11, 1433-1443.

Chapter 2

Scour below pipelines

Pipelines are installed in marine environments for transportation of gas and crude oil from offshore platforms, and for the disposal of industrial and municipal waste water into the sea.

The development in the offshore oil industry in the past 30 years or so has lead to tens of thousands of kilometers of pipeline networks laid in the North Sea, in the Gulf of Mexico, and in many other places across the globe, and these networks have become "lifelines" of the oil industry.

Typically, the pipe size may be from 20-30 cm to more than 1.0 m in diameter; the pipeline length may be from hundreds of meters to thousands of meters; and the water depth may be from tens of meters to hundreds of meters. The pipelines may be laid on the bed surface, they may be buried, or they may be trenched.

When a pipeline is exposed to direct flow action (in the case when it is laid on the seabed, or when it is trenched), and when the seabed is erodible, scour may occur around the pipe under the flow (waves/current) action, which may lead to suspended free spans of the pipeline. The pipeline along the length of the suspended span may or may not sag in the generated scour hole. In the case of a sagging pipeline, the pipeline may reach the bottom of the scour hole, which may be followed by backfilling and eventual self-burial of the pipeline, as will be detailed later in the chapter. There are several, highly complex, processes in this interesting pipeline/seabed interaction.

The purpose of this chapter is to study the scour processes around a pipeline, namely the onset of scour, the so-called tunnel and lee-wake erosion processes, the three-dimensional scour processes including the scour, self-burial, and backfilling processes along the free span areas and at span shoulders. The chapter also includes a review of scour protection measures, and a detailed account of the mathematical modelling studies.

2.1 Onset of scour

Consider a pipeline laid on an erodible bed. If the initial embedment of the pipeline in the bed is not very large, and the flow (induced by currents/waves) is sufficiently strong, the bed may be washed away underneath the pipe, the **onset of scour**. The onset of scour is basically related to the seepage flow in the sand beneath the pipeline, which is driven by the pressure difference between the upstream and downstream sides of the pipe.

The critical conditions for the onset of scour have been studied by Mao (1986), Chiew (1990), Sumer and Fredsøe (1991), Klomp et al. (1995) and Sumer, Truelsen, Sichmann and Fredsøe (2001 a).

Mao (1986) has described the role of vortices that form in front and at the rear of the pipe. He has also discussed the seepage flow underneath the pipe in relation to the onset of scour. The latter has been further elaborated by Chiew (1990). The latter author has also linked the onset of scour to the process of piping. Fredsøe and Sumer (1991) conducted experiments to determine the critical conditions in the case of waves, and expressed it in terms of two parameters, namely the Keulegan-Carpenter number, KC, and the initial embedment-to-diameter ratio, e/D. Klomp et al. (1995) later extended Sumer and Fredsøe's (1991) study to the case of combined waves and current. Subsequently, Sumer et al. (2001 a) studied the onset of scour in both currents and waves. They measured the pressure gradient underneath the pipe, and showed that the excessive seepage flow (driven by this pressure gradient) and the resulting piping is the major factor causing the onset of scour below the pipe. In the same study, the critical condition corresponding to the onset of scour has been determined. The following description is mainly based on Sumer et al.'s (2001 a) work.

2.1.1 Mechanism of onset of scour. Seepage flow and piping underneath the pipe

When a pipeline is laid on a sediment bed, and is subject to a current, the pressure difference between the upstream and the downstream of the pipe (Fig. 2.1) will induce a seepage flow in the sand bed underneath the pipe, as

2.1. ONSET OF SCOUR

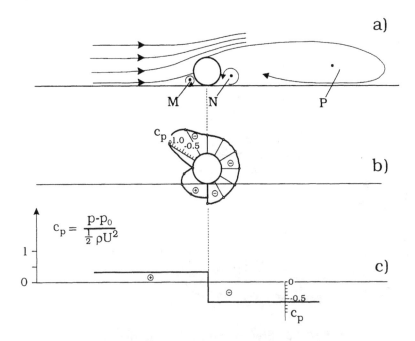

Figure 2.1: Pressure distributions for bottom-seated pipe. Bearman and Zdravkovich (1978).

sketched in Fig. 2.2. When the current velocity is increased, a critical point is reached where the discharge of the seepage flow will be increased more rapidly than the driving pressure difference dictates, and simultaneously the surface of the sand at the immediate downstream of the pipe will rise, and eventually a mixture of sand and water will break through the space underneath the pipe. This process is called **piping**, and is well-known in soil mechanics in conjunction with the so-called piping failures at hydraulic structures such as dams, cofferdams, etc. (Terzaghi, 1948).

Now, let us consider the critical condition for the piping for a cohesionless granular material. There are basically two forces: one is the agitating force (i.e., the seepage force), and the other is the resisting force (i.e., the submerged weight of the sand). The seepage force at the point where the sand-water mixture is expelled from the bed is directed vertically upwards (considering the bed as a potential line, Fig. 2.2, Detail A), and can be

CHAPTER 2. SCOUR BELOW PIPELINES

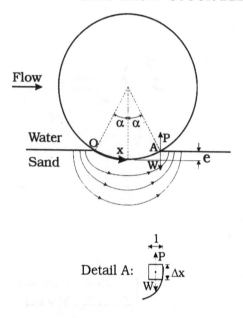

Figure 2.2: Seepage flow underneath the pipe.

written as

$$P = \frac{\partial p}{\partial x} \Delta x \tag{2.1}$$

in which p is the pressure, x is the distance along the perimeter of the pipe, measured from the junction between the upstream side of the pipe and the bed (Fig. 2.2), $\partial p/\partial x$ is the pressure gradient driving the seepage flow, and P is the force on a small element of sand (the size $\Delta x \times 1 \times 1$, Fig. 2.2) at the point where the mixture of sand and water breaks through. The submerged weight of the sand, W (Fig. 2.2), on the other hand, is

$$W = (\gamma_s - \gamma) \, \Delta x \, (1-n) = \gamma(s-1)(1-n) \, \Delta x \tag{2.2}$$

in which $s = \gamma_s/\gamma$ is the specific gravity of sand grains, γ is the specific weight of water, γ_s is the specific weight of sand grains, and n is the porosity. The critical condition will then occur when the seepage force P exceeds the submerged weight W (the friction forces are practically zero at the instant of failure):

2.1. ONSET OF SCOUR

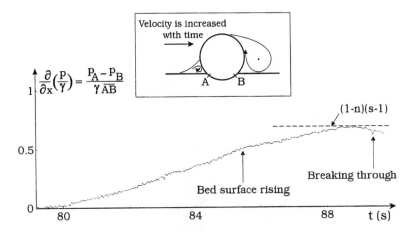

Figure 2.3: Time series of pressure gradient underneath the pipe. Current. Sumer et al. (2001 a).

$$P \geq W \tag{2.3}$$

Thus, from Eqs. 2.1 and 2.2, the critical condition is expressed by the following equation

$$\frac{\partial}{\partial x}(\frac{p}{\gamma}) \geq (s-1)(1-n) \tag{2.4}$$

i.e., the critical condition occurs when the pressure gradient $\frac{\partial}{\partial x}(\frac{p}{\gamma})$ exceeds the floatation gradient $(s-1)(1-n)$.

Current case

Fig. 2.3 shows the time series of the pressure gradient $\frac{\partial}{\partial x}(\frac{p}{\gamma})$ measured by Sumer et al. (2001 a) in the case of a steady current. In this test, the flow velocity is increased gradually until the critical point is reached. In the same test, the junction between the downstream side of the pipe and the bed was videotaped (simultaneously with the pressure measurements) with a mini underwater camera (Fig. 2.4 a). From these measurements, the authors have made the following observations:

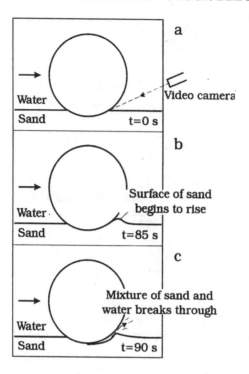

Figure 2.4: Piping. Times correspond to those in the previous figure. Sumer et al. (2001 a).

1. With increasing velocity, the pressure gradient $\frac{\partial}{\partial x}(\frac{p}{\gamma})$ is increased, Fig. 2.3. (This is because $p \sim U^2$, see Fig. 2.1 c).

2. As the pressure gradient increases, a point is reached where the surface of the sand at the immediate downstream of the pipe begins to rise (Figs. 2.3 and 2.4 b), consistent with the description of the piping process described in conjunction with dams in Terzaghi (1948). (Sumer et al. (2001 a) note that the video recording showed clearly that this change in the bed level was not in the form of piling-up of the sand due to the lee-wake vortex, but rather in the form of rise of the bed en masse).

3. This stage continues for some period of time (about 5 s, Figs. 2.3 and

2.1. ONSET OF SCOUR

2.4 b-c), and is subsequently followed by the process where a mixture of sand and water breaks through (Fig. 2.4 c). The instant when the surface downstream starts the rise marks the instant when the pressure gradient exceeds the floatation gradient. Subsequently, grains are progressively removed and a breakthrough develops. The process will depend on the porosity, internal friction, and length of flow path (the longer the path, the longer it takes for the breakthrough to develop).

4. The onset of scour never occurred concurrently along the length of the pipe in a two-dimensional fashion, but rather, it occurred locally, in a three-dimensional fashion.

5. For the piping condition to occur, the pressure gradient $\frac{\partial}{\partial x}(\frac{p}{\gamma})$ has to reach the value equal to $(1-n)(s-1)$, as seen from Fig. 2.3 (cf. Eq. 2.4). (Note that the mean value of the pressure gradient $\frac{\partial}{\partial x}(\frac{p}{\gamma})$ was found to be 0.74 with a standard deviation of $\sigma = 0.14$; The slight variation of the pressure gradient from one test to the other, characterized by $\sigma = 0.14$, was attributed to the turbulent wake behind the pipeline).

It should be noted that visual observations made in Sumer et al.'s work (2001 a) showed that, contrary to the generally accepted view (Mao, 1986, Chiew, 1990, and Sumer and Fredsøe, 1991), the vortices generated at the downstream and upstream parts of the pipe (see the small box in Fig. 2.3) did not undermine the pipe prior to the onset of scour (which would otherwise lead to a slight reduction in the length of the streamline of the seepage flow, presumably resulting in larger pressure-gradient forces).

Wave case

Fig. 2.5 shows the time series of the pressure gradient $\frac{\partial}{\partial x}(\frac{p}{\gamma})$ obtained in the experiments of Sumer et al. (2001 a) in the case of waves. In these experiments, the wave height is increased gradually until the critical point is reached. The crests in the time series correspond to the crest half periods in the surface elevation, while the troughs correspond to the trough half periods. As seen, the onset of scour takes place in the crest half period (see the instant of "breaking through" in Fig. 2.5). The pressure gradient in the trough half period is not large enough to cause piping.

As seen from Fig. 2.5, the onset of scour occurs when the pressure gradient $\frac{\partial}{\partial x}(\frac{p}{\gamma})$ reaches the value $(s-1)(1-n)$, or even exceeds it. This result is

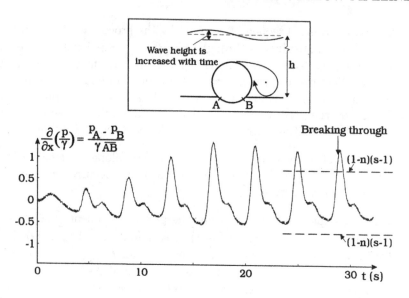

Figure 2.5: Time series of pressure gradient underneath the pipe. Wave. Sumer et al. (2001 a).

somewhat different from that obtained for the current case (Fig. 2.3). This difference may be attributed to the time over which the sand is exposed to the critical pressure-gradient force. In the case of the current, this period is quite long, namely in the order of magnitude of 5 s (Fig. 2.3), the mixture of sand and water breaks through only after $O(5\text{ s})$ upon the application of the critical pressure gradient force. By contrast, in the case of the waves, the pressure gradient necessary for the onset of scour is available only for a very short period of time ($O(0.5\text{ s})$) for each crest half period (Fig. 2.5). Apparently this small exposure to the critical pressure gradient is not long enough for the piping to occur. It is only when the pressure gradient is increased further, and after some number of exposures that the piping takes place, resulting in the onset of scour. It may be added that the breakthrough is a progressive process; each wave loosens some grains on the exit side.

Sumer et al. (2001 a) note that simultaneous measurements of the surface elevation η and the pressure gradient $\frac{\partial}{\partial x}(\frac{p}{\gamma})$ indicate that there is a phase difference between η and $\frac{\partial}{\partial x}(\frac{p}{\gamma})$; the pressure gradient (Fig. 2.5) lags about

2.1. ONSET OF SCOUR

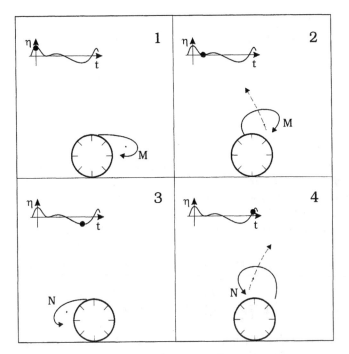

Figure 2.6: Sequence of flow pictures over one wave cycle. Sumer et al. (2001 a).

20-25⁰ behind the surface elevation. Fig. 2.6 shows a sequence of flow pictures over one wave period (corresponding to the flow conditions in Fig. 2.5). Sumer et al. point out that the moment where the onset of scour occurs coincides almost with the passage of the wave crest (Frame 1 in Fig. 2.6) where the flow is in the direction of wave propagation, and the lee-wake (with Vortex M) is well established. This observation is consistent with the flow pattern in the case of the steady-current (see the small sketch in Fig. 2.3).

2.1.2 Criterion for the onset of scour

In steady current

The criterion for the onset of scour (Eq. 2.4) can be written in the following nondimensional form. Onset of scour occurs if

$$\left\{ \frac{\partial p^\star}{\partial x^\star} \frac{U^2}{gD(1-n)(s-1)} + R \right\}_{cr} \geq 1 \quad (2.5)$$

in which

$$p^\star = \frac{p}{\rho U^2}, \quad x^\star = \frac{x}{D} \quad (2.6)$$

ρ is the water density, U is the undisturbed flow velocity at the top of the pipeline (the top velocity rather than the center-line velocity is adopted here, considering the cases where the pipeline may be buried with e/D larger than 0.5, e being the burial depth, Fig. 2.2), and g is the acceleration due to gravity. The term R is a small, nondimensional term, and is included here to represent the effects other than the pressure gradient force (mainly the effect of the vortices forming in front of the pipe and in the lee wake). Both $\partial p^\star/\partial x^\star$ and R are essentially a function of the burial-depth-to-diameter ratio, e/D. Therefore, the criterion for the onset of scour can be written in the following form.

$$\left[\frac{U^2}{gD(1-n)(s-1)} \right]_{cr} \geq f\left(\frac{e}{D}\right) \quad (2.7)$$

where the function $f(e/D)$ is to be determined from experiments. It may be noted that f is actually a function of not only e/D, but also the pipe Reynolds number, $Re = UD/\nu$, and the relative roughness k_s/D in which ν is the kinematic viscosity and k_s is the surface roughness of the pipe. However, it is expected that the influence of these latter parameters will not be very significant, if there is no significant change in the flow regime, i.e., if the flow around the pipe does not change from the subcritical regime to the supercritical regime, or from the supercritical regime to the transcritical regime (see, e.g., Sumer and Fredsøe, 1997). This issue will be further elaborated in the next subsection.

Also, it may be mentioned that cohesionless granular material is considered in the present analysis. Otherwise, soil properties (including permeability) will also influence the onset of scour (clearly, in the case when the permeability $\rightarrow 0$, the breakthrough will never occur).

The focus in the present subsection will be on the variation with e/D.

Sumer et al. (2001 a) determined the critical condition experimentally, and plotted the data in the form of Eq. 2.7. Fig. 2.7 displays this data. The figure shows that the larger the burial depth, the higher the critical velocity for the onset of scour. This is because, as the burial depth increases, the

2.1. ONSET OF SCOUR

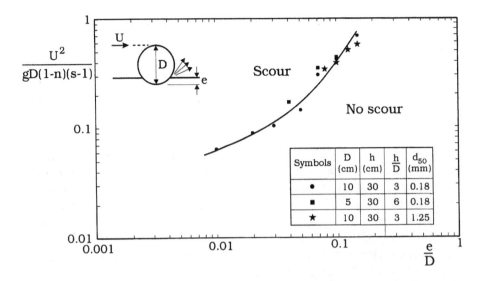

Figure 2.7: Onset of scour. Current. Sumer et al. (2001 a).

pressure gradient will be decreased; therefore relatively higher velocities will be required to cause piping.

Fig. 2.7 further shows that

1. the results for two different pipe diameters, namely $D = 5$ and 10 cm, coincide when plotted in terms of the nondimensional quantities in Eq. 2.7, and

2. likewise, the results for two different sand sizes, namely $d_{50} = 0.18$ and 1.25 mm, collapse on a single curve, revealing that the results are unaffected by the sand size. (Note that the experiments for coarser sand in Sumer et al.'s (2001 a) study were not conducted for burial depths smaller than $e/D = 0.08$ on grounds that, for such small values of e/D, the sand will no longer act as a continuous medium; therefore, the results will not make sense).

The data in Fig. 2.7 can be represented by the following empirical expression

$$\frac{U_{cr}^2}{gD(1-n)(s-1)} = 0.025 \exp[9(\frac{e}{D})^{0.5}] \tag{2.8}$$

in which U_{cr} is the critical undisturbed flow velocity (measured at the level of the top of the pipeline) for the onset of scour.

Finally, it may be noted that the time required for the flow to remove the grains and open a "breach" will be appreciably longer for larger diameter pipes than for those used in Sumer et al.'s (2001 a) study.

Effect of change in flow regime

Sumer et al. (2001 a) have also studied the effect of the change in the flow regime. To this end, the 10 cm diameter pipe in their experiment was coated with 0.3 cm cylindrically shaped plastic grains (0.3 cm in diameter and 0.3 cm in height). The burial depth tested in this experiment was $e/D = 0.1$. The grains were glued (in a densely packed manner) to the cylinder, and the roughness height (measured from the base pipe surface to the top of the roughness elements) was 0.3 cm, or alternatively Nikuradse's equivalent sand roughness $k_s \cong 2 \times 0.3 = 0.6$ cm, giving a relative roughness of $k_s/D = 6 \times 10^{-2}$. To keep the boundary condition in the sand the same as in the case of the smooth pipe, the holes between the roughness elements were filled with plastic for the portion of the pipe that stays in the sand bed. The only difference between the rough-pipe test and the smooth-pipe test was that, in the smooth-pipe case, the flow was in the subcritical regime ($Re = 6 \times 10^4$), whereas, in the rough-pipe case, it was in the transcritical regime ($Re = 6 \times 10^4$, $k_s/D = 6 \times 10^{-2}$) (Sumer and Fredsøe, 1997). The result of this experiment is compared with its smooth-pipe counterpart in the following table:

Pipe	$Re = UD/\nu$	k_s/D	Flow regime	$\frac{U_{cr}^2}{gD(1-n)(s-1)}$
Smooth	6×10^4	-	Subcritical	0.42
Rough	6×10^4	6×10^{-2}	Transcritical	0.72

As seen, the critical value of the parameter $U_{cr}^2/(gD(1-n)(s-1))$ is, in the case of the rough pipe, a factor of 1.7 larger. This is because the flow in this case is in the transcritical regime; therefore the pressure gradient will be smaller (due to the relatively larger wake pressure, see e.g., Sumer and Fredsøe, 1997, p. 41), and hence, relatively larger velocities will be required for the onset of scour. This result suggests that, for extremely large pipelines with smooth surface ($Re > O(10^5)$), or for medium/large size pipelines with very large roughness ($Re > O(10^4)$, $k_s/D > O(10^{-2})$) (i.e., in the case where

2.1. ONSET OF SCOUR

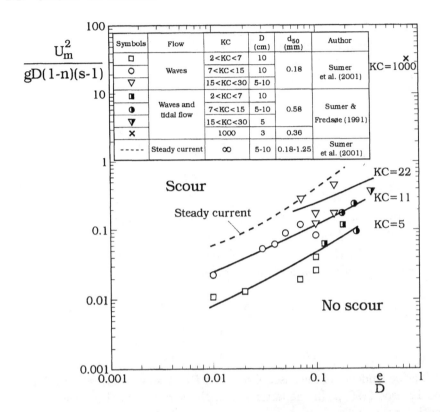

Figure 2.8: Onset of scour. Steady-current result taken from the previous figure. Sumer et al. (2001 a).

the flow regime is transcritical, see e.g. Sumer and Fredsøe, 1997), the critical curve for the onset of scour displayed in Fig. 2.7 may be shifted upwards so that the critical value of the parameter $U_{cr}^2/(gD(1-n)(s-1))$ would be a factor of 1.5-2 larger than depicted in Fig. 2.7.

In waves

In the case of waves, the criterion given in Eq. 2.7 can be adopted provided that

1. U is replaced by U_m, the maximum value of the orbital velocity of water particles at the bed, and

2. there will be an additional parameter regarding the function f. namely f will be a function of e/D and KC. This is because, in this case, the terms $\partial p^*/\partial x^*$ and R in Eq. 2.5 are also governed by KC, the Keulegan-Carpenter number (see, e.g., Sumer and Fredsøe, 1997), defined by

$$KC = \frac{U_m T_w}{D} \quad (2.9)$$

in which T_w = the wave period. KC can be written as

$$KC = \frac{2\pi a}{D} \quad (2.10)$$

if the orbital velocity is assumed to vary sinusoidally. Here a = the amplitude of the orbital motion of the water particles at the bed, $a = U_m T_w/(2\pi)$. (The role of the KC number in wave scour will be discussed in greater details later in the chapter).

So, the critical condition will be

$$\left[\frac{U_m^2}{gD(1-n)(s-1)}\right]_{cr} \geq f(\frac{e}{D}, KC) \quad (2.11)$$

The dependence of the onset of scour on KC has been discussed by Sumer and Fredsøe (1991). In the latter study, the variation of the critical burial depth for the onset of scour with KC was obtained; however, the role of the parameter $U^2/(gD(1-n)(s-1))$ was not recognized. The data obtained by Sumer and Fredsøe (1991) and subsequently by Sumer et al. (2001 a) are displayed in Fig. 2.8 in the format of Eq. 2.11. It is seen that both parameters, namely KC and $U_m^2/(gD(1-n)(s-1))$, are equally significant.

For a given value of KC, the critical value of the parameter $U_m^2/(gD(1-n)(s-1))$ increases with increasing e/D. This can be explained in the same way as in the case of the steady current. Likewise, for a given value of e/D, the critical value of the parameter $U_m^2/(gD(1-n)(s-1))$ increases with increasing KC. This is because the pressure gradient decreases with increasing KC (cf. the pressure diagram given in Sumer and Fredsøe, 1991, Fig. 2.3, and that in Bearman and Zdravkovich, 1978, Fig. 2.1); therefore, larger and larger velocities will be needed for the onset of scour with increasing KC, meaning that the critical value of $U_m^2/(gD(1-n)(s-1))$ will increase with KC.

2.1. ONSET OF SCOUR

Fig. 2.8 indicates that, as the Keulegan-Carpenter number increases, the critical value of $U_m^2/(gD(1-n)(s-1))$ approaches to that obtained in the case of the steady current. For example, for $e/D = O(0.05)$, the critical value of $U_m^2/(gD(1-n)(s-1))$ approaches the value for the steady current for $KC > O(20)$. This is linked to the fact that the pressure gradient in the case of the waves approaches to the pressure gradient experienced in the case of the steady current.

Sumer et al.'s (2001 a) study focused on the variations with respect to KC and e/D. The variation with the number of waves (or the time) required for the piping to occur has not been studied. Similar to the case of steady current, the time required for the piping to develop will be appreciably longer for larger diameter pipes than for those used in Sumer et al.'s (2001 a) study.

Example 1 *Critical velocity for the onset of scour. Numerical example*

1. Given $D = 1$ m, $e = 0.05$ m, $n = 0.43$ and $s = 2.65$. What is the critical velocity (measured at the top of the pipeline) that causes the onset of scour below the pipeline in the case of a steady current?

From Eq. 2.8

$$U_{cr}^2 = 0.025 \exp[9(\frac{e}{D})^{0.5}]gD(1-n)(s-1) =$$

$$= 0.025 \times \exp\left[9(\frac{0.05}{1})^{0.5}\right] \times 9.81 \times 1 \times (1-0.43) \times (2.65-1) = 1.73$$

or $U_{cr} = 1.3$ m/s.

2. Suppose that the pipeline is exposed to waves, with a maximum velocity (at the bed) of the same magnitude, namely $U_m = 1.3$ m/s, and with a wave period of $T_w = 10$ s. Find whether or not the onset of scour occurs.

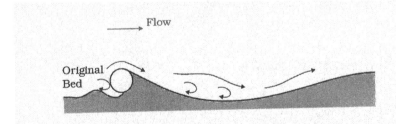

Figure 2.9: Scour profile when scour does not break under the pipe. Chiew (1990).

The Keulegan-Carpenter number will be

$$KC = \frac{U_m T_w}{D} = \frac{1.3 \times 10}{1} = 13$$

and for $KC = 13$ and $e/D = 0.05/1 = 0.05$, the critical velocity from Fig. 2.8 is found

$$\frac{U_{m,cr}^2}{gD(1-n)(s-1)} = 0.08$$

or $U_{m,cr} = 0.86$ m/s. Since $U_m = 1.3$ m/s $> U_{m,cr} = 0.86$ m/s, the onset of scour will occur for this wave climate.

3. Now, suppose that the pipeline is exposed to waves with $U_m = 1.1$ m/s and $T_w = 18$ s. In this case,

$$KC = \frac{U_m T_w}{D} = \frac{1.1 \times 18}{1} \simeq 20$$

and for $KC = 20$ and $e/D = 0.05$, the critical velocity from Fig. 2.8 is found

$$\frac{U_{m,cr}^2}{gD(1-n)(s-1)} \simeq 0.2$$

or $U_{m,cr} = 1.4$ m/s. This velocity is larger than $U_m = 1.1$ m/s; therefore, the onset of scour will not occur, meaning that the scour will not break underneath the pipe.

However, the action of the lee-wake vortices may cause scour at the two sides of the pipe. This has also been observed in the case of currents; see Fig. 2.9 for example, taken from Chiew (1990), where, although the scour does not break underneath the pipe, a certain scour pattern forms, as illustrated in the figure. In the case of waves, similar scour patterns may be experienced (symmetric or asymmetric between the two sides of the pipe, depending on the symmetry/asymmetry in the wave pattern).

2.2 Tunnel erosion

The onset of scour is followed by the stage called **tunnel erosion**.

In this initial stage, the gap between the pipe and the bed, e, remains small, i.e., $e \ll D$ in which D is the pipe diameter. During this stage, a substantial amount of water is diverted to the gap, as sketched in Fig. 2.10,

2.2. TUNNEL EROSION

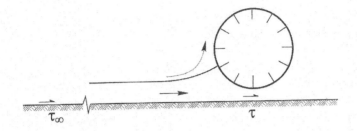

Figure 2.10: Definition sketch of approach flow.

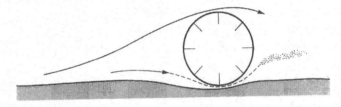

Figure 2.11: Tunnel erosion below a pipeline.

leading to very large velocities in the gap and presumably resulting in very large shear stresses on the bed just below the pipeline. The measurements of Sumer et al. (1990) indicate that the amplification factor in the bed shear stress, α, just below the pipe for the value of the gap-to-diameter ratio $e/D = 0.05$ is $\alpha = O(4)$ in which α is defined as in Eq. 1.1. (The previously mentioned measurements were undertaken for the case where the pipe is exposed to an oscillatory flow with the Keulegan-Carpenter number ranging from $KC = 10$ to 100). The velocity measurements reported in Jensen, Sumer, Jensen and Fredsøe (1990) where the pipe was exposed to steady currents show that, here too, the amplification α (corresponding to the initial stage of the scour process) is $\alpha = O(3-5)$. The potential-flow solution of Müller (1929) gives the corresponding α values a factor of 2 larger. (It may be noted that Müller's theory was later modified by Fredsøe and Hansen (1987), based on the observation that the velocities at the top and the bottom of the pipe are nearly identical).

The large increase in the bed shear stress below the pipe results in a tremendous increase in the sediment transport; a factor of 4 increase in the bed shear stress causes a factor of 8 increase in the sediment transport, considering $q_b \sim \tau^{3/2}$. This suggests that, immediately after the onset of scour, the scour under the pipeline will occur very violently; a mixture of sand and water flows in the form of a violent "jet" (Fig. 2.11), as revealed by visualization observations, see for example Mao (1986).

This scour process is termed the **tunnel erosion** (Leeuwenstein, Bijker, Peerbolte and Wind (1985), Hansen, Fredsøe and Mao (1986)).

In practice, the tunnel erosion can occur in various other situations as well, such as that below a protection mattress covering a pipeline, that below a solid subsea structure supported by piles, and that below the cone-shaped protection structure at the bottom of a pile, to give but a few examples.

The tunnel erosion is "relieved" by the decrease of the gap-flow velocity, as the gap becomes larger and larger due to the scour. This stage is followed by the stage called the lee-wake erosion, which is described in the following section.

2.3 Two-dimensional scour

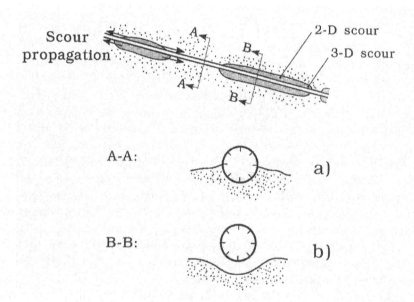

Figure 2.12: General scour picture around a pipeline.

2.3. TWO-DIMENSIONAL SCOUR

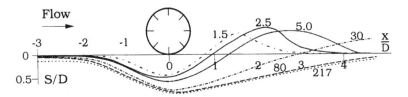

Figure 2.13: Scour development. Times in minutes. $\theta = 0.098$. Current. Mao (1986).

Scour below pipelines in the field occurs in a three-dimensional fashion: As pointed out in the previous section, the scour breaks out underneath the pipe locally, and it propagates along the length of the pipeline in both directions, as sketched in Fig. 2.12. The scour holes formed in this way are interrupted by stretches, called span shoulders, where the pipe obtains its support, Section A-A in Fig. 2.12.

However, after the process has reached a reasonably developed stage, the scour in the middle part of a scour hole can be considered as a two-dimensional process. It may be noted that a truly two-dimensional scour process can be achieved in flume tests where a rigidly fixed pipe (placed across the total width of the flume) is used to simulate a unit segment of a pipeline (Kjeldsen, Gjorsvik, Bringaker and Jacobsen (1973), Lucassen (1984), Mao (1986) and Sumer and Fredsøe (1990) among others).

2.3.1 Lee-wake erosion

In the case of the two-dimensional scour below the pipeline, the previously described tunnel erosion is followed by the stage called the **lee-wake erosion**, as mentioned in the preceding paragraphs.

Fig. 2.13 presents the results of a typical scour test where the pipe is rigidly fixed with initially a zero gap, and exposed to a steady current. The figure illustrates how the scour process evolves with time. The dotted line in the figure represents the equilibrium scour profile attained.

As seen, the scour occurs extremely fast at the beginning (tunnel erosion). As a result, a dune begins to form at the downstream side of the pipe. However, this dune gradually migrates downstream, and eventually may disappear as the scour progresses. Apparently, from the equilibrium profile in

Fig. 2.13, there will be more scour at the downstream side of the pipe than at the upstream side of it, resulting presumably in a steep upstream slope and a more gentle downstream slope.

Basically the scour at the stage of the lee-wake erosion is governed by the vortex shedding (Figs. 2.14 and 2.15), and the scour characteristics are controlled by the lee-wake of the pipe eventually (Sumer, Jensen, Mao and Fredsøe, 1988 a): When the gap between the pipeline and the bed reaches a certain value (due to scour), the vortex shedding will begin to occur (Sumer et al., 1988 a and Jensen, Sumer, Jensen and Fredsøe, 1990). The vortices shed from the bed side of the pipe sweep the bed, as they are convected downstream (Figs. 2.14 and 2.15). Bed shear stress measurements show that the Shields parameter can easily be raised up to $O(4)$ times momentarily during these periods (see Fig. 2.16 taken from Sumer, Chua, Cheng and Fredsøe, 2002), indicating that the sediment transport at the lee side of the cylinder will increase tremendously due to this action. This will presumably result in the lee-wake erosion.

Fig. 2.17 presents the results of a test where the spectral density distributions of the streamwise component of the velocity were measured near the upper edge of the pipe, as the scour below the pipe develops. In the figure f = the vortex shedding frequency, and U = the depth-averaged mean flow velocity.

As seen from this figure, vortex shedding, which is characterized by a dominant peak in the spectral distribution with a Strouhal number $St(= fD/U) = 0.2$, is present in the wake of the pipe, starting from rather early stages of the scour process. Fig. 2.17 indicates that vortex shedding is established within the first 15 minutes. Subsequently, the scour downstream of the pipe occurs under the action of the organized wake flow, namely an agglomeration of separation vortices that are shed from the pipe and steadily

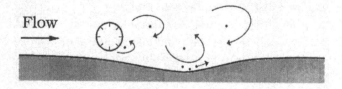

Figure 2.14: Sediment motion caused by vortex passing overhead. Sumer et al. (1988 a).

2.3. TWO-DIMENSIONAL SCOUR

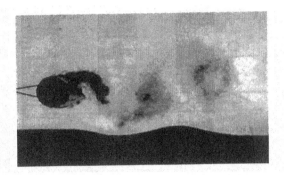

Figure 2.15: Organized lee-wake flow with aggregate of shed vortices. Sumer et al. (1988 a).

convected downstream. (A detailed account of vortex shedding and the flow around a circular cylinder placed near a wall is given in Sumer and Fredsøe, 1997).

In the test presented in Fig. 2.17, the value of the undisturbed-flow Shields parameter was maintained at $\theta = 0.05$, just above the critical value for incipient sediment transport, For larger values of θ, the vortex shedding would be established at even earlier stages of the scour process. Therefore, in such situations, the full development of the downstream scour can, for all practical purposes, be considered to occur under the action of the organized wake flow.

From the preceding measurements, and from several other lines of evidence, Sumer et al. (1988 a) concluded that

1. the vortex shedding is present in the lee-wake from rather early stages of the scouring process; and

2. the scour downstream of the pipe (the lee-wake erosion) should be governed by this organized flow eventually.

As mentioned earlier, the scour process finally reaches a steady state, the **equilibrium stage** (Fig. 2.13, the dotted-line profile). The equilibrium stage is reached when the bed shear stress along the bed underneath the pipe becomes constant and equal to its undisturbed value, namely

$$\tau = \tau_\infty \tag{2.12}$$

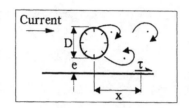

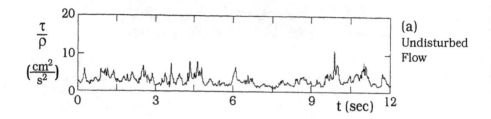

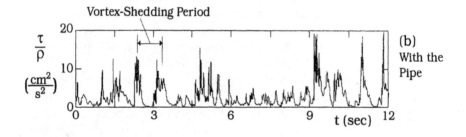

Figure 2.16: Time series of bed shear stress (a) without and (b) with the pipe, $e/D = 0.95$ and $x/D = 1.6$. Sumer et al. (2002).

in which the first term also includes the effect of a large, local, bed slope. Obviously, the sediment transport will be the same at all sections over the reach of the scour hole, and therefore the amount of sediment which enters the scour hole will be identical to that leaving the scour hole, when this stage is reached.

In this context, the velocity measurements of Jensen et al. (1990) (where the "frozen" scoured bed profiles were used to represent different stages of the scour process) are quite indicative:

Fig. 2.18 shows the results of Jensen et al.'s (1990) velocity measurements very close to the bed. The figure shows the following.

2.3. TWO-DIMENSIONAL SCOUR

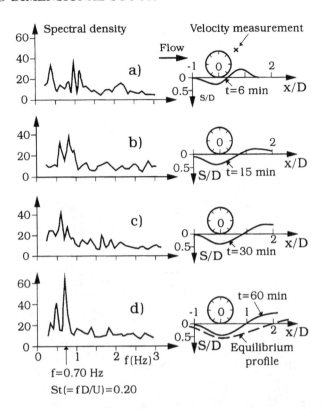

Figure 2.17: Spectral distribution of near-pipe velocity. $\theta = 0.05$. Sumer et al. (1988 a).

1. While the velocity below the pipe is increased tremendously at the initial stage of the scour process (Profile II), it eventually becomes practically identical to the undisturbed flow velocity, as the scour approaches towards its equilibrium stage (Profile V),

2. This suggests that, at the equilibrium stage, the sediment transport at all sections over the bed would occur at the same rate (disregarding the effect of gravity at large, local, slopes to a first approximation).

3. This in turn implies that the scour process would stop.

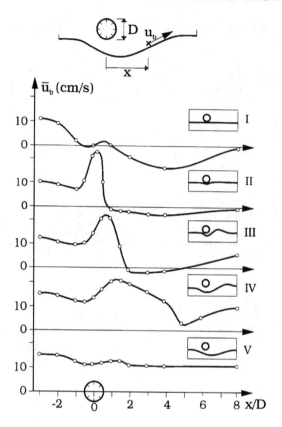

Figure 2.18: Near-bed velocity as scour develops. Jensen et al. (1990).

2.3.2 Scour depth

The scour depth develops towards the equilibrium stage through a transition period, as illustrated in Fig. 2.19 for a pipe rigidly placed on a bed with initially a zero gap. The depth corresponding to the fully-developed stage is called the **equilibrium scour depth**.

This section will focus on the equilibrium scour depth.

Scour depth in steady currents

Scour depth in this case has been studied extensively (Chao and Hennessy

2.3. TWO-DIMENSIONAL SCOUR

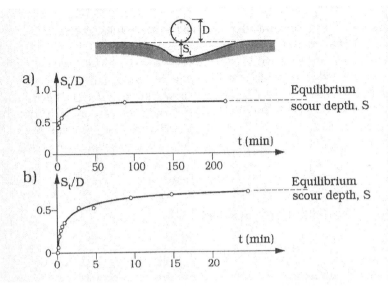

Figure 2.19: Time development of scour depth. (a) Current, $\theta = 0.098$ (Mao, 1986). (b) Wave, $\theta = 0.035$, $KC = 27$ (Fredsøe et al., 1992).

(1972), Kjeldsen et al. (1973), Littlejohns (1977), Herbich (1981), Bijker and Leeuwenstein (1984), Lucassen (1984), Leeuwestein et al. (1985), Herbich (1985), Herbich, Schiller, Watanabe, Dunlap (1984), Bijker (1986), Ibrahim and Nalluri (1986), Mao (1986), Kristiansen (1988) and Kristiansen and Tørum (1989)).

Kjeldsen et al. (1973) were the first to establish an empirical relation between the equilibrium scour depth, S, the pipe diameter, D, and the flow velocity, V

$$S = 0.972 \, (\frac{V^2}{2g})^{0.2} \, D^{0.8} \tag{2.13}$$

This is a dimensionally homogeneous equation. The relation in the preceding equation suggests that the nondimensional scour depth S/D is proportional to $\theta^{0.2}$:

$$\frac{S}{D} \propto \theta^{0.2} \tag{2.14}$$

in which θ = the Shields parameter, defined by Eq. 2.18. It may be noted that the scour in Kjeldsen et al.'s study was for the live-bed situation ($\theta > \theta_{cr}$) in which θ_{cr} = the critical value of the Shields parameter corresponding to the initiation of the motion at the bed.

The exact flow picture created by the presence of the pipe actually depends on the following quantities: the pipe diameter, D, the flow velocity, V (often taken as the undisturbed flow velocity at the center of the pipe), the kinematic viscosity of the fluid, ν, the pipe roughness, k_s, and the grain diameter of the bed material, d. From dimensional analysis, the nondimensional scour depth S/D can be found to depend on the following parameters:

$$\frac{S}{D} = f(k^*, Re, \theta) \qquad (2.15)$$

in which, k^* is the relative roughness,

$$k^* = \frac{k_s}{D} \qquad (2.16)$$

Re the Reynolds number,

$$Re = \frac{VD}{\nu} \qquad (2.17)$$

and θ the Shields parameter,

$$\theta = \frac{U_f^2}{g(s-1)d} \qquad (2.18)$$

in which U_f, the undisturbed bed shear velocity, may be calculated by the Colebrook-White formula

$$\sqrt{\frac{2}{f}} = \frac{V}{U_f} = 8.6 + 2.5 \ln\left(\frac{D}{2k_b}\right) \qquad (2.19)$$

in which V is the undisturbed flow velocity at the center of the pipe. The bed roughness k_b is usually taken as $2.5d$.

Of the three parameters in Eq. 2.15, the influence of k^* and Re appears through their effects on the downstream flow of the pipe. If the pipe is hydraulically rough, the wake flow is almost unaffected by Re, while for a hydraulically smooth pipe, some influence of Re is expected in the downstream vortex shedding pattern. Fig. 2.20 displays a plot of the data by Kjeldsen et al. (1973), Lucassen (1984), Mao (1986), and Kristiansen (1988) on the scour depth. It is seen in the figure that there is some weak influence of Re on the scour depth; a slight decrease in S occurs for Re around $10^5 - 3 \times 10^5$. For

2.3. TWO-DIMENSIONAL SCOUR

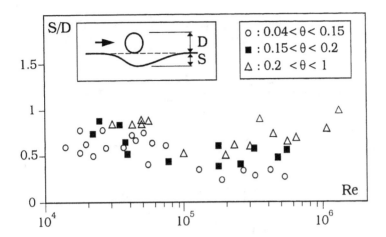

Figure 2.20: Data for equilibrium scour depth. Current. Live bed ($\theta > \theta_{cr}$). Compiled by Sumer and Fredsøe (1990).

a free circular cylinder, this coincides with the transition from subcritical to supercritical flow (Sumer and Fredsøe, 1997, p. 10). In this transition region, the vortex shedding becomes less pronounced, which might lead to a smaller lee-wake erosion and hence less scour depth.

As far as the influence of θ is concerned, this must be examined in two different categories: the clear-water case ($\theta < \theta_{cr}$), and the live-bed case ($\theta > \theta_{cr}$). In the clear-water case, the variation in scour depth with θ is more pronounced: as S/D increases from zero at very small θ values up to values of 0.4-1.0 when θ approaches the live-bed case. However, when the live-bed case is reached, very small variation with θ is observed (see the discussion in Section 1.4). Fig. 2.20 also reveals this. As already pointed out, Kjeldsen et al.'s study indicates that S/D increases with θ by a power of 0.2 (Eq. 2.14), while others simply disregard this weak variation.

Finally, from the data shown in Fig. 2.20, the mean value of the normalized scour depth and its standard deviation are found to be as follows:

$$\frac{S}{D} = 0.6 \text{ with } \frac{\sigma}{D} = 0.2 \tag{2.20}$$

which is valid for the case of live bed ($\theta > \theta_{cr}$).

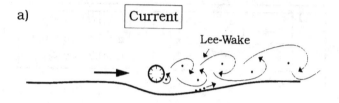

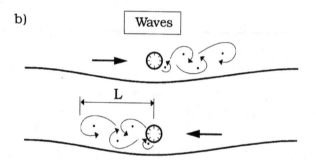

Figure 2.21: Lee-wake effect.

Scour depth in waves and tidal flows

This section considers the case where the flow attacks the pipe from both sides due to the near-bed flow induced by waves, or by slowly varying unsteady current conditions like a tidal current. The major difference between this case and the steady current case is that the downstream-wake system now occurs on both sides of the pipeline, as illustrated in Fig. 2.21.

The formation and extension of the wake pattern in oscillatory motion are governed by the Keulegan-Carpenter number, KC, defined earlier in Eqs. 2.9 and 2.10. As seen from the latter equation, small KC numbers mean that the orbital motion of the water particles is small relative to the total width of the pipe. When KC is very small, separation behind the pipe may not even occur. Large KC numbers, on the other hand, mean that the water particles travel quite large distances relative to the total width of the pipe, resulting in separation and probably vortex shedding. For very large KC

2.3. TWO-DIMENSIONAL SCOUR

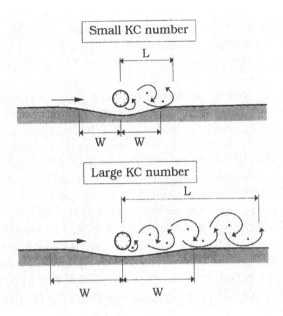

Figure 2.22: Streamwise extent of lee-wake corresponding to one half period of flow.

numbers ($KC \to \infty$), it may be expected that the flow for each half period of the motion resembles that experienced in a steady current.

For the case of the large KC numbers ($KC \gtrsim 6$), a vortex street is formed on the lee-side of the cylinder. The extension of this street, L (Fig. 2.21 b) is increased linearly with KC, corresponding to

$$\frac{L}{D} = 0.3 \, KC \qquad (2.21)$$

This equation is based on the flow visualization study described in Jensen, Jensen, Sumer and Fredsøe (1989). A detailed account of the vortex-flow regimes around a cylinder placed near a wall, and subject to an oscillatory flow is given by Sumer et al. (1991); see also Sumer and Fredsøe (1997), and, for wall-free cylinders, Williamson (1985).

It turns out from the preceding equation that the length of the part of the bed that is exposed to the lee-wake is dependent on KC. The larger the KC number, the larger the streamwise extent of the area affected by the lee-wake during one half-period of the flow, as sketched in Fig. 2.22. Because the flow erodes more heavily in the wake region, as described previously, we now obtain more gentle slopes with increasing KC in the scour hole on both sides of the pipe.

When the slopes of the scour holes become more gentle, the flow region below the pipe will be less protected against the outer flow, so the flow velocities below the pipe will increase, resulting in more scour below the pipe. Even a simple model like that of Hansen, Fredsøe and Mao (1986) predicts this increase in flow velocities with increasing width of scour hole.

From the preceding considerations, it is expected that the scour depth will increase significantly with increasing KC number. For large values of KC, the equilibrium scour depth is expected to approach a constant value, considering the finite lifetime of the lee-wake vortices. Experiments show that the organized structures in the wake can be detected at downstream distances of ~100 diameters from the pipe (Jensen, 1987; Antonia, Browne and Bisse, 1987). Taking $2a$ ~$100D$, then the corresponding KC number (Eq. 2.10) will be ~ 300. So, KC ~ 300 can be considered to be the Keulegan-Carpenter number beyond which the nondimensional scour depth S/D remains practically constant. For marine pipelines exposed to waves, KC is below ~ 100. However, higher values of KC help illustrate the basic scour mechanism, and include the tidal current case.

The variation of the scour depth with KC has been investigated by Sumer and Fredsøe (1990). Their result is shown in Fig. 2.23. The figure includes two sets of data: Lucassen's (1984) data, and Sumer and Fredsøe's (1990) data. Lucassen's data, given in dimensional form in the original reference and not investigated systematically in dimensionless terms, were recast in terms of KC and θ by Sumer and Fredsøe (1990), and plotted along with their own data. The ranges of various quantities in Lucassen's experiments are $KC = 2 - 25$; $Re = 2.5 \times 10^3 - 4 \times 10^4$; $D = 25 - 180$ mm with $d_{50} = 0.1$ mm in the main experiments, and 0.22 mm in their supplementary experiments. Here Re is the Reynolds number defined by

$$Re = \frac{U_m D}{\nu} \qquad (2.22)$$

The corresponding ranges in Sumer and Fredsøe's experiments are $KC = 2 - 1000$; $Re = 2.5 \times 10^3 - 1.3 \times 10^4$; $D = 10 - 50$ mm with $d_{50} = 0.18$, 0.36 and 0.58 mm. Both experiments involve only the live-bed conditions ($\theta > \theta_{cr}$). The Shields parameter, θ, is calculated by Eq. 2.18 where U_f is

2.3. TWO-DIMENSIONAL SCOUR

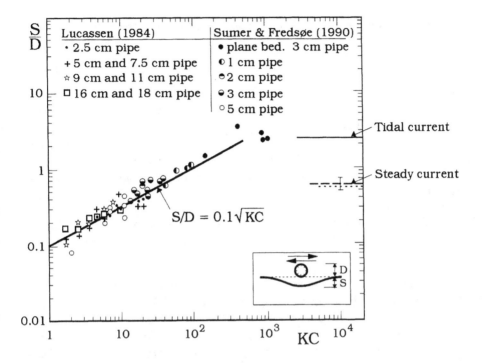

Figure 2.23: Equilibrium scour depth. Wave. Live bed ($\theta > \theta_{cr}$). Sumer and Fredsøe (1990).

now replaced by U_{fm}, the maximum friction velocity, calculated from

$$U_{fm} = \sqrt{\frac{f_w}{2}} U_m \qquad (2.23)$$

in which f_w = the wave friction coefficient (see, for example, Fredsøe, 1984).

Fig. 2.23 shows the variation of scour depth (normalized by the pipe diameter) with respect to KC. The figure also includes the steady current result. The latter is calculated from the data depicted in Fig. 2.20, and the corresponding dashed line and the dotted line are Hansen et al.'s (1986) theoretical solutions for $\theta = 0.1$ and 0.2, respectively.

Fig. 2.23 clearly demonstrates that the non-dimensional equilibrium scour depth correlates remarkably well with the KC number. The results show that the scour depth increases with KC, as argued in the preceding paragraph.

The scour data in the preceding figure can be represented by the following least-square fit

$$\frac{S}{D} = 0.1\sqrt{KC} \qquad (2.24)$$

(for the case of live bed $\theta > \theta_{cr}$). The preceding relationship has been confirmed later by Gokce and Gunbak's (1991) and Cevik and Yuksel's (1999) experimental data (although Cevik and Yuksel's study has indicated somewhat smaller scour depths for smaller KC, probably due to smaller values of the Shields parameter) and Hansen's (1992) numerical results.

Sumer and Fredsøe (1990) have further discussed the following aspects of the problem:

1. The scour in the case of the live bed is a weakly varying function of θ. The existing data seems to indicate a rather weak increase in S/D with θ.

2. They argued that, similar to the case of the steady current, a small decrease in the scour depth might be expected for the critical range of Re ($Re \sim 3 \times 10^5$). However, it was not possible to verify this by the available data, which covered only the range $Re = 2.5 \times 10^3 - 4 \times 10^4$. (Cevik and Yuksel (1999) have plotted their data as a function of the so-called period parameter $\beta = Re\ /KC$, and they have found that S/D decreases with increasing β. This variation in S/D may be due to the variation in KC rather than in β. Unless the influence of β is investigated in a systematic way in which KC is maintained constant while Re is changed, any results would be inconclusive. It may be noted that the definition of β in the latter study is written inadvertently as the inverse of $\beta = Re\ /KC$).

3. The effect of the presence of bed ripples was also discussed. They obtained exactly the same relative scour depth for different pipe diameters but for the same KC numbers; whereas, in these tests, the dimensions of the ripples with respect to the pipe diameter changed considerably. This confirms that the effect of ripples is not essential for the scour process. One reason for the ripple effect to be small is that presumably no ripples are present in the neighbourhood of the pipeline, where the flow field disturbed by the pipe prevents ripple formation.

Example 2 *A numerical example for the prediction of scour depth in waves*

2.3. TWO-DIMENSIONAL SCOUR

Consider a 30 cm diameter pipeline laid on the seabed. Predict the scour depth when this pipeline is exposed to waves with a period of $T_w = 10$ s and a wave height of $H = 2$ m. The water depth is $h = 10$ m. The sand size is 0.2 mm.

1. Calculate the quantity L_0 (the deep-water wave length)

$$L_0 = \frac{gT_w^2}{2\pi} = \frac{9.81 \times 10^2}{2\pi} = 156 \text{ m}$$

2. Calculate the parameter h/L_0

$$\frac{h}{L_0} = \frac{10}{156} = 0.064$$

3. From the wave tables for sinusoidal waves

$$\sinh(kh) = 0.733 \text{ for } \frac{h}{L_0} = 0.064$$

4. Calculate the amplitude of the orbital motion of water particles at the seabed, assuming that the small-amplitude sinusoidal wave theory is applicable (Appendix A)

$$a = \frac{H}{2}\frac{\cosh(k(z+h))}{\sinh(kh)} = \frac{H}{2}\frac{\cosh(k(-h+h))}{\sinh(kh)} = \frac{H}{2}\frac{1}{\sin(kh)} = \frac{2}{2}\frac{1}{0.733} =$$

$$= 1.36 \text{ m}$$

Here z = the vertical distance measured from the mean water level which is put equal to $z = -h$ (the seabed). The maximum value of the velocity at the bed is (Appendix A)

$$U_m = \frac{\pi H}{T_w}\frac{\cosh(k(z+h))}{\sinh(kh)} = \frac{\pi \times 2}{10}\frac{1}{0.733} = 0.86 \text{ m/s}$$

5. Check if the sinusoidal theory is applicable

$$U \text{ (the Ursell parameter)} = \frac{HL^2}{h^3} < 15$$

in which L, the wave length, is $L = L_0 \tanh(kh)$. (Otherwise, use the cnoidal theory). From the wave tables

$$\tanh(kh) = 0.591 \text{ for } \frac{h}{L_0} = 0.064$$

Therefore

$$L = 156 \times 0.591 = 92 \text{ m}$$

and then

$$U = \frac{HL^2}{h^3} = \frac{2 \times 92^2}{10^3} = 17$$

which is only slightly larger than 15. Therefore we may assume that the sinusoidal theory still is applicable.

6. Calculate the Keulegan-Carpenter number at the seabed

$$KC = \frac{2\pi a}{D} = \frac{2 \times \pi \times 1.36}{0.3} = 28.5$$

7. Predict the scour depth from Eq. 2.24

$$\frac{S}{D} = 0.1\sqrt{KC} = 0.1 \times \sqrt{28.5} = 0.53$$

or

$$S = 0.53 \times 0.30 = 0.16 \text{ m}$$

8. Calculate the Shields parameter

$$\theta = \frac{U_{fm}^2}{g(s-1)d} = \frac{\frac{f_w}{2}U_m^2}{g(s-1)d} = \frac{\frac{0.004}{2} \times 0.86^2}{9.81 \times (2.65-1) \times 0.0002} = 0.46$$

where f_w is calculated from

$$f_w = 0.035 RE^{-0.16}$$

as 0.004 (Fredsøe and Deigaard, 1992, p. 29), assuming that the bed is acting as a smooth wall ($dU_{fm}/\nu \lesssim 10$). Here, $RE = aU_m/\nu$, the wave-boundary-layer Reynolds number. As seen, the Shields parameter, $\theta = 0.46$, is larger than the critical value, $\theta_{cr} \simeq O(0.05)$, i.e., the bed is live, therefore the equation used to calculate the scour depth (Eq. 2.24) is valid.

2.3. TWO-DIMENSIONAL SCOUR

Scour depth in irregular waves. Sumer and Fredsøe (1996) have studied experimentally the influence of irregular waves on scour. A measured in-situ water elevation spectrum for the North Sea storm conditions was used as the control spectrum to produce the wave-generator signal. (This spectrum is well described by the JONSWAP wave spectrum).

The KC number in the case of irregular waves may be defined in several ways, such as $KC = U_m T_z/D$; $U_m T_s/D$; $U_m T_p/D$; $U_s T_z/D$; $U_s T_s/D$; $U_s T_z/D$. Here U_m is defined by

$$U_m = \sqrt{2}\sigma_U \quad (2.25)$$

in which σ_U = the r.m.s. value of the orbital velocity U at the bed

$$\sigma_U^2 = \int_0^\infty S(f) df \quad (2.26)$$

in which $S(f)$ = the power spectrum of U, and f = the frequency. The quantity U_s is

$$U_s = 2\sigma_U \quad (2.27)$$

and may be interpreted as the "significant" velocity amplitude, analogous to the half of the significant wave height. The periods, T_z; T_s; and T_p, on the other hand, are the mean zero-upcrossing period, the significant wave period, and the peak period ($= 1/f_p$), respectively (see, for example Sumer and Fredsøe, 1997, p. Chapter 7).

The authors have measured the scour depth, $(S/D)_{irregular}$, and compared it with that calculated from Eq. 2.24, namely $(S/D)_{regular} = 0.1\sqrt{KC}$ where KC was calculated in six different ways as described in the preceding paragraph. Apparently, the definition $KC = U_m T_p/D$ has given the best representation. Fig. 2.24 displays the results where KC is calculated in this way, i.e., $KC = U_m T_p/D$. It may be noted that the Keulegan-Carpenter number defined in this way reduces to the ordinary KC number in the case of regular waves, because $\sqrt{2}\sigma_U \to U_m$, and $T_p \to T_w$ in this latter case.

To conclude, the scour depth in the case of irregular waves can be predicted using the empirical expression given in Eq. 2.24 provided that KC is calculated by $KC = U_m T_p/D$ in which $U_m = \sqrt{2}\sigma_U$.

Scour depth in combined waves and current. Scour depth in the case of combined waves and current has been studied by Lucassen (1984), Bernetti, Bruschi, Valentini and Venturi (1990), Hansen (1992) and Sumer and Fredsøe (1996). Lucassen carried out a series of tests in combined waves

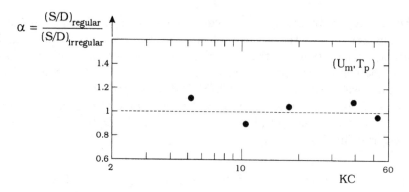

Figure 2.24: Comparison of measured $\frac{S}{D}$ in irregular waves with $\frac{S}{D} = 0.1\sqrt{KC}$. Sumer and Fredsøe (1996).

and currents (but with rather limited ranges of the governing parameters), while Bernetti et al. and Hansen studied the problem numerically. We shall return to these studies later in this subsection.

In Sumer and Fredsøe's work (1996), experimental data have been obtained, covering a rather wide range of KC (from 5 to about 50) and the full range of $U_c/(U_c + U_m)$ (namely, from 0 to 1) in which U_c = the undisturbed current velocity at the center of the pipe. Sumer and Fredsøe's (1996) work indicated that the scour depth may increase or decrease when a current is superimposed on waves, depending on the values of KC and $U_c/(U_c + U_m)$; see Fig. 2.25.

First of all, obviously S/D should go to $0.1\sqrt{KC}$ as $U_c/(U_c + U_m) \to 0$ (the waves-alone case, Eq. 2.24), and it should approach the value 0.6 ± 0.1 as $U_c/(U_c + U_m) \to 1$ (the current-alone case, Eq. 2.20). The data in Fig. 2.25 reveals this.

Secondly, there is a general trend that, irrespective of the KC number, the scour depth first decreases (slightly) as the current velocity is increased from zero. This is linked to the slight displacement of the upstream part of the scour hole in the direction of flow when a current is superimposed on waves so that the flow attack below the pipe will be "weaker". The aforementioned displacement is due to the smaller streamwise extent of the upstream lee wake when the current is superimposed.

Thirdly, Fig. 2.25 shows that the scour depth approaches its steady-

2.3. TWO-DIMENSIONAL SCOUR

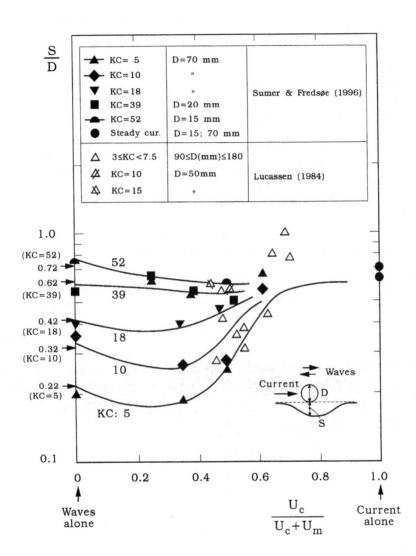

Figure 2.25: Equilibrium scour depth. Combined wave and current. Live bed ($\theta > \theta_{cr}$). Arrows on vertical axis from $\frac{S}{D} = 0.1\sqrt{KC}$ (waves alone). Sumer and Fredsøe (1996).

current value only after $U_c/(U_c + U_m)$ reaches the value of about 0.7. This is because, only after that point is reached, the lee-wake on the upstream side of the pipe completely disappears, and hence its influence on the scour process ceases to exist.

The data in Fig. 2.25 can explain the conflicting results of the previous authors, namely while Lucassen's experiments showed that the scour depth increased with increasing current, Hansen's numerical results indicated the opposite; this is because the range of $U_c/(U_c + U_m)$ in Hansen's study was from 0 to about 0.3, and therefore the scour depth decreased with increasing current, as revealed by Fig. 2.25, whereas the parameter $U_c/(U_c + U_m)$ in Lucassen's study ranged from about 0.4 to 0.7, and hence the scour depth increased with increasing current, Fig. 2.25.

Finally, although no direct comparison was made in the paper by Sumer and Fredsøe (1996), the numerical results of Bernetti et al. (1990) are in qualitative agreement with the experimental study of Sumer and Fredsøe (1996).

Based on the data in Fig. 2.25, the following empirical expressions can be worked out (Sumer and Fredsøe, 1996)

$$\frac{S}{D} = \frac{S_{cur}}{D} F \qquad (2.28)$$

in which S_{cur} = the scour depth in the current-alone case (Eq. 2.20), and F = a function of KC and $U_c/(U_c+U_m)$, and given by the following empirical equations:

1. When $0 < U_c/(U_c + U_m) \leq 0.7$:

$$F = \frac{5}{3}(KC)^a \exp(2.3b) \qquad (2.29)$$

2. When $0.7 < U_c/(U_c + U_m) \leq 1$:

$$F = 1 \qquad (2.30)$$

The coefficients a and b in Eq. 2.29, on the other hand, are given as follows:

1. For $0 \leq U_c/(U_c + U_m) \leq 0.4$:

$$a = 0.557 - 0.912(\frac{U_c}{U_c + U_m} - 0.25)^2 \qquad (2.31)$$

2.3. TWO-DIMENSIONAL SCOUR

$$b = -1.14 + 2.24(\frac{U_c}{U_c + U_m} - 0.25)^2 \tag{2.32}$$

2. For $0.4 < U_c/(U_c + U_m) \leq 0.7$:

$$a = -2.14\frac{U_c}{U_c + U_m} + 1.46 \tag{2.33}$$

$$b = 3.3\frac{U_c}{U_c + U_m} - 2.5 \tag{2.34}$$

Caution must be exercised in the implementation of the preceding empirical equations when KC is larger than the upper boundary of the KC range tested in the study, namely when $KC \gtrsim 50$. Also, it may be noted that the preceding results have been obtained from experiments where the live-bed $(\theta > \theta_{cr})$ conditions prevailed.

Scour depth in shoaling conditions. Cevik and Yuksel (1999) have studied the effect of shoaling conditions on the scour depth. They have employed three kinds of bottom slopes; the horizontal bottom, and two beach profiles with slopes 1/5 and 1/10.

In shoaling conditions, waves break. The breaking process creates strong downward directed flows that erode the bed. For example, in the case of plunging wave breaking, the plunging breaker may penetrate down to the bed, and mobilize sediment at the bed. This will presumably lead to scour at the pipe if the breaker occurs in the vicinity of the pipe.

There are basically three kinds of breakers: spilling breakers, plunging breakers and surging breakers (Fig. 8.4 in Chapter 8) (see, for example, Fredsøe and Deigaard, 1992). Obviously, the type of breaker may be an essential factor, influencing the scour. The parameters which govern the breaker type and its characteristics (such as the wave height, H_b, and water depth, h_b, at the breaking point) are

$$\frac{H_0}{L_0}, m \tag{2.35}$$

in which H_0 is the deep-water wave height, and L_0 is the deep-water wave length of the incident waves, and m is the beach slope (see, for example, Fredsøe and Deigaard, 1992, p. 86). This is when there is no pipe present. When there is a pipe present, the broken waves will be reflected to some extent from the pipe. Therefore, an additional parameter, x/L_0, may be involved with regard to the breaker type and its characteristics. Here, x is the distance of the breaking point from the pipe.

The mechanism regarding the scour processes around a pipeline under shoaling conditions is still unknown today. Cevik and Yuksel's (1999) study essentially views the entire process as a black box. Nevertheless, the latter authors have found that the scour depth in the case of the shoaling conditions is always larger than that in the case of the horizontal bottom under the same incident wave conditions, namely $S/S_0 > 1$. (In some individual cases, S/S_0 can be as much as $O(4)$). This is an important result in the sense that caution should be exercised for scour protection in areas where shoaling conditions prevail. Although no information has been given in Cevik and Yuksel (1999) about the width of the scour hole, an increase, to the same degree as in the scour depth, may be expected. Also, the location of the maximum scour depth is another important parameter. Cevik and Yuksel's analysis indicates that the maximum scour depth occurs always at the offshore side of the breaking depth.

Effect of other factors on scour depth

1. Effect of pipe roughness. In typical field situations, the pipe surface may be covered by marine growth to such an extent that the pipe surface acts as a rough wall ($k_s/D > 3 \times 10^{-3}$). In this case, the effect of the Reynolds number on vortex shedding practically disappears (Sumer and Fredsøe, 1997, p. 13).

In order to see that the scour process is not essentially affected by the pipe roughness, Sumer and Fredsøe (1990) repeated some of the tests with a pipe with a relative roughness of $k/D = 0.1$ (or, alternatively, $k_s/D = 0.2$, taking $k_s = 2k$, a roughness much larger than 3×10^{-3}). Their results showed that there hardly was any difference between the scour process on the case of the smooth pipe and that in the case of the rough pipe.

2. Effect of the Shields parameter. As stated earlier (Section 1.4), the effect of the Shields parameter, θ, is expected to be quite weak when the live-bed scour is considered. However, it becomes increasingly important when θ is reduced below its critical value. Fig. 2.26 illustrates this effect very clearly (The data is from Mao, 1986, and Hansen, 1992; see Section 2.7.1 for the details regarding Hansen's numerical model). As seen, the effect

2.3. TWO-DIMENSIONAL SCOUR

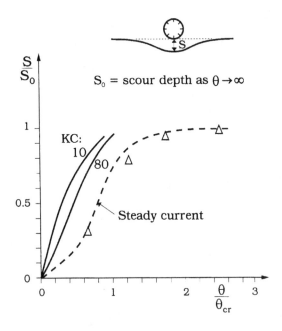

Figure 2.26: Effect of θ. Solid line: Numerical model (Hansen, 1992). Dashed line: Experiment (Mao, 1986).

of θ should be taken into account when θ is smaller than the critical value of the Shields parameter, θ_{cr} (i.e., in the case of the clear-water scour).

3. Effect of pipe position in vertical. The pipe is seldom in continuous contact with the bed, but is at some locations placed above the bed due to bed irregularities, and at other locations pressed down into the bed. Furthermore, the pipe will sag into the scour hole during the self-burial process, as will be discussed in the next section.

Fig. 2.27 shows the results of wave experiments reported in Sumer and Fredsøe (1990). Hansen et al. (1986) report similar results for steady currents. The quantity e in Fig. 2.27 is the clearance between the pipe and the bed. In the tests corresponding to the negative values of e (buried pipe), no external disturbance is applied to initiate the scour process.

The following conclusion is drawn from the figure. The smaller the value of e, the larger the influence of the presence of the pipe on the equilibrium

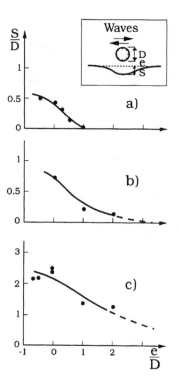

Figure 2.27: Effect of pipe position. (a) $KC = 11$, $\theta = 0.19$. (b) $KC = 27$, $\theta = 0.035$. (c) $KC = 900$, $\theta = 2.4$. Sumer and Fredsøe (1990).

scour depth.

It is interesting to note that no scour occurs if the pipe is placed above $e/D = 1$ for moderate KC numbers. For increasing KC values, however, scour can be generated for even larger values of e/D (approximately up to $e/D \sim 3$). This can be explained in terms of the increased extent of the lee-wake at larger KC numbers.

Finally, from the data reported in Hansen et al. (1986) and that displayed in Fig. 2.27, the scour depth can, to a first approximation, be approximated by

$$\frac{S}{D} = 0.625 \exp(-0.6\frac{e}{D}); \qquad -0.25 \lesssim \frac{e}{D} \lesssim 1.2 \qquad (2.36)$$

2.3. TWO-DIMENSIONAL SCOUR

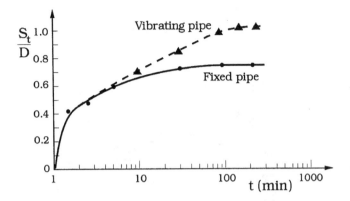

Figure 2.28: Scour development with time. $\theta = 0.1$. V_r (reduced velocity) = 7.2. Sumer et al. (1988 b).

for the case of steady currents, and by

$$\frac{S}{D} = 0.1\sqrt{KC}\exp(-0.6\frac{e}{D}); \quad 0 \lesssim \frac{e}{D} \lesssim 2 \qquad (2.37)$$

for the case of waves. The above equations are valid for the case of live bed ($\theta > \theta_{cr}$).

4. Effect of vibrations. When a pipeline is laid on the seabed, suspended spans of the pipeline will develop along the length of the pipeline, as will be discussed in greater detail in the next section under Three-Dimensional Scour.

These suspended spans may undergo flow-induced vibrations (Sumer and Fredsøe, 1997, p. 454). In this case, it is interesting to know to what extent the vibrations influence the scour. This problem has been studied by Sumer, Mao and Fredsøe (1988 b) and Kristiansen (1988) (see also Kristiansen and Tørum, 1989).

In these studies, a flexibly-mounted rigid pipe is employed to simulate a unit length of the pipe in the middle of the suspended span. The pipe is initially held stationary on or near the bed. Scour begins to occur below the pipe, as the flow develops (Fig. 2.28). Shortly after, a point is reached where the vibrations in the cross-flow directions begin to emerge (Fig. 2.29), and eventually reach a fully-developed stage with very large amplitudes. V_r, the

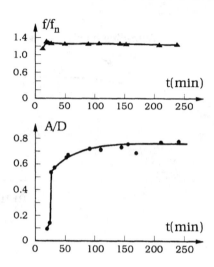

Figure 2.29: Frequency (f) and amplitude (A) of vibrations as scour develops. Sumer et al. (1988 b).

reduced velocity, in the caption of Fig. 2.28 is defined by $V_r = V/(Df_n)$ in which f_n is the natural frequency of the pipe.

Sumer et al.'s (1988 b) experiments showed the following.

There are three kinds of erosion. The first two are shown in Fig. 2.30, and are in addition to that experienced at a fixed pipe; one involves the erosion of the upstream face of the scour hole where the sand transport is, in the case of small values of the Shields parameter, in the form of bed load (Fig. 2.30 a), while the other involves the erosion of the bed just below the pipe where the entrainment of sand into the flow is caused by suspension (Fig. 2.30 b). Both types of erosion occur periodically, but in different phases. In the latter, the motion of the pipe towards the bed (not necessarily impinging on it) mobilizes the sand grains there, and then the sand grains are brought into suspension as the pipe motion reverses. In addition to these two kinds of erosion, the wake-induced erosion was observed to occur more violently in the case of the vibrating pipe.

Fig. 2.31 plots the normalized scour depth, S/D, against the initial gap ratio e/D, for both the fixed and the vibrating pipes. Note that the experiments with the negative values of e/D correspond to a pipeline which is in

2.3. TWO-DIMENSIONAL SCOUR

the process of sagging in a scour hole.

The figure shows that the scour is always larger in the case of the vibrating pipe than in the case of the fixed one. Also, the smaller the gap ratio e/D, the larger the effect of the pipe vibrations.

5. Effect of angle of attack. The angle of attack is also one of the influencing factors. Both experiments and theory indicate that the scour depth is reduced with decreasing angle of attack (Fig. 2.32). This may be attributed to the fact that the vortex shedding becomes less pronounced with decreasing angle of attack. Kozakiewicz, Fredsøe and Sumer (1995) report that, for a free cylinder, the lift force spectrum becomes broader, as the angle of attack is decreased, an indication of less organized vortex shedding with decreasing angle of attack.

6. Effect of multiple pipelines. When more than one pipeline are laid on the seabed (Fig. 2.33), the scour pattern as well as the maximum scour depth may change, depending on the number of pipelines and the spacing between them. This aspect of the problem has been studied by Westerhortmann, Machemehl and Jo (1992).

Fig. 2.34 shows the equilibrium scour profiles obtained in single and multiple pipeline situations for two selected cases. From the scour profiles, one can deduce that the maximum scour depth is not radically different in the three situations displayed in the figure, namely, in the case of a single pipe, in the case of two pipes, and in the case of three pipes with spacing equal to the pipe diameter.

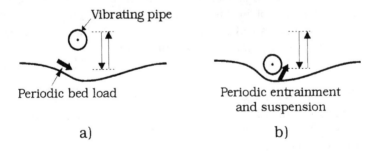

Figure 2.30: Two forms of vibration-induced erosion.

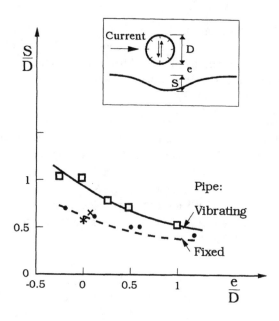

Figure 2.31: Scour depth. Vibrating pipe: V_r (reduced velocity) = 7.2. Sumer et al. (1988 b).

However, Westerhortmann et al.'s experiments with the pipes where the spacing between the pipes was equal to half the pipe diameter indicate that the maximum scour depth is generally reduced (between 5% and 35%) with respect to the case where the pipe spacing was equal to the pipe diameter. (Almost live-bed conditions prevailed in the experiments). The reduction in the scour depth can be explained in terms of the lee-wake erosion; the smaller spacing between the pipes will partially inhibit the shedding, and therefore reduce the effect of the lee wake on the scour depth.

7. Effect of armouring. In the case when the seabed is covered with sand containing shell fragments, the armouring effect may be expected, resulting in the reduction of the scour depth. This has been investigated by Sidek and Ibrahim (1992).

Fig. 2.35 shows the results of a typical experiment. The sand contained 10% shell fragments by weight. In the experiments, the mean grain diameter

2.3. TWO-DIMENSIONAL SCOUR

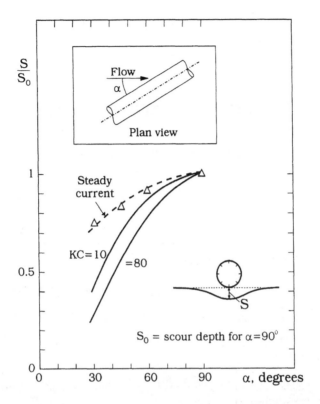

Figure 2.32: Effect of angle of attack. Solid line: Numerical (Hansen, 1992). Dashed line: Experiment (Mao, 1988).

was maintained unchanged with respect to the no-shell-fragment experiment so that comparison could be made on the same basis. Clearly, the effect is present; the scour is substantially reduced by the armouring effect.

8. Effect of cohesive sediment. Pluim-van der Velden and Bijker (1992) have studied the effect of cohesive sediment on scour in steady current. To this end, they have used three kinds of sediment: (1) sand; (2) natural sand-silt mixture; and (3) artificial sand-silt mixture (the sand was mixed with kaolinite in this latter case). The development of scour hole has been measured for these three kinds of sediment. Fig. 2.36 displays the results.

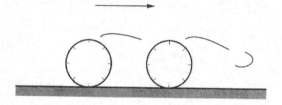

Figure 2.33: Multiple pipelines.

Pluim-van der Velden and Bijker (1992) draw the following conclusions:

1. In the case of the sand-silt mixture, an initial gap between the pipe and the bed was necessary to get scour started (cf., the mechanism of the onset of scour in the case of cohesionless sediment in Section 2.1).

2. The critical shear stress for a sand-silt mixture is higher than that for the sand bed.

3. Very high velocities are needed to start erosion in the case of the sand-silt mixture.

4. In the case of the sand-silt mixture, a W-shaped scour hole is obtained (Figs. 2.36 b and c).

5. Likewise in the case of the sand-silt mixture, no deposition occurs at the downstream side of the pipe.

6. Generally, in the case of the sand-silt mixture, the scour depth is smaller than in the case of the sand alone.

9. Effect of water depth. The effect of water depth on scour may be important when the water depth becomes comparatively small (in the case of river crossings, for example). This is basically because of the "blockage" effect. As the water depth decreases, more and more water will pass under the pipe, as revealed by the experiments of Chiew (1991 a, 1991 b). For example, the gap discharge can be as large as 60% of the incoming discharge when the water-depth-to-pipe-diameter ratio, h/D, becomes 2 for the value

2.3. TWO-DIMENSIONAL SCOUR

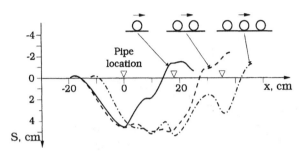

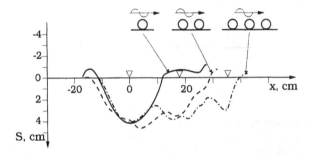

Figure 2.34: Effect of pipe spacing. Equilibrium profiles. Spacing between pipes is equal to pipe diameter. Westerhorstmann et al. (1992).

of the gap-to-diameter ratio, $e/D = 0.63$, whereas this figure is only about 10% for $h/D = 6$ (and for the same gap ratio). (Chiew, 1991 a).

This change in the gap discharge, and therefore in the gap velocity, has significant consequences for the scour process. This aspect of the problem has been investigated by Moncada-M. and Aguirre-Pe (1999). Fig. 2.37 displays the data obtained by the latter authors together with the data corresponding to the large-water-depth case. Moncada-M. and Aguirre-Pe studied the effect of water depth in terms of the Froude number

$$Fr = \frac{V}{\sqrt{gh}} \qquad (2.38)$$

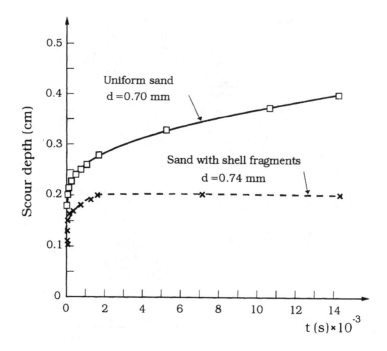

Figure 2.35: Effect of shell fragments in the sand. Sidek and Ibrahim (1992).

The Froude number may be involved because Fr governs the actual water depth at the pipe section, therefore the gap-flow discharge. (The depression in the water surface in an open channel with a structure (such as a step at the bottom) is controlled by Fr, see Henderson, 1966).

The following conclusions can be deduced from Fig. 2.37.

1. As seen from the figure, the scour depth increases with increasing Fr. This is because the flow velocity in the gap will increase with increasing Fr. This increased gap velocity will essentially generate a somewhat larger tunnel erosion (simply because the bed shear stress below the pipe becomes equal to that in the far field only with a larger scour depth), presumably leading to larger equilibrium scour depths with increasing Fr.

2. For very large Froude numbers, S/D can be a factor of 2-3 larger than that experienced in the case of large water depths, an increase in the

2.3. TWO-DIMENSIONAL SCOUR

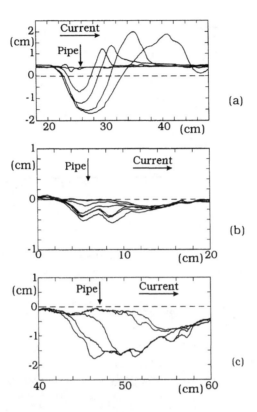

Figure 2.36: Scour development. (a) Sand. (b) Artificial silty sand. (c) Natural silty sand. Pluim-van der Velden and Bijker (1992).

scour depth which may prove very significant with regard to practical applications such as in the case of river crossings.

3. Finally, for large water depths, the scour depth approaches the value given in Eq. 2.20, namely

$$\frac{S}{D} \to 0.6 \text{ as } Fr \to 0 \qquad (2.39)$$

The effect of water depth on scour has been considered also by Chiew (1991 a, 1991 b). Chiew has proposed a calculation procedure for maximum

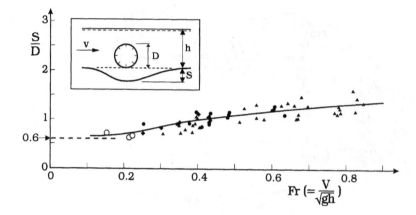

Figure 2.37: Effect of water depth. Moncada-M. and Aguirre-Pe (1999). o : Sumer and Fredsøe (1990).

scour depth, a procedure somewhat similar to that in Chao and Hennessy (1972) (see Section 2.7.1 below). The key element in Chiew's calculation procedure is an empirical relation between the discharge flow and the water-depth-to-pipe-diameter ratio. The scour depths predicted by Chiew using his calculation procedure have apparently been in agreement with the measurements.

2.3.3 Width of scour hole

Fig. 2.38 reproduces three equilibrium scour profiles reported in Sumer and Fredsøe (1990) for a broad range of the Keulegan-Carpenter number ($10 \lesssim KC \lesssim 1000$). As seen, the width of the scour hole increases with increasing KC, as anticipated (see the discussion in the previous section). The variation of the scour width with KC can, from the data in Fig. 2.38, be approximated by the following empirical relation:

$$\frac{W}{D} = 0.35\, KC^{0.65} \qquad (2.40)$$

in which W = the width measured from the center of the pipe to the end of the scour hole (see Fig. 2.38).

2.3. TWO-DIMENSIONAL SCOUR

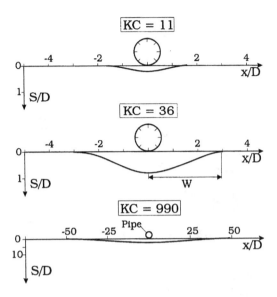

Figure 2.38: Equilibrium scour profiles. Sumer and Fredsøe (1990).

In a follow-up study, Sumer and Fredsøe (1996) have presented some further data; the focus was to study the influence of current in combined-waves-and-current environments, Fig. 2.39.

The data in Fig. 2.39 indicates the following.

1. The scour widths W_1 and W_2 (the upstream and downstream widths, respectively) reduce to that predicted by Eq. 2.40 as $U_c/(U_c+U_m) \to 0$ (the waves-alone case).

2. Furthermore, from Fig. 2.39, the net effect of superimposing a current on waves is to make the downstream width of the scour hole larger and the upstream width smaller (slightly), apparently due to the effect of the lee-wake.

3. It appears that the scour widths approach constant values for $U_c/(U_c+U_m)$ larger than 0.5-0.7, namely

$$\frac{W_1}{D} \to \sim 2; \text{ and } \frac{W_2}{D} \to \sim 4, \text{ as } U_c/(U_c+U_m) \to 1 \text{ (the current-alone case)}$$
(2.41)

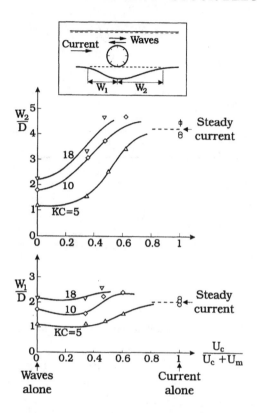

Figure 2.39: Equilibrium scour width. Combined wave and current. Sumer and Fredsøe (1996).

4. The above result is consistent with the corresponding findings related to the scour depth (see the previous subsection, and Fig. 2.25).

(It may be noted that the results summarized above correspond to the live-bed conditions, $\theta > \theta_{cr}$).

Finally, the scour width is influenced by various other factors such as the Shields parameter, the pipe position in the vertical, the vibrations of the pipeline, the angle of attack, the armouring, and the water depth.

Fig. 2.40 shows how the water depth influences the scour width, plotted in the same format as in Fig. 2.37. The general trend is that the scour width increases with increasing Froude number. The figure is reproduced

2.3. TWO-DIMENSIONAL SCOUR

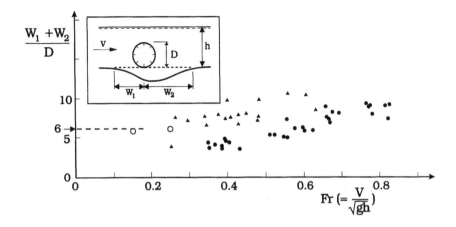

Figure 2.40: Effect of water depth. Moncada-M. and Aguirre-Pe (1999). o : Sumer and Fredsøe (1990).

from the work of Moncada-M. and Aguirre-Pe (1999). In the latter work, Sumer and Fredsøe's (1996) data has been plotted erroneously with the value $(W_1 + W_2)/D = 1$ (as the asymptotic value) when $Fr \to 0$. This has been corrected in Fig. 2.40. Moncada-M. and Aguirre-Pe (1999) also study the influence of the pipe position in the vertical on the scour width.

2.3.4 Time scale

The scour depth develops towards its equilibrium stage through a transitional period, as depicted in Fig 2.19. It is seen from the figure that the time variation of the scour depth can be approximately represented by the following relation

$$S_t = S(1 - \exp(-\frac{t}{T})) \qquad (2.42)$$

in which S = the equilibrium scour depth. The quantity T may be defined as the **time scale** of the scour process, representing the time period during which a substantial amount of scour develops (Section 1.3).

The time scale can be predicted from the scour-depth-versus-time information, either by calculating the slope of the line tangent to the $S_t(t)$ curve at $t = 0$ (see Fig. 1.1), or by integrating $S_t(t)$ over time.

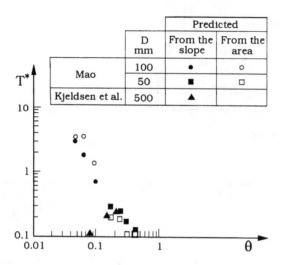

Figure 2.41: Time scale. Current. Live bed ($\theta > \theta_{cr}$). Fredsøe et al. (1992).

Fredsøe, Sumer and Arnskov (1992) investigated the time scale both in steady currents and in waves, employing the previously mentioned methods. The following paragraphs will summarize the results of this study.

Time scale in steady currents

On dimensional grounds, the time scale can be written as in the following nondimensional functional form

$$T^* = f(\theta) \qquad (2.43)$$

in which θ = the Shields parameter (Eq. 2.18), and T^* = the normalized time scale defined by

$$T^* = \frac{(g(s-1)d^3)^{1/2}}{D^2} T \qquad (2.44)$$

The preceding nondimensional formulation can also be obtained by normalizing the equation of sediment continuity.

The time scale obtained from the scour-depth-versus-time information given by Kjeldsen et al. (1973) and Mao (1986) is plotted in the preceding nondimensional form in Fig. 2.41. The ranges of various parameters are as

2.3. TWO-DIMENSIONAL SCOUR

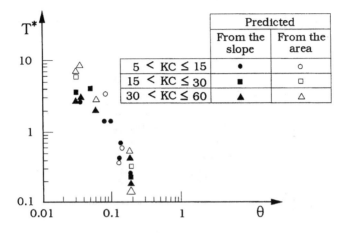

Figure 2.42: Time scale. Wave. Live bed ($\theta > \theta_{cr}$). Fredsøe et al. (1992).

follows. In Kjeldsen et al.'s (1973) study: $Re = 1.2 \times 10^5 - 2.2 \times 10^5$, $D = 500$ mm, and $d = 0.074$ mm, while in Mao's (1986) study: $Re = 3.1 \times 10^4 - 5 \times 10^4$, $D = 50$ mm, and 100 mm and $d = 0.360$ mm.

Fig. 2.41 shows that the normalized time scale T^* correlates quite well with the Shields parameter θ.

Furthermore, the figure shows that the larger the Shields parameter, the smaller the time scale. This is because the larger the Shields parameter, the larger the sediment transport due to scouring, and therefore the shorter the time period during which a substantial change in the scour depth will occur.

Time scale in waves

In the case of waves, one additional parameter, namely the Keulegan-Carpenter number, may be involved in the nondimensional formulation of the time scale (Eq. 2.43):

$$T^* = f(\theta, KC) \tag{2.45}$$

Fredsøe et al. (1992) analyzed the data obtained in the work of Sumer and Fredsøe (1990). The result is reproduced in Fig. 2.42 where the normalized time-scale data are plotted against the Shields parameter for different groups of the KC number. The Shields parameter, θ, is calculated by Eq. 2.18 where

CHAPTER 2. SCOUR BELOW PIPELINES

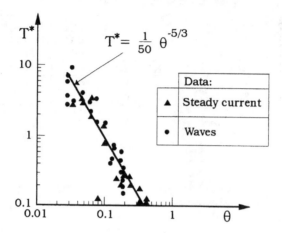

Figure 2.43: Time scale. All data (current/wave). Live bed ($\theta > \theta_{cr}$). Fredsøe et al. (1992).

U_f is replaced by U_{fm}, the maximum bed shear velocity, calculated from Eq. 2.23.

While the scour depth mainly depends on KC (Fig. 2.23), Fig. 2.42 indicates that the time scale is governed by only the Shields parameter.

In Fig. 2.43 are plotted the wave data (Fig.∞ 2.42) together with the steady-current data (Fig. 2.41). As seen, the two data sets collapse fairly well, indicating that the time scale does not "differentiate" whether the flow is a steady current or a wave. This may be attributed to the fact that the lee-wake erosion, the key element in the wave-induced scour, is insignificant at the initial stage of the scour process.

Finally, Fredsøe et al. (1992) give the following empirical expression representing the data in Fig. 2.43:

$$T^* = \frac{1}{50}\theta^{-5/3} \qquad (2.46)$$

Example 3 *A numerical example for the prediction of time scale in steady currents*

Given that $D = 30$ cm, $d_{50} = 0.5$ mm, what is the time scale of the scour process below a pipeline which is exposed to a mean current of $V = 0.6$ m/s perpendicular to the pipeline. The water depth is $h = 10$ m.

2.3. TWO-DIMENSIONAL SCOUR

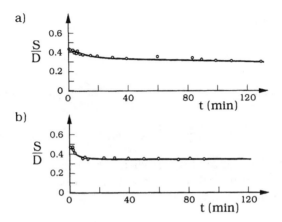

Figure 2.44: Time development of scour with sudden change in wave climate. Fredsøe et al. (1992).

1. Calculate the undisturbed bed friction velocity from

$$U_f = \frac{V}{2.5 \left[\ln(\frac{30h}{k_s}) - 1\right]}$$

in which $k_s = 2.5 d_{50}$:

$$U_f = \frac{0.6}{2.5 \times \left[\ln(\frac{30 \times 10}{2.5 \times (0.5 \times 10^{-3})}) - 1\right]} = 0.021 \text{ m/s}$$

2. Calculate the undisturbed Shields parameter:

$$\theta = \frac{U_f^2}{g(s-1)d_{50}} = \frac{0.021^2}{9.81 \times (2.65-1) \times (0.5 \times 10^{-3})} = 0.055$$

3. Calculate T^* from Eq. 2.46:

$$T^* = \frac{1}{50}\theta^{-5/3} = \frac{1}{50} \times (0.055)^{-5/3} = 2.51$$

4. Find T from Eq. 2.44:

$$T = \frac{D^2}{(g(s-1)d_{50}^3)^{1/2}}T^* =$$

$$= \frac{0.30^2}{(9.81 \times (2.65-1) \times (0.5 \times 10^{-3})^3)^{1/2}} \times 2.51 = 5020 \text{ s} \simeq 1.4 \text{ h}$$

Example 4 *A numerical example for the prediction of time scale in waves*

Calculate the time scale of scour when the pipeline in the previous example is exposed to waves with a period of $T_w = 10$ s and a wave height of $H = 2$ m. The water depth is $h = 10$ m.

1. From Example 2 in Section 2.3.2, the amplitude a and the maximum value of the velocity U_m of the orbital motion of water particles at the seabed are

$$a = 1.36 \text{ m, and } U_m = 2\pi a/T_w = 0.86 \text{ m/s}$$

2. Find the friction factor for the wave boundary layer, assuming that the bed is a rough boundary (from, for example, Fredsøe and Deigaard, 1992, p. 25):

$$f_w = 0.04\left(\frac{a}{k_s}\right)^{-\frac{1}{4}}, \quad \frac{a}{k_s} > 50$$

$$f_w = 0.007 \text{ for } \frac{a}{k_s} = \frac{a}{2.5d_{50}} = \frac{1.36}{2.5 \times (0.5 \times 10^{-3})} = 1088$$

3. Calculate the maximum bed friction velocity for the wave boundary layer:

$$U_{fm} = \sqrt{\frac{f_w}{2}}U_m = \sqrt{\frac{0.007}{2}}0.86 = 0.051 \text{ m/s}$$

4. Check if the bed is acting as a rough wall: $dU_f/\nu = 0.05 \times 5.1/0.01 = 26$, larger than 10; therefore, the calculation of the friction velocity assuming that the bed is a rough wall may be justified.

5. Calculate the undisturbed Shields parameter:

$$\theta = \frac{U_{fm}^2}{g(s-1)d_{50}} = \frac{0.051^2}{9.81 \times (2.65-1) \times (0.5 \times 10^{-3})} = 0.32$$

2.3. TWO-DIMENSIONAL SCOUR

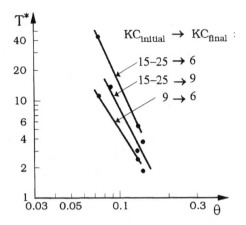

Figure 2.45: Time scale when wave climate is subject to sudden change. Fredsøe et al. (1992).

6. Calculate T^* from Eq. 2.46

$$T^* = \frac{1}{50}\theta^{-5/3} = \frac{1}{50} \times (0.32)^{-5/3} = 0.134$$

7. Find T from Eq. 2.44:

$$T = \frac{D^2}{(g(s-1)d_{50}^3)^{1/2}}T^* =$$
$$= \frac{0.30^2}{(9.81 \times (2.65-1) \times (0.5 \times 10^{-3})^3)^{1/2}} \times 0.134 = 268 \text{ s} \simeq 5 \text{ min}$$

2.3.5 Change in wave climate

In practice, it may be interesting to know the scour depth and time scale in transient cases involving changes in the wave climate. Fredsøe et al. (1992) examined this problem. Fig. 2.44 gives two examples of scour development in such a situation; in the first one, the wave climate changes so that the KC number changes from 19.6 to 5.6, and θ from 0.092 to 0.13, and in the second case, KC changes from 19.6 to 9.4, and θ from 0.092 to 0.13.

In a typical test, the waves are run with a particular wave period and a particular wave height, and this is done until the scour process reaches an equilibrium situation. The KC number and θ corresponding to this particular portion of the test are denoted by $KC_{initial}$, and $\theta_{initial}$. Subsequently, the wave climate is changed suddenly by changing the wave period and the wave height. The KC number and θ values corresponding to this part of the test are denoted by KC_{final}, and θ_{final}.

Fredsøe et al. compared the equilibrium scour depths obtained in the tests with those predicted from the expression given in Eq. 2.24, $S/D = 0.1\sqrt{KC}$ in which KC is taken to be KC_{final}. The agreement was found to be quite good; hence they concluded that, *in a transient situation where the waves change from one climate to another, the equilibrium scour depth is always determined by* KC_{final}.

Fig. 2.45 presents the nondimensional time scale data. The figure clearly shows that the time scale is a function of not only the Shields parameter but also the initial and the final values of the KC number. The Shields parameter referred to above is θ_{final}.

The way in which the time scale varies with θ_{final} is exactly the same as in Fig. 2.43. It is also evident that the closer the initial and final values of KC are, the smaller the time scale will be.

2.4 Three-dimensional scour

As has been pointed out at the beginning of the previous section, the pipeline scour in the field occurs in a three-dimensional fashion. Although the scour in the middle part of the suspended span of a pipeline can be considered as a two-dimensional process, scour around the span shoulders is definitely a three-dimensional one (Fig. 2.12).

In addition to the flow effects studied in conjunction with the two-dimensional scour (namely, the tunnel erosion and the lee-wake effects), a spiral type of vortex may form in front of the pipe. This vortex is caused by the three-dimensional separation under the adverse pressure gradient produced by the pipe in which the separated boundary layer rolls up to form a spiral vortex, as sketched in Fig. 2.46, a process similar to the formation of a horseshoe vortex around a pile (see Section 3.1.1). No detailed study is yet available, investigating the three-dimensional flow at the junction of a pipe and the span shoulder.

As for the 3-D scour itself, various scenarios may occur in a real life situation, depending on the flow, the soil, and the pipe stiffness. One scenario

2.4. THREE-DIMENSIONAL SCOUR

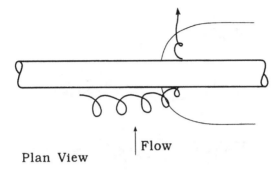

Figure 2.46: Sketch of 3-D flow around pipeline at span shoulder.

involves scour, sagging, backfilling, and eventual self-burial of the pipeline in the free-span areas. In another scenario, the self-burial of the pipeline also occurs at span shoulders. In a third scenario, the soil supporting the pipeline over a pipeline stretch may fail due to liquefaction, resulting in the self-burial of the entire pipeline.

The following paragraphs will address the first two issues. The third issue, namely the self-burial of pipelines in a liquefied soil, will be examined in Chapter 10.

2.4.1 Scour, backfilling, and self-burial in the free-span areas

Following the onset of scour, the scour spreads along the length of the pipe (Fig. 2.12). When the length of the suspended span of the pipeline becomes sufficiently large, the pipeline begins to sag into the scour hole until it reaches (or comes in the neighborhood of) the bottom of the scour hole (Figs. 2.47 c-d and 2.48 c-d). As the pipe sags into its scour hole, it will obviously influence the scour process. When the pipe comes in the neighborhood of the bottom of the scour hole, it will more or less block the flow (Fig. 2.49 d); and as a result, the scour process will come to an end (Figs. 2.47 d and 2.48 d), and the so-called backfilling process will start. The latter may result in partial or complete burial of the pipe in the span areas (the self-burial of pipelines) (Fig. 2.47 e and 2.48 e).

Example 5 *Length of pipeline span which produces deflection*

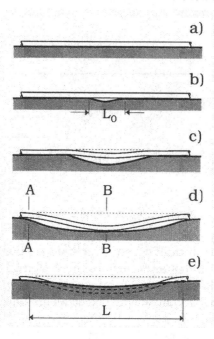

Figure 2.47: Sketch of 3-D scour process and self-burial.

This can be best described with the help of a numerical example.

Given the pipeline depicted in Fig. 2.50, work out the span lengths which can produce large deflections.

Consider that the deflection formula

$$U_{\max} = \alpha \frac{pL^4}{EI} \qquad (2.47)$$

can, to a first approximation, be applicable for the present case with $\alpha = 3/384$, an average value of α for two extreme end conditions, namely the fixed ends where $\alpha = 1/384$, and the hinged ends where $\alpha = 5/384$ (Fig. 2.51 b and c). Here, p = the intensity of the distributed load per unit length of the pipe, E = modulus of elasticity, and I = the inertia moment.

The modulus of elasticity for concrete $E_{concrete} = 2 \times 10^{10}$ N/m², and that for steel $E_{steel} = 2.1 \times 10^{11}$ N/m².

The product EI :

$$EI = E_{concrete}I_{concrete} + E_{steel}I_{steel}$$

2.4. THREE-DIMENSIONAL SCOUR

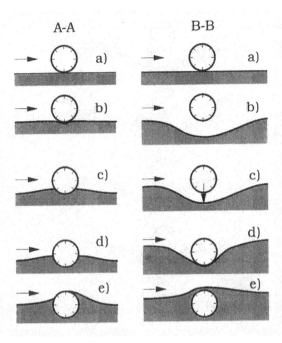

Figure 2.48: Time development at Sections A-A and B-B in the previous figure.

$$\begin{aligned} &= E_{concrete}\frac{\pi}{64}(D_1^4 - D_2^4) + E_{steel}\frac{\pi}{64}(D_2^4 - D_3^4) \\ &= 2 \times 10^7 \text{ Nm}^2 \end{aligned}$$

The intensity of the distributed load :

$$\begin{aligned} p &= \left[\frac{\pi}{4}(D_2^2 - D_3^2)\gamma_{steel} + \frac{\pi}{4}(D_1^2 - D_2^2)\gamma_{concrete} + \frac{\pi}{4}D_3^2\gamma_{oil} - \frac{\pi}{4}D_1^2\gamma_{water}\right] \times 1 \\ &= 1243 \text{ N/m} \end{aligned}$$

in which the densities of steel, concrete, oil, and water are taken as $\rho_{steel} = 7.8 \times 10^3$ kg/m^3, $\rho_{concrete} = 3 \times 10^3$ kg/m^3, $\rho_{oil} = 0.8 \times 10^3$ kg/m^3 and $\rho_{water} = 1 \times 10^3$ kg/m^3, respectively.

The maximum deflection is calculated from Eq. 2.47, and plotted versus the span length in Fig. 2.52. As seen, while the deflection is almost nil for a span length of 10-15 m (or alternatively, $L/D = 30-45$), it grows explosively for span lengths larger than 20 m (or alternatively for $L/D > 60$).

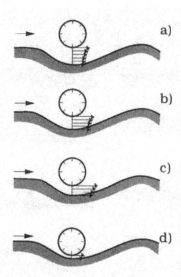

Figure 2.49: Time development of velocity in the gap between sagging pipe and bed.

Rate of spread of scour along the pipeline

The scour spreads along the length of the pipeline after it breaks out (Fig. 2.12). The velocity of this spreading may be termed the propagation velocity of the 3-D scour process.

Research dealing with this quantity is not extensive. Gravesen and Fredsøe (1983) gave an account of how to deal with the problem when extending the results of model experiments to nature. Various accounts of the spreading process have been given in Leeuwestein (1985) and Bernetti et al. (1990). Hansen, Staub, Fredsøe and Sumer (1991), on the other hand, have presented a semi-empirical model of the process. The following paragraphs summarize this model.

The model assumes that the two-dimensional scour in the middle of the span (Fig. 2.53) in a progressive scour hole has attained its equilibrium stage. Therefore the bed shear stress in this area can be put equal to its undisturbed value, namely τ_∞.

The bed shear stress will increase as we move towards the ends of the span, i.e., towards the corners. This is due to the contraction of streamlines.

2.4. THREE-DIMENSIONAL SCOUR

Let α be the amplification in the bed shear stress, defined by

$$\alpha = \frac{\tau}{\tau_\infty} \qquad (2.48)$$

in which τ = the bed shear stress at any z location (Fig. 2.53).

The model further assumes that the previously mentioned increase in the bed shear stress takes place over a distance βD, as schematically illustrated in Fig. 2.53.

The model is based on the sediment conservation equation. The erosion rate in the corner at the span support is determined by the difference between the sediment transport out of the corner area and the sediment transport into the corner area. The rate of volume erosion from the corner is determined by

$$\frac{d(Vol)}{dt} = \frac{(q_{c0} - q_0)\,\beta D}{1 - n} \qquad (2.49)$$

in which n = the porosity, βD = the length, measured along the pipe, of the corner region (β is a constant scaling factor and D is the pipe diameter), q_{c0} = the sediment transport rate in the corner area, and q_0 = the sediment transport rate upstream of the pipe on the flat seabed.

Only the bed-load transport of sediment is included. The model by Meyer-Peter and Müller (1948) is used

$$q_B = 8\,\sqrt{(s-1)\,g\,d_{50}^3}\,(\theta - 0.047)^{3/2} \qquad (2.50)$$

The longitudinal erosion rate at the shoulder of the free span is determined from Eq. 2.49 by

$$c = \frac{1}{eD}\frac{d(Vol)}{dt} \qquad (2.51)$$

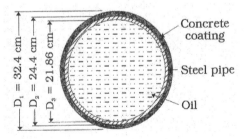

Figure 2.50: Pipeline considered in Example 5.

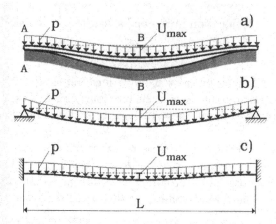

Figure 2.51: Comparison between pipeline and other girders (hinged ends and fixed ends).

in which e is the embedment of the pipe at the span shoulder (see the small sketch in Fig. 2.54). The quantity c is the velocity at which the scour propagates along the length of the pipeline (Fig. 2.53)

Thus, the longitudinal erosion rate is determined by sediment properties, the bed shear stress on the flat bed, the embedment of the pipe, and the two coefficients, α and β. Presently, the latter coefficients can only be evaluated from experiments. In the analysis of the experiment done by the authors, it has been assumed that $\beta = 1$ or in geometrical terms: The length of the corner area (measured along the pipeline) is one pipe diameter. Therefore, from Eqs. 2.49 and 2.51

$$c = \frac{q_{c0} - q_o}{e(1-n)} \qquad (2.52)$$

The amplification factor of the bed shear stress, α, has been determined from the experiments. For this, the following procedure has been adopted in Hansen et al.'s study (1991):

1. Guess the value of α;

2. Calculate the bed shear stress at the corner through $\tau = \alpha \tau_\infty$, and then from Eq. 2.50 the sediment transport rate q_{c0};

3. Determine the propagation velocity c from Eq. 2.52;

2.4. THREE-DIMENSIONAL SCOUR

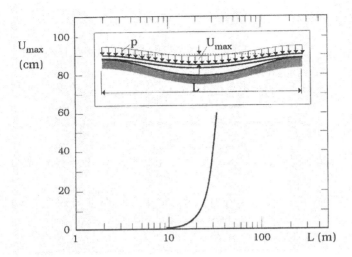

Figure 2.52: Maximum deflection as function of free-span length in the numerical example.

4. Repeat the procedure until an input value of α results in the measured value of the propagation velocity.

The results are plotted in Fig. 2.54. The figure shows clearly that the shear stress amplification decreases with increasing embedment of the pipe.

Example 6 *A numerical example for the prediction of propagation velocity of scour*

Given $D = 30$ cm, $e = 5$ cm, the water depth $h = 10$ m, the sand size $d_{50} = 0.5$ mm, the porosity $n = 0.4$, calculate the scour development rate for a mean current of 0.6 m/s (the current being perpendicular to the pipeline).

1. Calculate the undisturbed bed friction velocity

$$U_f = \frac{V}{2.5\left[\ln(\frac{30h}{k_s}) - 1\right]}$$

$$= \frac{0.6}{2.5 \times \left[\ln(\frac{30 \times 10}{2.5 \times (0.5 \times 10^{-3})}) - 1\right]} = 0.021 \text{ m/s}$$

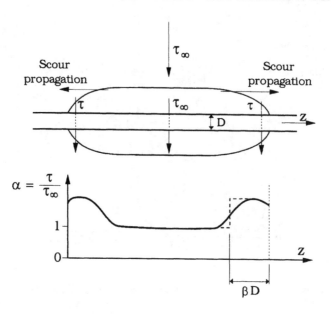

Figure 2.53: Amplification in the bed shear stress in 3-D scour.

2. Calculate the undisturbed Shields parameter

$$\theta = \frac{U_f^2}{g(s-1)d_{50}} = 0.055$$

3. Pick up the value of α from Fig. 2.54, corresponding to $e/D = 5/30 = 0.167$, namely $\alpha \simeq 1.8$.

4. Calculate the sediment transport rates on flat seabed and in the corners of the scour hole from Eq. 2.50:

Flat seabed : $\theta = 0.055$, resulting in $q_0 = 2.6 \times 10^{-7}$ m²/s
Corner area : $\theta_{c0} = \alpha\theta = 1.8 \times 0.055 = 0.099$, giving
q_{c0} = 4.3×10^{-6} m²/s

5. Calculate c from Eq. 2.52:

$$c = \frac{4.3 \times 10^{-6} - 2.6 \times 10^{-7}}{0.05 \times (1 - 0.4)} = 0.000135 \text{ m/s} = 0.48 \text{ m/h}$$

2.4. THREE-DIMENSIONAL SCOUR

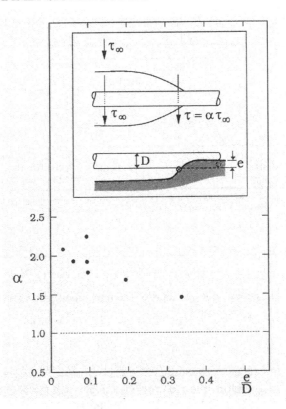

Figure 2.54: Amplification in the bed shear stress. Hansen et al. (1991).

6. Furthermore, it would be interesting to estimate the time period required for the development of $L = 20$ m long, say, span length. (This span length is a "critical" span length beyond which the maximum deflection of the pipe span reaches substantial values for a typical concrete-coated pipeline of about the same size, see Figs. 2.50 and 2.52). This will be

$$t = \frac{L/2}{c} = \frac{20/2}{0.48} = 21 \text{ h}$$

As seen, this time scale is large compared with the time scale of the two-dimensional scour predicted in the numerical example given in Example 3 in Section 2.3.4, namely $T = 1.4$ h.

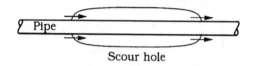

Figure 2.55: Migration of free span.

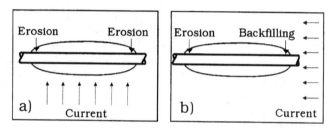

Figure 2.56: Current at oblique angle. Decomposition into two components (a) and (b).

When the current is perpendicular to the pipe with a free span, the flow conditions, and therefore the rate of scour development, will be exactly the same at the two ends of the scour hole. When the current approaches the pipe at an oblique angle, however, the flow conditions at the ends of the scour hole will be different.

It is assumed that when the deviation from the perpendicular incidence is larger than a certain angle, one end will tend to backfill while the other end will continue to erode. In such a situation, the free span will migrate (Hansen et al., 1991) (Fig. 2.55).

The components of sediment transport normal and parallel to the pipe, respectively, are

$$q_n = q_{tot} \sin(v) \tag{2.53}$$

$$q_t = q_{tot} \cos(v) \tag{2.54}$$

where v is the angle between the current and the pipe.

The sediment is assumed to be trapped at the upstream end of the scour hole (backfilling) and removed from the downstream end (erosion), see Fig.

2.4. THREE-DIMENSIONAL SCOUR

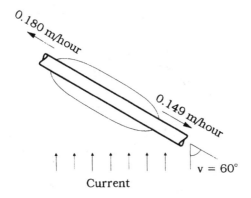

Figure 2.57: Calculation example. Hansen et al. (1991).

2.56 b. This is, of course, a very simplified description of the actual physical processes.

Also, the erosion due to the bed shear stress amplification in the corners is affected by the oblique angle. Only the component of the bed friction perpendicular to the pipe is assumed to contribute to the amplification of erosion capacity in the two corners, Fig. 2.56 b. The same amplification factor, α, is assumed, as for the current, perpendicular to the pipeline. Therefore, the erosion due to friction amplification will decrease when the angle between the current and the pipe decreases.

Regarding the effect of waves, Hansen et al. (1991) note that the bed shear stress amplification, α, and the longitudinal scaling coefficient, β, will be strongly dependent on the Keulegan-Carpenter number, wave-current velocity ratio, etc. Therefore, they point out that the model is not expected to be directly applicable to waves, or combined waves and current.

Example 7 *A numerical example for the prediction of propagation velocity of scour. Oblique pipe*

Suppose that the pipeline in the previous example is at an angle of 60^0 with the current direction. Predict the erosion rate.

The sediment transport in the corner is calculated to be

$$q_{co} = 1.6 \times 10^{-6} \text{ m}^2/\text{s}$$

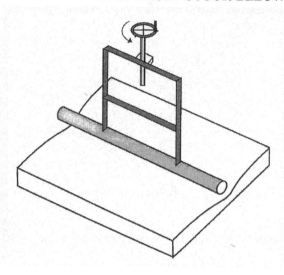

Figure 2.58: Apparatus for 2-D simulation of scour in the case of sagging pipe.

The migration velocity at the right end of the scour hole (Fig. 2.57) is calculated from Eqs. 2.49, 2.53, 2.54 and 2.51:

$$c = 0.149 \text{ m/h}$$

and at the left end of the scour hole

$$c = 0.180 \text{ m/h}$$

Effect of sagging on scour. 2-D laboratory simulation

The effect of sagging on scour can be studied by a two-dimensional laboratory simulation of the process in which a unit length of the pipeline at the middle of the span length is simulated by a rigid model pipe. The sagging of the pipeline itself is simulated by moving the model pipe vertically downwards at a certain velocity, Fig. 2.58 (Fredsøe, Hansen, Mao and Sumer, 1988). Gökçe and Günbak (1991) adopted a similar approach with a two-dimensional sagging pipe, with and without spoilers, exposed to waves.

2.4. THREE-DIMENSIONAL SCOUR

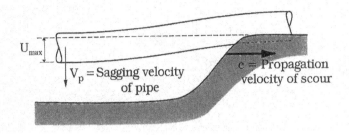

Figure 2.59: Definition sketch of sagging pipe.

Leeuwestein et al. (1985) adopted a different approach; each time the equilibrium scour depth was established, the pipe was placed down in the scour hole, and the procedure was resumed with the next test until the scour hole was backfilled, and the pipe was buried.

Regarding the first type of approach, there are two important parameters: One is the velocity of the pipe (the sagging velocity in the model), and the second is the initial scour profile.

The first parameter is determined in the field by

$$V_p = \frac{dU_{max}}{dt} \qquad (2.55)$$

and

$$c = \frac{dL}{dt} \qquad (2.56)$$

in which V_p = the sagging velocity of the pipe (Fig. 2.59), U_{max} = the deflection of the pipe span (Fig. 2.52), c = the propagation velocity of the 3-D scour at the ends of the span, L = the span length , and t = the time.

From Eq. 2.55 and 2.47

$$V_p = \frac{d}{dt}(\alpha\frac{pL^4}{EI}) = \frac{4\alpha pcL^3}{EI} \qquad (2.57)$$

As seen, V_p changes as the span develops. Furthermore, it is seen that the sagging velocity of the pipe is a function of not only L and c but also the stiffness of the pipe, namely EI/p, and hence V_p may vary over a rather broad range.

In the 2-D laboratory simulation of the process (Fredsøe et al., 1988), V_p is maintained constant during the course of a test, considering the impracticality of changing V_p with time. However, to study the influence of V_p on the

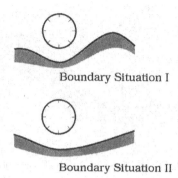

Figure 2.60: I: Initial scour profile when $T_v \sim T_h$; II: that when $T_v \ll T_h$. Fredsøe et al. (1988).

results, a reasonably wide range of V_p has been adopted, ranging from 0.62 mm/min to 12.4 mm/min.

There may be two situations as regards the occurrence of pipe sagging: $T_v \sim T_h$ and $T_v \ll T_h$ in which T_v = the time scale of the two-dimensional scour in the middle of the pipeline span (Section 2.3), and T_h = the time scale of the span development, representing the time period during which a substantial span length develops to produce measurable deflection in the middle of the pipe span.

Normally, $T_v \ll T_h$ (cf. Example 3 in Section 2.3.4 and Example 6 in Section in the present section), i.e., the 2-D scour reaches its equilibrium stage long before the sagging begins to occur. Hence, the 2-D simulation of the process in the laboratory may begin with an initial scour hole which is identical to that representing the equilibrium scour profile of the 2-D scour under the considered flow conditions.

Otherwise (i.e., when T_v is not small compared with T_h), the initial boundary condition of the 2-D laboratory simulation may be selected from scour profiles corresponding to the initial stage of the 2-D scour process. This aspect of the problem has been studied by Fredsøe et al. (1988), employing two different scour profiles, to see the influence of the initial scour profile. These profiles are reproduced in Fig. 2.60.

The results of Fredsøe et al.'s (1988) experiments are reproduced in Fig. 2.61. From the figure, the following conclusions are straightforward:

2.4. THREE-DIMENSIONAL SCOUR

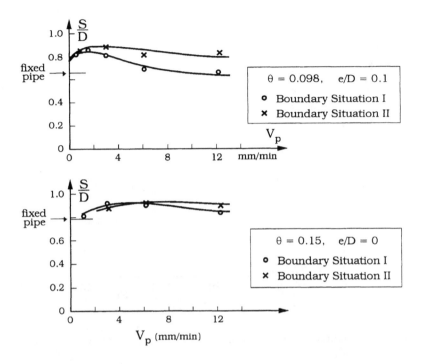

Figure 2.61: Equilibrium scour depth versus sagging velocity. Current. Fredsøe et al. (1988).

1. The scour hole in the case of the sagging pipe may be significantly deeper, depending on the "Boundary Situation" and the sagging velocity V_p. The increase in the scour depth is caused by the inevitable increase in the near-bed flow velocities due to sagging of the pipe at the early stage of the sagging process, as illustrated in Fig. 2.49 a-c. This increase in the scour depth can be as much as 40%, particularly for small sagging velocities, $V_p = O(3 \text{ mm/min})$, in combination with Boundary Situation II.

2. Boundary Situation II almost always results in deeper scour holes. This is not surprising, because the sagging process starts with a deeper scour hole. (However, this effect disappears as V_p decreases, because the influence of the initial boundary condition is not felt any longer for

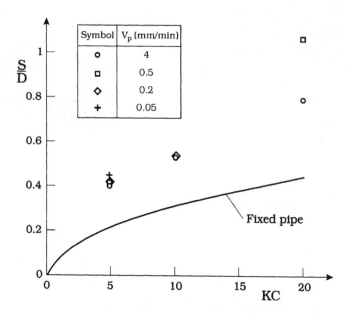

Figure 2.62: Equilibrium scour depth versus sagging velocity. Wave. Line: $\frac{S}{D} = 0.1\sqrt{KC}$. Roll (1989).

small V_p values, since the sagging process continues for a long period of time).

3. It can be deduced from Fig. 2.61 that the scour depth may remain practically unchanged (with respect to the case of the fixed pipe) if the sagging velocity is large, larger than $O(10$ mm/min$)$, and the pipe begins to sag in the early stages of the scour process.

Fig. 2.62 shows the results of Roll's (1989) investigation in the case of waves. In these experiments, no specific boundary conditions (such as Boundary Situations I and II of Fig. 2.60) were implemented at the beginning of the tests. The pipe was driven downwards with a specified velocity upon the onset of scour. As seen from the figure, the scour holes are always deeper than in the case of the fixed pipe (the scour depth can be increased by as much as a factor of 2). This may be due partly to the relatively small values of the sagging velocity ($V_p = 0.05 - 4$ mm/min, cf. Fig. 2.61), and partly to

2.4. THREE-DIMENSIONAL SCOUR

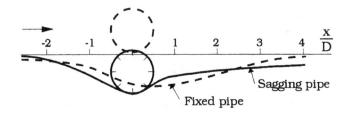

Figure 2.63: Scour profile when sagging pipe reaches seabed. Boundary Situation II. Fredsøe et al. (1988).

the "parallel" occurrence of the two processes, namely the sagging process and the scour process. Also, the lee-wake-dominated scour in the present case may also help develop deeper scour holes. The sagging velocities of the experiments presented in Fig. 2.62 are rather small (cf. Fig. 2.61). For higher velocities, the scour depth may be expected not to be affected much (not as much as in Fig. 2.62) (as implied by Fig. 2.61, higher velocities and Boundary Situation I).

Backfilling and self-burial

Fig. 2.63 compares two scour profiles,

1. one obtained with a sagging pipe, and

2. the other with a fixed pipe, exposed to the same current.

The sagging profile corresponds to the instant when the pipe reaches the bottom. Note the deposition of sand at the downstream side of the pipe, caused by the decrease in the flow velocity downstream the pipe, as the pipe approached the bed (Fig. 2.49 d).

As soon as the gap between the pipe and the bed is closed, the backfilling process starts. In a rather short period of time, the downstream side of the pipe is filled with the sediment coming not only from the upstream side of the pipe but also from the downstream side of it with the help of the lee-wake vortex.

Fig. 2.64 illustrates the scour and backfilling process in the case of waves where the previously mentioned backfilling process is present on both sides

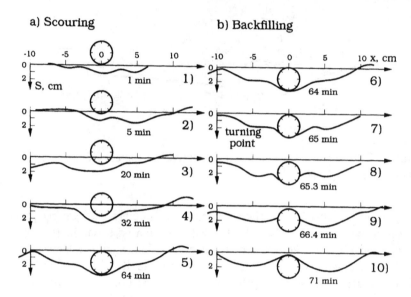

Figure 2.64: Time development of scour and backfilling. $KC = 20$. $\theta = 0.19$. $V_p = 0.5$ mm/s. Roll (1989).

of the pipe. At the end of the backfilling process, the pipe will be partially or completely buried in the bed, as revealed by Fig. 2.64 b (cf. Fig. 2.62).

Free-span length

The development of free spans is a rather complex process, and is still not very well understood. The key issue here is the prediction of span lengths.

Fig. 2.65 shows the probability density function of free-span length obtained from a site-specific field study (Orgill, Barbas, Crossley and Carter, 1992). The figure indicates that there is a considerable variation in L/D, ranging from 10 to as much as 100 with the mean value of about 20.

Another interesting piece of information is given by R.J. Brown in the compiled Technical Articles for State-of-the-Art of Marine Pipeline Engineering (Chapter 7, Part II, p. 12): Brown notes: "..., for a pipeline laid on the bottom, after five years of operation, it was found that, for every 15 feet installed, 1 foot was spanned. Initially, the value was, for every 35 feet installed, with 1 foot spanned".

2.4. THREE-DIMENSIONAL SCOUR

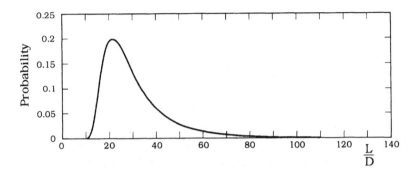

Figure 2.65: Probability density of free-span length. Site specific. Orgill et al. (1992).

The free-span length is governed by various effects (Fredsøe et al., 1988; Bruschi, Cimbali, Leopardi and Vincenzi, 1986; Bernetti, Bruschi, Valentini and Venturi, 1990; and Bijker, Staub, Silvis and Bruschi, 1991):

1. Changing flow conditions: A developing free span may stop growing when the flow velocity decreases below the threshold values for the onset of scour (see Section 2.1).

2. Changing soil conditions: The development of a free span will obviously stop when a support reaches a non-erodible bed area, an area where the soil contains a substantial amount of silt, or clay, or gravel.

3. Sinking of the pipeline at the span shoulders: Again, the development of a free span will stop when the pipe sinks in the span shoulder substantially, terminating the 3-D scour. This may be due to general shear failure (Section 2.4.2), or it may be due to liquefaction (Chapter 10). As for the latter, Bijker et al. (1991) note that the liquefaction potential strongly depends on the relative density of the soil, the permeability, the presence of the pipe, the influence of the stress history, adding that soil containing a significant percentage of silt (<63 μm) will show less permeability, and consequently higher liquefaction potential. (Chapter 10 will give a detailed analysis of the impact of liquefaction).

4. Sagging of the pipeline in the scour hole: As the span develops, the pipe may sag in the scour hole, as described in the preceding paragraphs.

When the pipe reaches the bed, the length of the free span will be cut in half.

Fredsøe et al. (1988) gave a simple model, to predict the free-span. The model simply assumes that the familiar deflection expression for a beam (Eq. 2.47) is applicable for the present case with $\alpha = 3/384$ (Example 5 in Section 2.4.1).

From the experimental information on the effect of sagging on the scour depth, the final scour depth is taken approximately equal to the pipe diameter. The consequence of this approximation will be discussed later in the section.

Now, if $S = D$ is substituted in Eq. 2.47, the span length can be obtained as follows

$$L = 3.35 \, D^{1/4} \, L_s^{3/4} \tag{2.58}$$

in which L_s, the stiffness length, is defined by

$$L_s = \left(\frac{EI}{p}\right)^{1/3} \tag{2.59}$$

The preceding analysis implies that the span length of a pipeline cannot be larger than that given in Eq. 2.58.

The analysis has been given for the case of the steady current. Nevertheless, the expression given in Eq. 2.58 can, to a first approximation, be used for the wave case as well (not for very small KC numbers, $KC \lesssim O(20)$, and not for very large KC numbers, $KC \gtrsim O(300)$, however).

A sensitivity analysis was carried out by Fredsøe et al. (1988) regarding the value of α, and it was found that L is not very sensitive to the value of α.

2.4.2 Scour, backfilling and self-burial at span shoulders

Pipe sinking at span shoulders is caused by soil failure. There are two kinds of soil failure: (1) the general shear failure, and (2) the failure due to liquefaction. This section will focus on the former. The failure due to liquefaction will be examined in Chapter 10.

The process of sinking of a pipeline due to shear failure and the resulting self-burial have been investigated by Sumer and Fredsøe (1994) for the case of steady currents, and by Sumer et al. (2001 a) for the case of waves.

2.4. THREE-DIMENSIONAL SCOUR

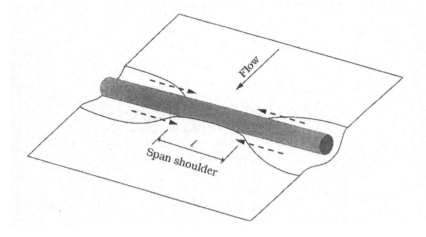

Figure 2.66: Definition sketch for the test in the next figure.

Process of sinking at span shoulders in steady current

The process could possibly be best described by reference to the experiment reported in Sumer and Fredsøe (1994). A rigid pipe was employed in this experiment. The pipe was free to move in the vertical direction, and it was initially sitting on the bed. The experimental conditions were such that the scour first emerged at the two ends of the pipe, and subsequently spread progressively towards the center of the pipe where the pipe was supported by a ridge (representing a span shoulder) in the middle of the flume (Fig. 2.66). (The scour first emerges at the two ends of the pipe because of the small gaps between the ends of the pipe and the side walls of the flume). Fig. 2.67 b presents the time series (the time development) of the length of the supporting ridge, ℓ, while Fig. 2.67 a presents the time series of the sinking of the pipe in the sand (at the supporting ridge). As seen from Fig. 2.67, there are three stages in this pipe/sand-bed interaction process, namely the scour, the sinking, and the backfilling. Now let us see these stages in detail:

1. During the first 0.1 h (after the scour breaks at the two ends of the pipe), the **scour** process continues to spread towards the center of the pipe until the supporting length ℓ is reduced to a value of about 50 cm.

2. When this point is reached, the pipe begins to sink in the sand at the

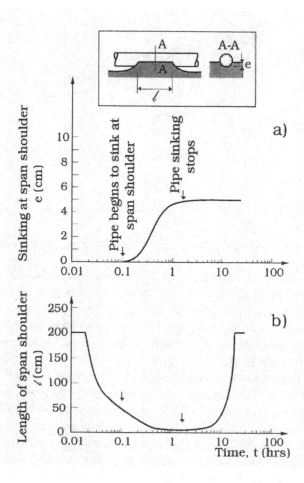

Figure 2.67: Time development of pipe sinking at span shoulder. Sumer and Fredsøe (1994).

2.4. THREE-DIMENSIONAL SCOUR

supporting ridge (Fig. 2.67 a). The **sinking** of the pipe is obviously due to the failure of the sand. Clearly, as the scour progresses, the load due to the weight of the pipe will be exerted on a smaller and smaller length of the supporting ridge. This will eventually lead to a situation where the bearing capacity of the soil will be exceeded, and therefore the pipe will consequently sink in the sand.

As seen from the figure, ℓ continues to decrease, i.e., the scour at the two ends of the pipe continues to occur, until the time $t \simeq 1.5$ h is reached. This means that, during this period, the bearing capacity of the sand at the supporting ridge is continuously exceeded, as the area of the ridge (which bears the weight of the pipe weight) becomes smaller and smaller with the continuing scour at the two ends of the span shoulder.

This continues until the scour at the two sides of the ridge comes to a complete stop. When this point is reached, the bearing area will no longer change, therefore the bearing capacity of the soil will no longer be exceeded, and hence the sinking of the pipe in the soil will terminate, as revealed in Fig. 2.67 a. (Regarding the termination of the scour process itself, this is related to the protection of the pipe in the scour hole; the flow velocities below the pipe at the junctions between the pipe and the scour hole are reduced so that the scour will no longer occur in these locations).

3. The third stage of the process begins with the time at which the scour stops. During this final stage, the space between the pipe and the scour hole is gradually filled with sand, ℓ (the length of the supporting ridge) this time being gradually increased (Fig. 2.67 b), the **backfilling** process. This continues until the entire length of the pipe is covered by sand. As seen from Fig. 2.67 b, the latter process is completed by $t = 20$ h.

This experiment clearly shows that the pipe may bury itself at span shoulders, the **self-burial** of pipelines at span shoulders. In this process, the key "ingredients" are apparently: (1) the three-dimensional scour at the two ends of the span shoulders, (2) the continuous shear failure at the span shoulders (resulting in sinking of the pipeline in the supporting soil stretch), and (3) the backfilling.

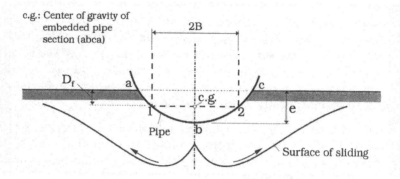

Figure 2.68: Sinking of pipeline at span shoulder due to general shear failure. Sumer and Fredsøe (1994).

Sumer and Fredsøe (1994) have made a theoretical assessment of the sinking of the pipe. Before the critical condition is reached, the soil beneath the pipe is in a state of elastic equilibrium. When the pipe begins to sink, however, the soil gradually passes into a state of plastic equilibrium. The failure occurs by sliding in the two outward directions (Fig. 2.68). This type of failure is known as the general shear failure in soil mechanics (Terzaghi, 1948).

Sumer and Fredsøe (1994) have adopted the formula given for a rectangular cross-section footing to predict the bearing capacity of the soil, an approximation considering that no information is available for a strip loading with the footing geometry illustrated in Fig. 2.68, namely the portion of the pipe *abca* in the figure. As indicated in Fig. 2.68, it is assumed that the bottom of the "equivalent" rectangular cross-section lies at the level of the center of gravity of the embedded portion of the pipe section, a usual practice in soil mechanics to make a quick assessment of the settlement of structures with cross-sections different from a rectangular shape. Another assumption in this theoretical assessment is that the width of the equivalent rectangular cross-section, $2B$, is taken as the length $\overline{12}$ (Fig. 2.68).

The formula estimating the bearing capacity for general shear failure in the case of rectangular, rough-base, infinitely long, strip footings for non-cohesive soil reads as follows

$$\frac{Q}{A} = \gamma' \, B \, N_\gamma + \gamma' \, D_f \, N_q \tag{2.60}$$

2.4. THREE-DIMENSIONAL SCOUR

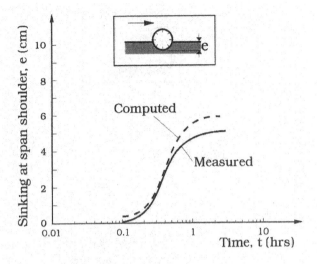

Figure 2.69: Sinking at span shoulder. Sumer and Fredsøe (1994).

in which Q = the bearing-capacity load, A = the bearing area, γ' = the specific weight of the soil in water, and N_γ and N_q = the bearing-capacity factors (Terzaghi, 1948).

When the footing is not an infinitely-long strip footing, but rather a rectangular one, the so-called shape factors are introduced in the preceding formula (DIF, 1984):

$$\frac{Q}{A} = \gamma' \, B \, N_\gamma \, s_\gamma + \gamma' \, D_f \, N_q \, s_q \qquad (2.61)$$

in which

$$s_\gamma = 1 - 0.4 \frac{2B}{\ell} \qquad (2.62)$$

and

$$s_q = 1 + 0.2 \frac{2B}{\ell} \qquad (2.63)$$

in which ℓ is the length of the rectangular bearing area, and A in this case will be

$$A = 2B\ell \qquad (2.64)$$

Given the measured values of the length ℓ at discrete times, the sinking of the pipe has been determined by Eqs. 2.61-2.64. The result of Sumer

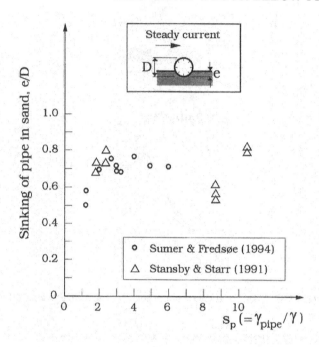

Figure 2.70: Self-burial depth at span shoulder. Current. Sumer and Fredsøe (1994).

and Fredsøe (1994) is reproduced in Fig. 2.69 together with the measured sinking. As seen, despite the approximate nature of the simple theoretical analysis, Eqs. 2.61-2.64 predict the sinking of the pipe fairly well.

Finally, Fig. 2.70 illustrates the influence of the pipe weight on the sinking of the pipe. From the figure, it appears that no drastic effect seems to exist (although the sinking seems to decrease with decreasing specific gravity of the pipe, $s_p = \gamma_{pipe}/\gamma$, below approximately $s_p = 2$). This finding (namely, the sinking depth is practically independent of the pipe weight) is not an entirely unexpected result. This is because basically the final displacement of the pipe is governed by the scour at the span shoulders; as the scour process stops, the sinking process will presumably come to an end according to the mechanism described in the previous paragraphs. Therefore, the pipe weight should not play any significant role in the sinking process.

2.4. THREE-DIMENSIONAL SCOUR

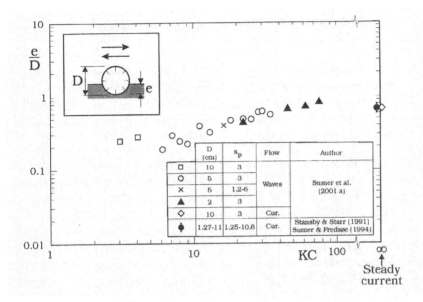

Figure 2.71: Self-burial depth at span shoulder. All data (current/wave). Live bed ($\theta > \theta_{cr}$). Sumer et al. (2001 a).

Sinking at span shoulders in waves

As mentioned in the preceding paragraphs, the scour at the two ends of the span shoulder is the key element in the process of pipe sinking. The scour process in the case of waves is essentially governed by the Keulegan-Carpenter number, KC. This is linked to the lee-wake, precisely in the same way as in the case of the two-dimensional scour process (see Section 2.3.2). This suggests that the sinking depth (the self-burial depth), e, should be a function of KC:

$$\frac{e}{D} = f(KC)$$

Fig. 2.71 displays the data regarding the equilibrium self-burial depth, reproduced from Sumer et al. (2001a). The steady current data from Stansby and Starr (1991) and Sumer and Fredsøe (1994) (Fig. 2.70) are also included in Fig. 2.71. The following conclusions can be deduced from Fig. 2.71.

1. The self-burial depth is apparently a function of the Keulegan-Carpenter

number KC, revealing the argument put forward in the preceding paragraph..

2. e/D increases with increasing KC. This is because the higher the Keulegan-Carpenter number, the longer the lee-wake (that forms behind the pipe in each half period of the motion), and the larger the scour. Since the self-burial depth e increases with increasing scour depth, then e should increase with increasing KC.

3. The influence of the specific gravity of the pipe, s_p, on the end results is insignificant. See the data point for $KC = 16$, the cross, and also the legend in Fig. 2.71; For this KC number, five tests were conducted with five different values of the pipe specific gravity in the range $s_p = 1.25$-6, and it was found that the sinking depth was practically unchanged (Sumer et al., 2001 a). It may be noted that Sumer and Fredsøe (1994) reached the same conclusion in the case of the steady current (Fig. 2.70). The sinking is uninfluenced because the key element in the process of sinking is the scour; when the scour stops, the sinking will stop, as discussed in the preceding subsection. Since the specific gravity of the pipe is not an influencing factor for the scour, it will therefore not affect the sinking, and hence the self-burial depth.

Fig. 2.72 presents an interesting comparison: it compares two sets of data (including the steady-current results): (1) the self-burial-depth data of Fig. 2.71; and (2) the scour-depth data (for a *fixed*, bottom-seated pipeline) of Fig. 2.23 .

As seen, these two sets of data agree quite well. This can be explained as follows.

Now, as the sand at the span shoulder fails progressively, the pipe sinks in the sand, and, at the same time, it falls in the scour holes at the two sides of the span shoulder. The scour process comes to an end when the pipe reaches the bottoms of these scour holes. The scour depth S in the latter case (Fig. 2.73 b) will be quite close to that obtained in the case of the fixed pipe (Fig. 2.73 a). This is because the fall velocity V_p is rather large ($O(10 \text{ mm/min})$), and the pipe begins to "sag" in the scour holes as the scour process continues, a situation similar to that displayed in Fig. 2.61 with $V_p = O(10 \text{ mm/min})$ and Boundary Situation I. This leads to the following important conclusion:

2.4. THREE-DIMENSIONAL SCOUR

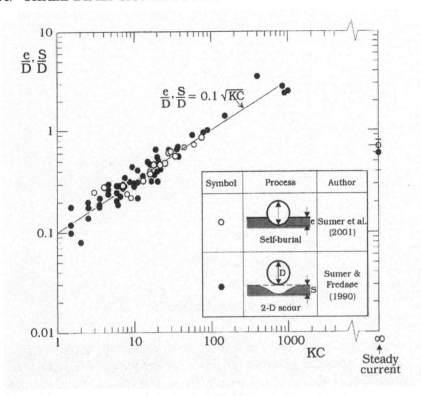

Figure 2.72: Self-burial depth at span shoulder and scour depth. Live bed ($\theta > \theta_{cr}$). Sumer et al. (2001 a).

the self-burial depth at span shoulders should be the same as the scour depth experienced in the case of a fixed pipe with an initial zero gap.

Although not tested directly in Sumer et al.'s (2001 a) study (due to the experimental constraints), the preceding results imply that

1. a pipeline may be self-buried completely for KC larger than $O(100)$; and

2. the self-burial depth of pipelines may reach values as high as $e/D \simeq 3$, for very large KC numbers such as $O(1000)$, representing the tidal flow conditions.

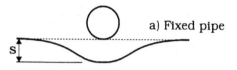

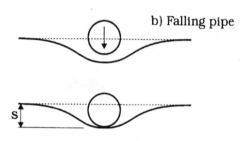

Figure 2.73: Definition sketch.

Finally, it may be noted that the following empirical expression, given originally by Sumer and Fredsøe (1990) for the equilibrium scour depth below a fixed pipeline (with an initial zero gap between the pipe and the bed), can be used to assess the self-burial depth of pipelines at span shoulders for $KC \lesssim 100$ (Fig. 2.72):

$$\frac{e}{D} = 0.1 \sqrt{KC} \qquad (2.65)$$

Caution must be exercised when extrapolating the preceding equation for KC larger than 100, as it has not been tested for the self-burial of pipelines for such large KC numbers.

Time scale of self-burial process

The sinking of the pipe develops towards the equilibrium stage through a transitional period (see the definition sketch in Fig. 2.74). The time scale of the process may be defined by the following equation

$$T = \frac{1}{e} \int_0^\infty (e - e_t)\, dt \qquad (2.66)$$

2.4. THREE-DIMENSIONAL SCOUR

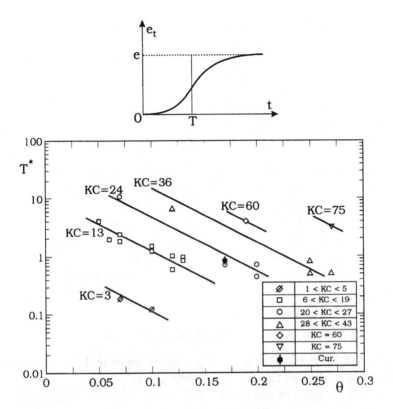

Figure 2.74: Time scale of self-burial process. Live bed ($\theta > \theta_{cr}$). Sumer et al. (2001 a).

Here, e is the equilibrium-sinking depth and e_t is the sinking depth at time t. The time scale T can be interpreted as the time over which a substantial amount of self-burial takes place.

The data regarding the time scale is reproduced from Sumer et al. (2001 a) in Fig. 2.74. Here, T^* is the normalized time scale defined by

$$T^* = \frac{[g(s-1)d^3]^{1/2}}{D^2} T \qquad (2.67)$$

(Fredsøe et al., 1992), considering that the time scale of the self-burial process is similar to that of the scour process that induces the self burial (see Section 2.3.4).

1. As seen from Fig. 2.74, the time scale decreases with increasing θ. This is because the larger the value of θ, the higher the sediment transport in the scour process, and the faster the sinking of the pipe in the sand; therefore, the time scale should increase with increasing θ. On the other hand, the figure shows that the larger the Keulegan-Carpenter number, the larger the time scale. This is because the larger the value of KC, the larger the scour depth, therefore the larger the volume of the sand undergoing scour, hence the larger the time scale.

2. Although there is only one data point for the case of the steady current, apparently the time scale of the self-burial process in the case of the steady current seems to be quite close to that for the case of waves with $KC = 24$. This may be linked to the self-burial depth. Fig. 2.72 suggests that the self-burial depths for the two cases (namely, the steady-current case and the case of $KC = 24$) are rather close. Therefore, the time scale of the self-burial process should also be close to each other.

2.4.3 Stimulated self-burial of pipelines

As has been mentioned in the preceding sections, pipelines laid on the seabed may bury themselves in the seabed by various mechanisms.

Fig. 2.75 depicts the results of a field study made by Delft Hydraulics Laboratory where a 12" (30.5 cm) pipeline, laid in 1980 on the North Sea bottom at a water depth of 30 m, was monitored. It appeared that, within a few months after the pipeline was laid on the seabed, it caused the formation of a trench, 3 pipe diameters deep and 50 diameters wide, and then the pipe gradually sank into this trench and buried itself, without any human interference, with a thickness of the covering sand layer of more than 30 cm (Kroezen, Vellinga, Lindenberg and Burger, 1982).

In order to exploit the self-burial potential in the circumstances where pipelines would not bury themselves fast enough (or they would not bury themselves at all), the Delft Hydraulics Laboratory in an extensive research has investigated the feasibility of a method called the stimulated self-burial, Hulsbergen (1984), Hulsbergen (1986), and Hulsbergen and Bijker (1989). The idea is to stimulate a controlled local scour by using fins (called spoilers) attached to the pipeline (Fig. 2.76).

The idea of using spoilers attached to a pipeline is not new. Brown

2.4. THREE-DIMENSIONAL SCOUR

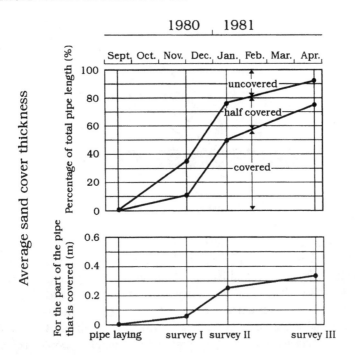

Figure 2.75: Behaviour of a 12" pipeline on the floor of the North Sea. Kroezen et al. (1982).

(1966) reports the results of a study where the pressure distribution around a pipeline model sitting on the bed was measured with and without spoilers. No mention was made, however, of possible applications of spoilers to enhance the self-burial of pipelines.

Different versions of the spoilers shown in Fig. 2.76, and many others, are known to have been used specially in wind engineering as vortex-shedding suppression devices (Blevins, 1977; and Sarpkaya and Isaacson, 1981).

Regarding the application of spoilers to self-burial, the Delft study revealed that the spoilers would indeed stimulate the self-burial process substantially. The major effects of a spoiler can be summarized as follows.

1. The spoiler reduces the time necessary to accomplish a given self-burial depth by a factor of ten with respect to a plain pipeline. As seen from Fig. 2.77, a self-burial depth of for example 0.1 m (about 30% of the

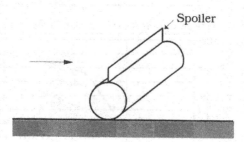

Figure 2.76: Spoiler arrangement to stimulate self-burial.

pipe diameter) is reached after some 30 minutes if a spoiler is applied, while it takes about 6 h to reach the same self-burial depth for a plain pipeline. Hulsbergen and Bijker (1989) note that especially in areas like the Southern North Sea, where low to moderate near-bed tidal current velocities are a common feature, this large difference in the burial speed implies that a pipeline with a spoiler will be well self-buried by the first tidal flow phase after laying, whereas a plain pipeline would stay on top of the bed for many cycles, maybe for many months.

2. The final depth of the self-burial in the spoiler case will be larger than in the case of plain pipelines. This trend is seen clearly in Fig. 2.77. The same effect was observed also by other researchers (Gökçe and Günbak, 1991; and Chiew, 1993).

The presence of the spoiler actually diverts more flow to pass under the pipe, which helps enhance the tunnel erosion, meaning that the burial at early stages occurs much faster and much deeper, as indicated in Fig. 2.77.

Hulsbergen and Bijker (1989) report also the results of their investigation with regard to the forces on a pipeline with a spoiler. The drag is increased and the lift is decreased with respect to a plain pipeline. The inertia coefficient on the other hand is found to increase by a factor of about 1.3 with respect to its plain-pipe value in the case of a spoiler which is 25% in height of the pipe diameter.

On the basis of the research and development undertaken by the Delft Hydraulics, the first 4 km of spoiler have been applied on a Placid International Oil Ltd. pipeline in the Dutch sector of the North Sea in June 1988

2.5. SCALE EFFECTS

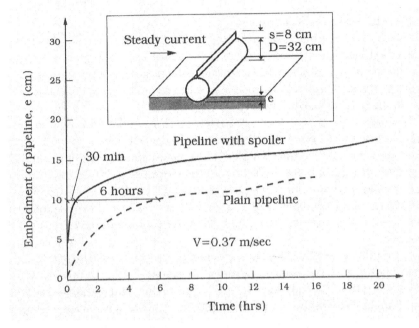

Figure 2.77: Effect of spoiler on scour. Hulsbergen and Bijker (1989).

(Hulsbergen and Bijker, 1989). The purpose of this project was to see the technical feasibility of the spoiler application and its performance in a real-life situation, to monitor the self-burial behaviour and to draw conclusions with regard to the use of spoilers in subsequent larger projects.

2.5 Scale effects

Scale effects must be considered when the results of small-scale laboratory experiments are extrapolated to real-life situations. There are several scale effects, such as the effect of the pipe Reynolds number; the effect of the pipe roughness; the effect of the bed ripples (the pipe-diameter-to-ripple-size ratio is much larger in the field than in the laboratory); the effect of wave/soil interaction; the effect of incoming-flow turbulence; the effect of the pipe stiffness (in the case of the 3-D scour).

Little is known about these scale effects; none of these issues have been studied in a systematic manner.

From the existing data for steady currents, Sumer and Fredsøe (1990) concluded that there is some weak influence of the pipe Reynolds number on the scour depth; however, this result is at best suggestive due to various other uncertainties involved, such as the Shields parameter dependence, the incoming-flow turbulence, and the bed ripples.

As regards the effect of the bed ripples, the wave data depicted in Sumer and Fredsøe (1990) showed that the normalized scour depth was basically the same for a wide range of the ripple-size-to-pipe-diameter ratio, implying that the effect of ripples was not essential in the scour process. The latter authors argue that one reason for this is that presumably no ripples are present in the neighborhood of the pipeline due to the very strong flow (θ values larger than $O(0.6 - 0.7)$). However, the subject has not been investigated in a detailed manner.

The influence of scale effects (in general) when investigating processes in the laboratory have been discussed by, amongst others, Hughes, (1993), Oumeraci (1994); Whitehouse (1998), Sutherland and Whitehouse (1998) and Sumer, Whitehouse and Tørum (2001 b). Clearly no one experimental set up can meet the scaling requirements for all situations and hence each scenario will have its own scaling solution. The following discussion addressing the situation for scour testing draws from Sumer et al. (2001 b).

Whilst the scaling of waves and currents in the laboratory is well understood in practice (Froude and Reynolds scaling) the many requirements of an experiment preclude an exact scaling exercise. Of course scale effects can be reduced to a minimum by running tests in suitably large facilities. In order to design the experiment at an appropriate scale the most important factor is to have an understanding of the important processes acting in the prototype situation. This means the model scaling can be optimized to address the influence of these processes.

Once the relevant processes are determined the next step will be to determine appropriate and meaningful non-dimensional quantities to represent these processes. Obviously, these nondimensional quantities in the laboratory experiments need to be maintained in the same range as that experienced in the field.

If necessary, the influence of scale effects can be examined by running tests at a number of scales to understand better the prototype situation.

The question of scale effects relates both to the geometric scale (i.e. the size of features with respect to the structure) and to the time scale of the process (i.e. how fast features develop).

Finally, there may also be model effects that need to be considered. The key features of the structure, and more specifically their influence on the processes, will need to be represented in the model to reduce the level of unwanted model effects in the experiments. With experience the model effects can be kept to a sufficiently low level of influence in the model design or, with a diverse data set, they can be factored out in the interpretation of the results.

Model effects may also relate to the fact that the scour in nature may be transient as the magnitude of the forcing changes with time, and experimental studies typically operates with a steady forcing. This may mean that in the laboratory the equilibrium scour development can be achieved whilst in the field the scour development may be less well pronounced simply because there is insufficient time for it to develop.

The great advantage of laboratory testing is that the key factors causing scour development can be investigated in a controlled fashion. The results can be represented by non-dimensional semi-empirical parameters that can be used in design work. However, the potential influence of scale effects or model effects must always be considered in the interpretation of results. The reader is referred to the references given above for more information on this topic.

2.6 Scour protection for pipelines

There are basically three kinds of protection measures against the scour: (1) the pipeline may be buried in a trench; (2) it may be covered with a stone protection layer; or (3) it may be covered with a protective mattress.

Regarding the **pipelines buried in a trench**, the key issue is to determine the burial depth. This depth depends on various variables such as the wave climate, the current climate, sediment properties, and the liquefaction potential. A trench is dredged, and the pipeline is placed in the trench, and usually backfilled with the excavated material. If the excavated material does not have the quality to protect the pipe against the risk of liquefaction (where the pipeline may float to the surface of the bed), it must be replaced with a coarser material, a material which is sufficiently permeable so that all pore pressures developed in it by cyclic stress applications (due to waves, or earthquakes, or other effects) will dissipate as rapidly as they develop (Seed

114   CHAPTER 2. SCOUR BELOW PIPELINES

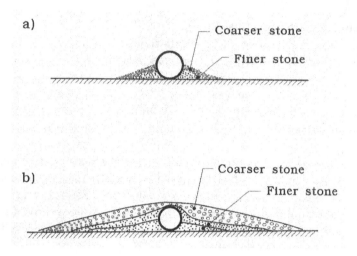

Figure 2.78: Stone protection for pipeline. (a): Hjorth (1975). (b): Herbich (1981).

and Rahman, 1978). Pipelines buried in the soil may also sink, depending on the density of the pipe and the density of the mixture of the liquefied soil and the water. The floatation and sinking of pipelines in liquefied soil will be studied in greater detail in Chapter 10.

In cases when the required burial depth is large, or when it is required that the pipeline should be easily accessible, or when the cost of the pipe burial is simply too high, the pipeline may be buried close to the surface (or it may be allowed to sit on the bed) with a stone cover or a protective mattress.

The **stone cover** is an effective measure for scour protection. It may also be employed in place of sediment backfill referred to in the preceding paragraphs, to provide a stronger and heavier cover for pipelines buried in areas of heavy traffic, or large wave forces.

The extent of the protection layer necessary to provide an adequate protection is usually smaller than the extent of the scour itself. Hjorth (1975) states that, to protect a pipeline against the undermining, it should be sufficient to stabilize the bed near the pipe by means of a stone cover, as illustrated in Fig. 2.78 a. He reports that all known experience indicates that if this cover is extended to the mid-height of the pipe, the bed will be safely

2.6. SCOUR PROTECTION FOR PIPELINES

protected. The stone cover may also be constructed so that it covers the pipeline completely (Fig. 2.78 b). It may be noted that the cover may be constructed in the form of a filter (as in Fig. 2.78 a and b), to prevent the stones from sinking in the bottom soil.

The conventional design strategy for a stone cover requires the stability of the top layer. For this, the Shields criterion is used; namely, the Shields parameter calculated for the stones must be smaller than θ_{cr}, the critical value of the Shields parameter corresponding to the initiation of motion at the top layer of the protection layer.

The stability of stone covers has recently been investigated by Klomp and Tonda (1995), adopting a different strategy: a certain damage is accepted for the design conditions (yet, the damage may not lead to failure of the structure), an approach similar to that used in the design of rubble-mound breakwaters. (Allowing a certain damage may avoid undue conservatism). An empirical relationship between the loading and the damage was established:

$$S_{1000} = 21.4 \tan(\alpha) \Psi^{2.25} \tag{2.68}$$

in which S_{1000} is the damage after 1000 waves, α is the side slope of the stone cover, and Ψ is the so-called mobility number in which S_{1000} and Ψ are, respectively, defined by

$$S_{1000} = A_e/D_{n50}^2 \tag{2.69}$$

and

$$\Psi = U_m^2/(gD_{n50}\Delta) \tag{2.70}$$

in which A_e is the erosion area, D_{n50} is the nominal diameter of stone ($= (M_{50}/\rho_s)^{1/3}$), U_m is the near-bed orbital velocity, and Δ is the relative density ($= (\rho_s - \rho)/\rho$). Van der Meer (1993) gives design values of S in the case of rubble mound breakwaters (for a two-diameter thick armour layer): $S = 2-3$ for the initial damage (corresponding to 0-5 % damage according to the familiar Hudson formula), $S = 3 - 12$ for the intermediate damage, and $S = 8 - 17$ for the failure. In Klomp and Tonda's work, the time-dependent damage, and the effect of wave-current interaction were also investigated.

A stone-protection layer may be subject to failure. This is caused by different mechanisms. A detailed account of the failure mechanisms are given in Chapter 3 in Section 3.4 in conjunction with the scour protection around piles.

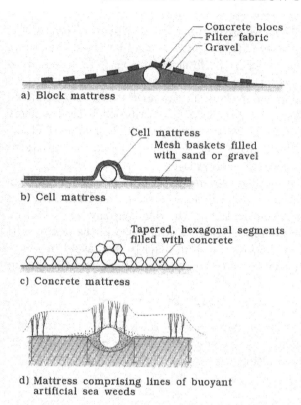

Figure 2.79: Mattress protection. (a)-(c): Herbich (1981). (d): Offshore, August Issue 1988, p.48.

The third method, namely the **protective mattress**, is an alternative to the stone cover. This method is employed in the case where rock is not easily available. It may also be employed when there is a risk of further scour due to the stone elements themselves, plus when there is a requirement for further maintenance work due to the settlement and sinking of the stone. Also, the mattresses may be preferred in favour of the stone protection when a certain reach or a certain point of the pipeline should be readily accessible.

There are various kinds of mattresses such as block mattresses (Fig. 2.79 a), cell mattresses filled with sand or gravel (Fig. 2.79 b), cell mattresses filled with concrete (Fig. 2.79 c), and mattresses comprising lines of buoy-

2.7. MATHEMATICAL MODELLING

ant artificial sea weeds (Fig. 2.79 d). The "edge" scour can be a problem when the mattresses with rigid edges are used (such as the concrete cell mattresses). Even the entire system comprising the pipeline and its protective mattress can be undermined due to the tunnel erosion (particularly in the areas where there are other disturbances such as that due to a pipeline bend), hampering the scour protection. This effect may lead to a suspended span of the pipeline with the protective mattress hanging from the pipe into the naturally generated trench, consequently with a very large scour depth.

An extensive review of the subject is given by Herbich (1981).

Finally, it proves necessary to monitor the scour development (if any) and the scour protection after the protection work is applied in the field. This obviously enables the operator to detect any scour at the bed developing at/around the protection layer, to locate if there is any local failure of the protection work itself, and to take remedial measures if and when necessary.

2.7 Mathematical modelling

The mathematical models on scour below pipelines may be divided into three categories: (1) the potential-flow models, (2) the advanced models, and (3) the integrated models. The first two models concern the flow and the resulting scour for a fixed pipeline, while the third one aims at predicting the time evolution of the interaction between a pipeline and the bed under a given time series of wave and current conditions.

2.7.1 Potential-flow models

Several authors have developed potential-flow models for flow and scour around pipelines, Chao and Hennessy (1972), Hansen, Fredsøe and Mao (1986), Hansen (1992), Bernetti, Bruschi, Valentini and Venturi (1990), Li and Cheng (1999 a).

Chao and Hennessy's (1972) model. von Muller (1929) has used the method of dipoles to develop a potential-flow model for a cylinder placed above a plane bed. Chao and Hennessy (1972), however, were the first to implement the potential-flow theory for scour calculations, whereby they expressed the discharge between the gap and the bed as (see the definition

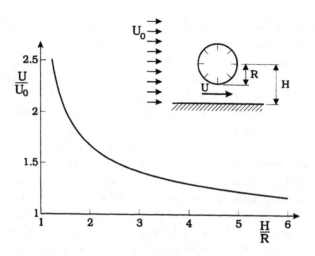

Figure 2.80: Average velocity in "scour hole". Chao and Hennessy's (1972) potential-flow theory.

sketch in Fig. 2.80)

$$q = \int_0^{H-R} u(y)\, dy = U_0 \left[H - R\left(\frac{1}{2\frac{H}{R} - 1}\right) \right] \quad (2.71)$$

The average velocity in the gap (normalized with the approach velocity U_0) was then calculated (from Eq. 2.71):

$$\frac{U}{U_0} = \frac{\frac{q}{H-R}}{U_0} = \frac{\frac{H}{R}}{\frac{H}{R} - 1} - \frac{1}{\frac{H}{R} - 1}\left(\frac{1}{2\frac{H}{R} - 1}\right) \quad (2.72)$$

Fig. 2.80 displays this velocity as a function of H/R.

Subsequently, the bed shear stress, τ, was calculated, based on this velocity (Eq. 2.72) and the friction factor (from the friction factor versus Reynolds number relationship reported by Lovera and Kennedy (1969)). Chao and Hennessy (1972) then argued that the scour continues until the bed shear stress attains the value of the critical shear stress, τ_{cr} (corresponding to the initiation of motion at the bed). The gap corresponding to this situation (where $\tau = \tau_{cr}$) was then designated as the equilibrium scour depth.

Three constraints with Chao and Hennessy's analysis are:

2.7. MATHEMATICAL MODELLING

1. The potential-flow theory can not adequately describe the flow in the gap, as revealed by the measurements reported in Fredsøe and Hansen (1987). For example, the potential-flow theory gives $U/U_0 = 1.9$ for $H/R = 1.6$, while this velocity has been measured to be $U/U_0 \cong 1.5$. The disagreement between the potential-flow theory and the experiments becomes larger and larger with decreasing values of the gap-to-diameter ratio.

2. Clearly, the potential-flow theory cannot handle the lee-wake flow; therefore it cannot handle the lee-wake erosion, an important stage in the scour process, as seen in the preceding sections.

3. The assumption that the bed shear stress eventually attains the critical value of the bed shear stress τ_{cr} may be valid only for the clear-water scour case, $\theta < \theta_{cr}$. For the live-bed situation, the bed shear stress eventually attains the level of the bed shear stress in the approach flow, which is necessarily larger than the critical shear stress, as discussed in conjunction with Fig. 2.18.

Chiew (1991 b) has circumvented this problem by calculating U from an empirical relation involving the directly measured flow discharge in the gap. It may be noted that the latter quantity in Chiew's study has been obtained as a function of the water-depth-to-pipe-diameter ratio, h/D (the latter being in the range of $1.5 \lesssim h/D \lesssim 6$). The scour depths predicted by Chiew using his calculation procedure have apparently been in agreement with the measured ones.

Hansen et al.'s (1986) model. These authors have extended Fredsøe and Hansen's (1987) modified potential theory to the case of the scoured bed. The principal idea in this study is to implement infinite series of pairs of dipoles to fulfill the requirement that the prescribed bed is actually a streamline. They assumed a symmetric bed form (see the definition sketch in Fig. 2.81) with

$$h(x) = e + \frac{D}{2} + \frac{S}{2}(1 + \cos(2\pi \frac{x}{W})), \quad |x| < \frac{W}{2}, \quad (2.73)$$
$$h(x) = e + \frac{D}{2}, \quad |x| > \frac{W}{2}$$

Their flow calculation gives the velocity at the bed as depicted in Fig. 2.81

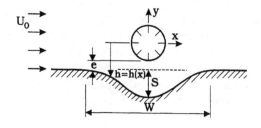

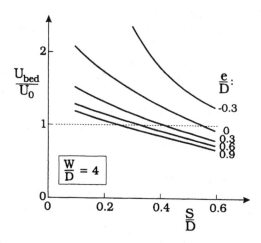

Figure 2.81: Variation of bed velocity with scour depth. Hansen et al.'s (1986) potential theory.

for $W/D = 4$ for different values of the gap-to-diameter ratio. In Fig. 2.81

$$\frac{U_{bed}}{U_0} = 1 \qquad (2.74)$$

is also plotted (the dashed line) which may, to a first approximation, characterize the equilibrium situation, the stage where the scour will stop. Here, U_{bed} is the velocity at the bed just under the center of the pipe.

For scour calculations, the equation of continuity for sediment is considered:

$$\frac{\partial q}{\partial x} = -(1-n)\frac{\partial h}{\partial t} \qquad (2.75)$$

2.7. MATHEMATICAL MODELLING

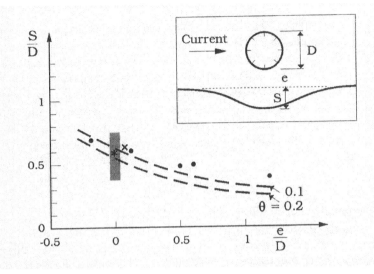

Figure 2.82: Dashed line: Numerical, Hansen et al. (1986). Experiment data compiled by Sumer et al. (1988 b).

in which h = the scoured bed level (Fig. 2.81), and q = the sediment transport rate per unit width (by volume), given by the Meyer-Peter Muller formula (Fredsøe and Deigaard, 1992, p. 214)

$$\Phi = \frac{q}{\sqrt{(s-1)gd^3}} = 8(\theta - \theta_{cr})^{3/2} \qquad (2.76)$$

where the Shields parameter is replaced by a new, modified Shields parameter

$$\theta^* = \theta - 0.1\frac{\partial h}{\partial x} \qquad (2.77)$$

the last term, $\frac{\partial h}{\partial x}$, accounting for the effect of a local bed slope (Fredsøe, 1974). The equilibrium scour profile is determined by

$$\frac{\partial q}{\partial x} = 0 \qquad (2.78)$$

meaning that the sediment transport must be constant along the bed.

Apparently, this is possible for various combinations of W and S. The authors have worked out the sediment transport q as a function of x for a number of combinations of W and S, and used the least square method to decide what combination of W and S gives the best result in conjunction with Eq. 2.78. (Note that the combinations of W and S in the calculations are chosen such that Eq. 2.74 is satisfied). The results obtained in this way are plotted together with the experimental data in Fig. 2.82 (for two different values of the Shields parameter, the dashed lines). As seen, the agreement with the experimental data is fairly good. It may also be noted that the scour width (the half width, $W/2$, Fig. 2.81) calculated by Hansen et al. is, for $e/D = 0$,

$$\frac{W/2}{D} = 2 - 3$$

and this compares fairly well with $W_1/D \sim 2$; and $W_2/D \sim 4$ (see Eq. 2.41).

Although the flow at the upstream side of the pipe is well represented by Hansen et al.'s theory, and the scour depth is well predicted, the theory is unable to predict the downstream part of the bed profile, the portion of the bed with the distinct feature of the mild slope. This is because, although the modified potential-flow theory takes into account the lee-wake to some degree by setting the velocities at the bottom and the top of the pipe equal, it cannot handle the lee-wake altogether (obviously, no vortex shedding). The latter is shown to be the key element with regard to the lee-wake erosion, as discussed in Section 2.3.1.

Hansen's (1992) model. Hansen (1992) has extended Hansen et al.'s (1986) theory (1) to the case of waves, (2) to the case of combined waves and current (wave-dominated), and (3) to the case of obliquely incident waves. Clearly, to be able to calculate the scour in the case of waves, the lee-wake needs to be taken into consideration. To this end, Hansen has adopted the following relation for the extension of the lee-wake

$$\frac{L}{D} = 0.8 + \frac{2a}{D} \tag{2.79}$$

from Jacobsen, Bryndum and Fredsøe (1984) (cf. Eq. 2.21). In the case of combined waves and current, the stroke $2a$ in Eq. 2.79 is replaced by

$$2a^- = \int_0^{T_w} U^-(t)\, dt \text{ and } 2a^+ = \int_0^{T_w} U^+(t)\, dt \tag{2.80}$$

where U^- and U^+ are the total velocities when the flow is in the negative and positive directions, respectively, during one wave period.

2.7. MATHEMATICAL MODELLING

Given e/D, W/D, and S/D, the flow (and therefore the bed velocity) can be calculated in the same way as in Hansen et al. (1986), presumably leading to a diagram similar to that illustrated in Fig. 2.81.

For waves, $W/2$ is taken to be equal to the extension of the wake given in Eq. 2.79. (This relates the scour width to the Keulegan-Carpenter number). Then, the equilibrium scour is found from the previously mentioned U_{bed}/U_0 versus S/D diagram (cf. Fig. 2.81), corresponding to the condition $U_{bed}/U_0 = 1$ in which U_0 is the velocity of the approaching flow (Fig. 2.81), and U_{bed} is the velocity at the bed (just under the center of the pipe).

The values of the scour depth calculated in this way in Hansen's (1992) study (for $e/D = 0$) agree fairly well with the experimental data (Eq. 2.24) for $O(20) < KC < O(80)$. For smaller values of KC ($KC = O(5)$), however, the theoretical predictions are a factor of 2 larger than the experimentally obtained values. For values of KC larger than $O(80)$, the theory begins to deviate from the experimental data (At $KC = 80$, the theory underpredicts the scour depth by about 10%). Hansen (1992) has also calculated scour depth in the case of waves for other values of the initial gap between the pipe and the bed, $e/D = -0.2$; and 0.1 for the same range of KC, $5 < KC < 80$.

In the case of combined waves and current, the scoured bed profile has been considered to have two widths, W^- and W^+, which are taken to be equal to those given by Eqs. 2.79 and 2.80, and the calculations have been performed accordingly. However, the flow was wave-dominated; therefore the variation illustrated in Fig. 2.25 could not be captured entirely.

Hansen (1992) with his model has also studied the influence of the Shields parameter and the influence of the angle of wave attacks on the scour depth. The results of Hansen's calculations regarding the influence of the Shields parameter for $KC = 10$ and 80 are plotted in Fig. 2.32.

Bernetti et al.'s (1990) model. Bernetti et al. (1990) have developed a numerical model where the flow is calculated from the potential-flow theory. Bernetti et al.'s model actually consists of four components: (1) the flow component; (2) the bed-morphology component; (3) the component calculating the propagation of the scour along the pipeline; and (4) the component calculating the pipe sagging into the scour hole.

The flow component calculates the flow in the same way as in Chao and Hennessy's (1972) study (Fig. 2.80). This enables the authors to work out the average velocity in the gap between the pipe and the bed.

The bed-morphology component essentially uses the continuity equation for the bed sediment (Eq. 2.75). The sediment transport is calculated from Bijker's (1971) formula. The latter enables the authors to calculate the scour also in the case of combined waves and current.

The component calculating the propagation of scour along the pipeline considers the geometry of the flow as sketched in Fig. 2.57 with $\alpha = \phi$ in which α = the angle of the slope of the bed at the span shoulder, and ϕ = the angle of internal friction of the soil. The propagation of the scour is computed with a procedure similar to the 2-D scour hole.

The final component of the model considers the sagging of the pipeline at the mid-span, based on the computed, free-span length in the previous component. Clearly, the sagging of the pipe will change the flow conditions, and therefore the scour.

The procedure used in the model is as follows: (1) For a given sea state (the current plus wave conditions), calculate the flow; (2) Calculate the bed morphology; (3) Calculate the scour propagation along the pipe; (4) Calculate the sagging of the pipe at the mid-span; (5) Update the sea state (usually every 1 or 3 hours); and (6) Stop the simulation if the pipe reaches the bottom of the scour hole. With this, the authors state, the free-span length has then reached its maximum allowable value, in agreement with the discussion in conjunction with Eq. 2.58.

Bernetti et al. (1990) have compared their model results for the case of 2-D scour below a fixed pipe with Mao's (1986) steady-current experimental data (the time variation of the scour depth, and the scour depth variation with the Shields parameter), and obtained fairly good agreement. Their results for the combined-wave-and-current situation are in qualitative agreement with the experimental data of Sumer and Fredsøe (1996) in Fig. 2.25. Bernetti et al. also give the results of their free-span-length calculations as a function of time.

Li and Cheng's (1999 a) model. Li and Cheng have numerically solved the Laplace equation

$$\frac{\partial^2 \Phi}{\partial x^2} + \frac{\partial^2 \Phi}{\partial y^2} = 0 \qquad (2.81)$$

under the boundary conditions that (1) the normal component of the velocity at the pipe boundary and at the bed is zero; and (2) the bed shear stress is equal to the critical bed shear stress corresponding to the initiation of motion at the bed (similar to the work of Chao and Hennessy, 1972)

$$\tau = \tau_{cr} \qquad (2.82)$$

2.7. MATHEMATICAL MODELLING

in which the last term, namely τ_{cr}, also includes the effect of a large, local bed slope. In Eq. 2.81, Φ is the velocity potential. No sediment transport calculations have been performed; instead, the bed has been adjusted in an iterative way so that the equilibrium profile has been obtained where the condition in Eq. 2.82 has been fulfilled. The procedure used in Li and Cheng's work is: (1) Assume an initial form for the bed; (2) Solve the flow for this boundary; (3) Check if the boundary condition in Eq. 2.82 is satisfied; (5) If not, modify the bed form; (6) Repeat the steps in (1)-(5), until the aforementioned boundary conditions are fully satisfied.

As discussed earlier, the condition $\tau = \tau_{cr}$ at the bed implies that the model results may be valid only for the clear-water scour case, $\theta < \theta_{cr}$. Indeed, the model apparently gives a fairly good comparison with the experimental results in the case of the clear water scour, but not for the case of the live-bed scour. Also, the model fails to simulate the scour process properly for the downstream portion of the scour hole for the same reasons as mentioned in conjunction with the work of Chao and Hennessy (1972) in the preceding paragraphs.

2.7.2 Advanced models

These models calculate the flow around the pipe, solving basically the Navier-Stokes (N.-S.) equations numerically.

Numerical treatment of flow around cylinders has improved significantly with the increasing capacity of computers. A detailed review of the subject (in the case of a free cylinder) can be found in Sumer and Fredsøe (1997, Chapter 5).

While there have been numerous numerical investigations of flow around free cylinders (Sumer and Fredsøe, 1997), there have been comparatively few numerical investigations of flow around pipelines (where the pipe is confined with a rigid or an erodible bed below) (Leeuwenstein and Wind, 1984; van Beek and Wind, 1990, and Brørs, 1999, involving $k - \varepsilon$ **simulations**; Li and Cheng, 1999 b, 2001 and 2002, involving **Large Eddy Simulations**; and Sumer et al. 1988 a, Jensen, Jensen, Sumer and Fredsøe, 1989; and Jensen et al., 1990, involving **discrete-vortex models**). These studies, in the case of the erodible bed, also involve the numerical calculation of the bed morphology.

$k - \varepsilon$ simulation

The approach could possibly be best described by reference to the work of Brørs (1999).

The flow configuration in Brør's study is as follows. A fixed pipe is sitting on an erodible bed. The flow is switched on, and the bed below the pipe begins to undergo a 2-D scour. The objective is to obtain the time evolution of the scour hole, as the flow around the pipeline continues.

Brør's model has two modules: (1) the flow module; and (2) the sediment module.

The flow module. In this module, basically the Reynolds-averaged N.-S. equations are solved. These equations are (1) the continuity equation

$$\frac{\partial u_i}{\partial x_i} = 0 \qquad (2.83)$$

(2) the equation of motion

$$\frac{\partial u_i}{\partial t} + u_j \frac{\partial u_i}{\partial x_j} = -\frac{1}{\rho_f}\frac{\partial p}{\partial x_i} + \nu \frac{\partial}{\partial x_j}\left(\frac{\partial u_i}{\partial x_j} + \frac{\partial u_j}{\partial x_i}\right) - \frac{\partial \overline{u'_i u'_j}}{\partial x_j} + \left(\frac{\rho_s}{\rho_f} - 1\right)cg\delta_{i3} \qquad (2.84)$$

and (3) the equation of conservation of mass for suspended sediment

$$\frac{\partial c}{\partial t} + (u_j - w_s \delta_{j3})\frac{\partial c}{\partial x_j} = -\frac{\partial \overline{c' u'_j}}{\partial x_j} \qquad (2.85)$$

(The latter equation can be "switched on and off", as required; this aspect of the model will be discussed later in the section. When Eq. 2.85 is switched off, the last term in Eq. 2.84 will drop). Here, u_i = the i th component of the time-averaged velocity, u'_i = i th component of the fluctuating velocity, x_i = the Cartesian coordinates (x_3 is directed vertically upwards), t = time, and p = the time-averaged pressure. The density of the mixture of water and sediment is expressed as $\rho = \rho_f + (\rho_s - \rho_f)c$ in which c = the concentration of the suspended sediment, and the subscripts f and s refer to the fluid and the sediment components, respectively. Also, in the above equations, c' = the fluctuating component of the concentration, w_s = the fall velocity of sediment grains in still water and δ_{ij} = the Kronecker delta, i.e., $\delta_{ij} = 1$ for $i = j$, and $\delta_{ij} = 0$ for $i \neq j$.

2.7. MATHEMATICAL MODELLING

The turbulent fluxes of momentum and sediment are related to the mean-flow gradients through Boussinesq approximations:

$$\overline{u'_i u'_j} = \nu_T \left(\frac{\partial u_i}{\partial x_j} + \frac{\partial u_j}{\partial x_i} \right) - \frac{2}{3} \delta_{ij} k \qquad (2.86)$$

and

$$\overline{c' u'_j} = \frac{\nu_T}{\sigma_c} \frac{\partial c}{\partial x_j} \qquad (2.87)$$

in which ν_T = the turbulence viscosity, σ_c = the turbulent Prandtl number for concentration, and k is the turbulent kinetic energy

$$k = \frac{1}{2} \overline{u'_k u'_k} \qquad (2.88)$$

To close the above system, the k–ε model is adopted. This model involves two equations, one for the quantity k, and the other for the quantity ε, the rate of dissipation of turbulent kinetic energy:

$$\frac{\partial k}{\partial t} + u_j \frac{\partial k}{\partial x_j} = -\frac{\partial}{\partial x_j} \left(\frac{\nu_T}{\sigma_c} \frac{\partial k}{\partial x_j} \right) + P + G - \varepsilon \qquad (2.89)$$

and

$$\frac{\partial \varepsilon}{\partial t} + u_j \frac{\partial \varepsilon}{\partial x_j} = -\frac{\partial}{\partial x_j} \left(\frac{\nu_T}{\sigma_c} \frac{\partial \varepsilon}{\partial x_j} \right) + (C_{\varepsilon 1} P - C_{\varepsilon 2} \varepsilon + C_{\varepsilon 3} \max(0, G)) \frac{\varepsilon}{k} \qquad (2.90)$$

in which

$$\nu_T = \frac{C_\mu k^2}{\varepsilon} \qquad (2.91)$$

and P and G represent the production terms related to the velocity gradient and concentration gradient, respectively,

$$P = \nu_T \frac{\partial u_i}{\partial x_j} \left(\frac{\partial u_i}{\partial x_j} + \frac{\partial u_j}{\partial x_i} \right) \qquad (2.92)$$

and

$$G = \left(\frac{\rho_s}{\rho_f} - 1 \right) c g \frac{\nu_T}{\sigma_c} \frac{\partial c}{\partial x_3} \qquad (2.93)$$

Here $C_{\varepsilon 1}$, $C_{\varepsilon 2}$, $C_{\varepsilon 3}$ = are constants of the model, equal to 1.44, 1.92, and 1.44, respectively; and C_μ = another constant equal to 0.09.

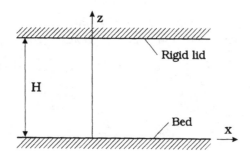

Figure 2.83: Computational domain.

A survey of turbulence models can be found in Rodi (1980).

The flow is simulated as a steady, pressure-driven water tunnel with a rigid lid at the top, the usual approach in the numerical simulation of free-surface flows for convenience, Fig. 2.83. 2-D calculations are carried out; i.e., $i = 1$, and 3 in the above equations.

At the inlet boundary, an equilibrium layer is specified below a chosen height δ. Above this level, a constant free-stream velocity $u = u_0$ is specified. Therefore the inlet conditions:

$$u(z) = \min\left\{\frac{U_f}{\kappa}\ln(\frac{30z}{k_s}),\ u_0\right\}; \quad v(z) = 0 \qquad (2.94)$$

$$k(z) = \max\left\{C_\mu^{-1/2}(1 - \frac{z}{\delta})^2 U_f^2,\ 0.0005u_0^2\right\} \qquad (2.95)$$

$$\varepsilon(z) = \frac{C_\mu^{3/4} k(z)^{3/2}}{\ell}, \quad \ell = \min\left\{\kappa z(1 + 1.5\frac{z}{\delta})^{-1},\ C_\mu \delta\right\} \qquad (2.96)$$

in which the friction velocity is evaluated as $U_f = \kappa u_0 / \ln(30\delta/k_s)$. For the bed roughness, $k_s = 2.5d = 2.5 \times (2 \times 10^{-4}\text{m}) = 5 \times 10^{-4}\text{m}$ is selected (corresponding to 0.2 mm sand) as a default value if nothing else is stated. The cylinder surface is also assumed rough with the same roughness. The quantity κ, the von Karman constant, is taken to be 0.42.

In the calculations with suspended sediment, the near-bed sediment concentration is specified as

$$c_{p1} = \max(c_0,\ c_{p2}) \qquad (2.97)$$

2.7. MATHEMATICAL MODELLING

in which

$$c_0 = 0.3 \text{ if } \theta > 0.75 \qquad (2.98)$$
$$= 0.3(\theta - \theta_s)(0.75 - \theta_s)^{-1} \text{ if } \theta_s < \theta < 0.75$$
$$= 0 \text{ if } \theta < \theta_s$$

in which c_{p1} = the concentration at the bed, c_{p2} = the concentration at the mesh point nearest the bed, and θ_s = the Shields parameter corresponding to the initiation of suspension from the bed, taken as 0.25 (see Example 20 in Section 7.1.1 in Chapter 7 for a detailed analysis for the latter).

At the seabed and at the pipe, the flow domain is put some distance away from the surface, and the gap is bridged using the logarithmic law of the wall.

At the top boundary,

$$\text{At } z = H : \quad \frac{\partial u}{\partial z} = 0, \text{ and } v = 0 \qquad (2.99)$$

and the free-stream values in Eqs. 2.95 and 2.96 are specified for the quantities k and ε.

At the outlet, the pressure is given a reference value $p(z) = 0$, while the other flow variables are allowed to adjust freely with zero x–gradient conditions.

The sediment module. This component of the model is concerned with the bed morphology. Basically, the equation of the continuity of sediment is solved numerically to continuously "update" the bed morphology. The latter equation reads

$$\frac{\partial h}{\partial t} = \frac{1}{1-n}(-\frac{\partial q}{\partial x} + D + E) \qquad (2.100)$$

in which h = the scoured bed level (see e.g., Fig. 2.81). This equation is a slightly different version of Eq. 2.75 with two new terms D and E in which D = the rate at which the sediment is deposited on the bed, and E = the rate at which it is entrained into the flow.

The sediment transport q is taken as

$$q = q_0 - C q_0 \frac{\partial h}{\partial x} \qquad (2.101)$$

q_0 = the bed-load sediment transport rate per unit width (by volume), and taken to be

$$q_0 = 12\sqrt{g(\rho_s/\rho_f - 1)d^3}(\theta - \theta_{cr})\theta^{1/2} \text{ if } \theta > \theta_{cr} \qquad (2.102)$$
$$= 0 \text{ otherwise.}$$

The quantities D and E are taken to be

$$D = w_s c_0 \qquad (2.103)$$

and

$$E = \frac{\nu_T}{\sigma_c} \frac{\partial c}{\partial x_3} \qquad (2.104)$$

Again, when the suspended load is switched off, then the preceding two equations will drop, therefore Eq. 2.100 will reduce to Eq. 2.75.

Brørs (1999) has compared his results for the case of a pipe placed above a rigid, plane wall (with a gap-to-diameter ratio $e/D = 0.6$) with the results of experiments reported in Mao (1986) and Sumer et al. (1988). Vortex shedding has been resolved in the calculations (cf. Fig. 2.14), and various gross features of the vortex shedding including the Strouhal period and the downstream extent of the organized lee-wake flow (with the aggregate of shed vortices) apparently have been found to be in fairly good agreement with the experimental data. The increase in the bed shear stress experienced as the vortices pass overhead (Fig. 2.16) has been captured in the calculations. The instantaneous bed shear stress at the downstream of the pipe (corresponding to the passage of the vortex shed from the lower edge of the pipe) can be a factor of 3-4 larger than the bed shear stress in the approach flow, in agreement with the discussion in the previous sections in conjunction with Fig. 2.16.

As for the scour calculations, the following procedure has been used:

1. Start the simulation with an initially scoured bed (in the example given in Brørs' paper, this was a sine-shaped scour hole with a depth of $0.1D$).

2. Calculate the flow, using the flow module. (The time step, Δt, in the flow calculations should be sufficiently small, so small that it enables the vortex shedding to be resolved in the case when there is vortex shedding. After a grid update, the flow model should be run sufficiently long so that transients created by the grid update are allowed to die out. Brørs gives the latter time as D/u_0).

3. Calculate the rate at which the bed morphology changes, namely $\partial h/\partial t$, solving Eq. 2.100 at every grid point along the length of the bed profile. (In these calculations, use the bed shear stress, turbulent viscosities, and sediment concentrations obtained in the previous step).

2.7. MATHEMATICAL MODELLING

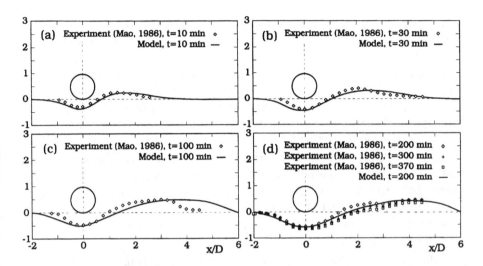

Figure 2.84: Time development of scour. Model: Brør's (1999) k-ε model.

4. Find the bed-morphology changes by multiplying the bed-change rates ($\partial h/\partial t$) with an appropriate morphological time step Δt_b, and "update" the bed accordingly. (Δt_b is chosen such that the maximum depth change along the bed is $0.01D$).

5. Adapt the grid to this new, updated bed profile.

6. Repeat Steps 2-5.

Regarding the flow time-steps and the morphological time-steps, Brørs indicates that these are typically $\Delta t = 10^{-3}$ s (or less), and $\Delta t_b = 2-4$ s for the flow conditions used in the simulations, namely, $D = 10$ cm, and $u_0 = O(10 \text{ cm/s})$. A Galerkin finite-element method has been used in the flow calculations in Step 2, while the finite differences have been used in solving Eq. 2.100 for the morphological calculations.

Fig. 2.84 compares the results obtained by Brørs with the experimental data from Mao (1986). The pipe diameter $D = 10$ cm, and $\theta = 0.048$ with $d = 0.36$ mm.

Brørs (1999) reports that, although vortex shedding has been obtained in the case of the plane rigid bed (as discussed above), no vortex shedding has been predicted in the case of the sediment bed, even during the later stages of the scour process. He attributes the early appearance of the steady state in the scour simulations ($t = 200$ min, Fig. 2.84 d) to the absence of vortex shedding. No clear explanation has been given why the vortex shedding does not occur in the numerical simulations while the experiments (corresponding to similar conditions) do show that vortex shedding occurs from the very early stages of the scour process (Sumer et al., 1988, and Jensen et al., 1990; see also discussion in Section 2.3.1).

It is interesting to note the following: When the calculations are started with a deeper initial scour hole and a smooth downstream dune, a simulation where the suspended load is switched off has produced a final bed shape, similar to that presented in Fig. 2.84. However, with the initial bed shape adopted in the calculations, the bed profile has become unstable in the initial stage when the suspended load is switched off.

Finally, Brørs gives the following information regarding the computational time for a Sun 10 workstation for the entire scour process (until the equilibrium scour has been reached). Number of bed updates is 6000; number of time steps between two successive bed updates is 200. Hence the total number of time steps is 6000×200 = 1.200.000. Computing time for each time step is 0.4 s. Therefore, the total computing time is 1.200.000×0.4 s or 130 h. Brørs (personal communication, 2000) has also provided the authors the following information: The preceding figures are for the year 1994. The same calculations on a 700 MHz PC of today (2000) take approximately 20 h.

Leeuwenstein and Wind (1984) and later van Beek and Wind (1990) have adopted a similar approach for numerical computation of scour and deposition around pipelines. In the former study, computations have been carried out for the time-averaged bed shear stress for a given scoured bed profile, using the $k - \varepsilon$ model. In the follow-up study (van Beek and Wind, 1990), the development of scour holes below a pipeline has been predicted with and without an attached spoiler. In the calculations, again, a $k - \varepsilon$ model has been used, along with a transport equation for suspended sediment to generate changes in the bed morphology.

Large Eddy Simulation (LES)

The principal idea in the method of LES is to average the N.-S. equations over a small volume of space. (Recall that the N.-S. equations are

2.7. MATHEMATICAL MODELLING

averaged over time to obtain the Reynolds equations, i.e., Eqs. 2.83 and 2.84). Normally, the size of the averaging volume is chosen such that the small-scale component of turbulence is, for convenience, not resolved. The aforementioned space-averaging gives the following set of equations: (1) the continuity equation,

$$\frac{\partial u_i}{\partial x_i} = 0 \qquad (2.105)$$

and (2) the equation of motion

$$\frac{\partial u_i}{\partial t} + u_j \frac{\partial u_i}{\partial x_j} = -\frac{1}{\rho}\frac{\partial p}{\partial x_i} + \nu \frac{\partial}{\partial x_j}(\frac{\partial u_i}{\partial x_j} + \frac{\partial u_j}{\partial x_i}) - \frac{\partial \overline{u'_i u'_j}}{\partial x_j} \qquad (2.106)$$

in which u_i = the space-averaged (the resolved) component of the velocity; p = the space-averaged (the resolved) component of the pressure; and u'_i = the fluctuating component of the velocity over the averaging volume (called the subgrid-scale component of the velocity, to differentiate it from the term "fluctuation" in the conventional Reynolds decomposition), and $\overline{u'_i u'_j}$ = the subgrid-scale stresses, analogous to the Reynolds stresses in the case of the Reynolds decomposition.

(Compare Eqs. 2.105 and 2.106 with Eqs. 2.83 and 2.84 (2.84 without the concentration term)).

The subgrid-scale stresses (SGS) are written (by analogy to the Boussinesq approximation)

$$\overline{u'_i u'_j} = \nu_t (\frac{\partial u_i}{\partial x_j} + \frac{\partial u_j}{\partial x_i}) \qquad (2.107)$$

in which ν_t = the turbulence viscosity associated with the unresolved component of turbulence. Hence, Eq. 2.106 will be

$$\frac{\partial u_i}{\partial t} + u_j \frac{\partial u_i}{\partial x_j} = -\frac{1}{\rho}\frac{\partial p}{\partial x_i} + \frac{\partial}{\partial x_j}\left[(\nu + \nu_t)(\frac{\partial u_i}{\partial x_j} + \frac{\partial u_j}{\partial x_i})\right] \qquad (2.108)$$

(It may be noted that the turbulence viscosity ν_t here is different from that in the $k - \varepsilon$ model defined by Eq. 2.86, ν_T. As emphasized in the present definition, Eq. 2.107, ν_t is associated with the unresolved small-scale turbulence averaged over the "averaging" volume, while ν_T is associated with the turbulence over the entire spectrum).

On dimensional grounds, the turbulence viscosity is written as

$$\nu_t = C_s^2 \Delta^2 s \qquad (2.109)$$

in which Δ is a length scale and C_s is a nondimensional constant. (By making an analogy to Prandtl's mixing-length hypothesis, it may be seen that s must be a quantity like $\partial u/\partial y$). We shall return to this point later in the section.

Clearly, the resolved quantities, $u_i(x_i,t)$, $p(x_i,t)$ and $\nu_t(x_i,t)$ in Eqs. 2.107-2.109 should be dependent on the size of the averaging volume. The smaller the size of this volume, the better the resolution (both in space and time), and therefore the more information on space and time variations of $u_i(x_i,t)$, $p(x_i,t)$ and $\nu_t(x_i,t)$, implying that these quantities are dependent on the size of the averaging volume. This feature in LES is handled through ν_t as follows.

As seen in Eq. 2.109, ν_t involves a length scale, Δ. This quantity is taken as the length scale representing the size of the averaging volume

$$\Delta = (\Delta x_1 \Delta x_2 \Delta x_3)^{1/3} \tag{2.110}$$

in which Δx_1, Δx_2, Δx_3 are the grid sizes in the x_1-, x_2-, x_3- directions, respectively. (The averaging volume being considered to be $\Delta x_1 \Delta x_2 \Delta x_3$). Furthermore, the quantity s in Eq. 2.109 is taken as

$$s = (2S_{ij}S_{ij})^{1/2} \tag{2.111}$$

in which S_{ij}

$$S_{ij} = \frac{1}{2}(\frac{\partial u_i}{\partial x_j} + \frac{\partial u_j}{\partial x_i}) \tag{2.112}$$

It is easy to see that s can also be written in the following form

$$s = \left[(\frac{\partial u_i}{\partial x_j} + \frac{\partial u_j}{\partial x_i})\frac{\partial u_i}{\partial x_j}\right]^{1/2} \tag{2.113}$$

The above model (Eqs. 2.105-2.111) is known as the Smagorinsky (1963) model for LES. As seen, only one "closure" coefficient is required, namely C_s, the Smagorinsky constant. The value of the Smagorinsky constant is determined from experiments.

To sum up, the solution of the above set of equations, namely Eqs. 2.105-2.107, and Eqs. 2.109-2.112, basically gives the time series of the hydrodynamic quantities (the velocity components and the pressure) in space, resolved with a resolution equal to the grid size.

A detailed account of turbulence models for LES has been given by Meneveau and Katz (2000) in a recent review paper.

2.7. MATHEMATICAL MODELLING

As has already been mentioned, Li and Cheng (1999 b, 2001) have investigated the flow and scour around a pipeline, using LES. These authors have adopted the 2-D LES equations (Eqs. 2.105 and 2.106) with the Smagorinsky model (Eqs. 2.105-2.111) with $\Delta = (\Delta x_1 \Delta x_2)^{1/2}$, and $C_s = 0.1$. They have run several simulation tests with a rigid bed with the pipe placed close to the bed. This has enabled the authors to validate their results against the experimental data from Sumer et al. (1988 a) and Jensen (1987). Their results (Li and Cheng, 1999 b and 2001) have confirmed the significance of the fluctuating component of the bed shear stress (induced by the vortex shedding) with regard to the scour at the downstream of a pipeline (Sumer et al., 1988 a, see also Section 2.3.1). Also, their simulation data (Li and Cheng, 2002) regarding the mean- and turbulent-flow quantities agree fairly well with the experimental data of Jensen (1987).

In Li and Cheng (1999 b and 2001), no sediment transport calculations have been performed; instead, the bed has been adjusted in an iterative way so that the equilibrium profile has been obtained, in exactly the same fashion as in the potential-flow work of the same authors (Li and Cheng, 1999 a, see Section 2.7.1). In Li and Cheng (1999 b), the equilibrium bed profile is defined such that Eq. 2.82 is fulfilled (the clear-water scour), while, in Li and Cheng (2001), the critical shear stress in Eq. 2.82 is replaced by the bed shear stress in the far field, i.e., the condition in Eq. 2.12 (the live-bed scour). As mentioned previously in conjunction with the potential-flow models, the downside of the bed-adjustment technique is that only the equilibrium profile of the scour hole is obtained (no information about the time evolution of the scour hole).

Li and Cheng (2002), in their subsequent study, have adopted the equation of the continuity of sediment (Eq. 2.100) to calculate the time development of the bed morphology. The simulated scour profiles have been found to be in accord with the corresponding experimental data. Particularly, the lee-wake erosion has been well captured.

Discrete-Vortex models

The discrete-vortex methods have been developed in the past to circumvent the problem of the large number of grid points required in the finite-difference/finite-element solutions of the N.-S. equations, a problem which can become prohibitive at large Reynolds numbers. A simple method offering an alternative (as a grid-free, or almost grid-free numerical method) is the discrete-vortex method.

A detailed account of the vortex methods (as implemented for the investigation of flow around free cylinders) can be found in Sumer and Fredsøe (1997).

In the following paragraphs, the method will be described briefly as applied to the flow around a pipeline placed near a bed.

Discrete vortices are released steadily into the flow from the boundaries, namely the pipe surface and the bed. The strength of these vortices are calculated in such a way that the zero normal velocity and zero slip conditions are satisfied together on the pipe surface and also that the zero normal velocity condition is satisfied on the bed. The vortices released into the flow are convected according to a Lagrangian scheme, which includes a random-walk element in it, to take care of the diffusion of vorticity. This scheme is actually equivalent to solving the vorticity transport equation

$$\frac{\partial \omega}{\partial t} + u\frac{\partial \omega}{\partial x} + v\frac{\partial \omega}{\partial y} = \nu(\frac{\partial^2 \omega}{\partial x^2} + \frac{\partial^2 \omega}{\partial y^2}) \tag{2.114}$$

in which $\omega =$ the vorticity

$$\omega = \frac{\partial v}{\partial x} - \frac{\partial u}{\partial y} \tag{2.115}$$

The vorticity ω can be calculated at any point in the flow field, and then the solution of the Poisson equation

$$\frac{\partial^2 \psi}{\partial x^2} + \frac{\partial^2 \psi}{\partial y^2} = -\omega \tag{2.116}$$

gives the stream function ψ from which the velocity is obtained through

$$(u, v) = (\frac{\partial}{\partial y}, \frac{\partial}{\partial x})\psi \tag{2.117}$$

where u and $v =$ the $x-$ and $y-$ components of the velocity. It may be noted that the vorticity transport equation above (Eq. 2.114) is obtained from the N.-S. equations by eliminating the pressure.

Sumer et al. (1988 a), Jensen et al. (1989) and Jensen et al. (1990) have made simulations of the flow around a pipe placed near a bed using a discrete-vortex model (the cloud-in-cell (CIC) model). Fig. 2.85 displays two snap-shots of the flow from Sumer et al.'s (1988 a) calculations, one for the case of a plane bed, and the other for the case of a scoured bed, for the same value of the clearance between the pipe and the bed.

2.7. MATHEMATICAL MODELLING

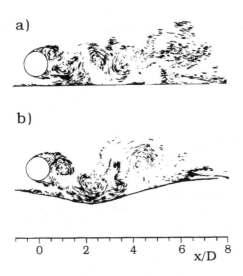

Figure 2.85: Vorticity field from the cloud-in-cell vortex method. Sumer et al. (1988 a).

2.7.3 Integrated models

The principal idea here is to develop a computer model which would enable the engineer to predict the emergence and disappearance of scour along a pipeline, the pipeline self-burial, the trench backfilling, migrating sand wave exposure, and undermining of pipelines. Obviously, such a model needs to accommodate all the possible processes such as the onset of scour, the tunnel erosion, the wake erosion, the 2-D scour, the 3-D scour, the sagging of the pipeline, the self-burial (in free span areas and at span shoulders), and other sediment transport processes. The input parameters will be the time series of the wave and current climate, the geometrical and structural properties of the pipeline plus the soil properties. The output will be the time series of the position of the pipe (with respect to the bed) at every point along the length of the pipeline. Papers by Staub and Bijker (1990), and Hansen, Klomp, Smed, Chen, Bijker and Bryndum (1995) present the results of a work where such an approach has been adopted. It may also be noted that the work by Bernetti et al. (1990) (described in the preceding paragraphs under Potential Flow Models) may be considered as one of the first examples of these types of models although it does not include the processes such as the onset of scour, the wake erosion, and the self-burial (in free span areas and at span shoulders). Finally, the paper by Bruschi, Drago, Venturi, Jiao and Sotberg (1998) gives a detailed account of the existing models to help estimate integrity of free spans.

2.8 References

1. Antonia, R.A., Browne, L.W.B. and Bisset, D.K. (1987): Topology of organized structures in a turbulent plane wake. In: Advances in Turbulence. Ed. G. Comte-Bellot and J. Matthieu. Springer Verlag, New York, N.Y., 337-345.

2. Bearman, P.W. and Zdravkovich, M.M. (1978): "Flow around a circular cylinder near a plane boundary", J. Fluid Mech., vol. 109, 33-48.

3. Bernetti, R., Bruschi, R., Valentini, V. and Venturi, M. (1990): Pipelines placed on erodible seabeds. Proc. 9th International Conference on Offshore Mechanics and Arctic Engineering, ASME, Houston, TX, vol. V, 155-164.

4. Bijker, E.W. (1971): Longshore transport computations. J. Waterways, Harbours and Coastal Engineering Division, ASCE, vol. 97, No. WW4, 687-701.

5. Bijker, E.W. (1986): Scour around marine structures. Proc. 20th Coastal Engineering Conference, Taipei, Taiwan, vol. 2, 1754-1768.

6. Bijker, E.W. and Leeuwenstein, N. (1984): Interaction between pipelines and the seabed under the influence of waves and currents. In: Seabed Mechanics, B. Denness, ed., Graham and Trotman, Gettysburg, Md., 235-242.

7. Bijker, R., Staub, C., Silvis, S. and Bruschi, R. (1991): Scour induced free spans. Proc. 23rd Offshore Technology Conference, Houston, TX, Paper No. OTC 6762, 583-588.

8. Blevins, R.D. (1977): Flow-Induced Vibrations. van Nostrand.

2.8. REFERENCES

9. Brown, R.J. (1976): Technical Articles for State-of-the-Art of Marine Pipeline Engineering. Compiled by R.J. Brown and Associates. Zug, Houston, The Hague, Singapore, Montreal.

10. Brown, R.J. (1966): Drag and lift forces on a submarine pipeline subjected to a transverse horizontal current. Society of Petroleum Engineering Journal, September 1966, vol. 6, 254-260.

11. Bruschi, R., Cimbali, W., Leopardi, G. and Vincenzi, M. (1986). Scour induced free span analysis. Proc. 5th International Conference on Offshore Mechanics and Arctic Engineering, ASME, Tokyo, Japan, vol. III, 656-669.

12. Bruschi, R., Drago, M., Venturi, V., Jiao, G. and Sotberg, T. (1998): Models bid to help estimate integrity of subsea free spans. Offshore Pipeline Technology, April 1998, 48-54.

13. Brørs, B. (1999): Numerical modelling of flow and scour at pipelines. J. Hydraulic Engineering, ASCE, vol. 125, No. 5, 511-523.

14. Cevik, E. and Yuksel, Y. (1999): Scour under submarine pipelines in waves in shoaling conditions. J. Hydraulic Engineering, ASCE, vol. 125, No. 1, 1-11.

15. Chao, J.L. and Hennessy, P.V. (1972): Local scour under ocean outfall pipelines. J. Water Pollution Control Fed., vol. 44, No. 7, 1443-1447.

16. Chiew, Y.-M. (1990): Mechanics of local scour around submarine pipeline J. Hydraulic Engineering, ASCE, vol. 116, No. 4, 515-529.

17. Chiew, Y.-M. (1991 a): Flow around horizontal circular cylinder in shallow flows. J. Waterway, Port, Coast and Ocean Engineering, ASCE, vol. 117, No. 2, 120-135.

18. Chiew, Y.-M. (1991 b): Prediction of maximum scour depth at submarine pipelines. J. Hydraulic Engineering, ASCE, vol. 117, No. 4, 452-466.

19. Chiew, Y.-M. (1992): Effect of spoilers on scour at submarine pipelines. J. Hydraulic Engineering, ASCE, vol. 118, No. 9, 1311-1317.

20. DIF (1984): Dansk Ingeniørforenings Code of Practice for Foundation Engineering. DS 415, 3rd ed., Teknisk Forlag, Normstyrelsens Publikationer, NP-168-N.

21. Fredsøe, J. (1974): On the development of dunes in erodible channels. J. Fluid Mech., vol. 64, 1-16.

22. Fredsøe, J. (1984): Turbulent boundary layers in wave-current motion. J. Hydraulic Engineering, ASCE, vol. 110, No. 8, 1103-1120.

23. Fredsøe, J. and Hansen, E.A. (1987): Lift forces on pipelines in steady flow. J. Waterway, Port, Coastal and Ocean Engineering, ASCE, vol. 113, No. 2, 139-155.

24. Fredsøe, J. and Deigaard, R. (1992): Mechanics of Coastal Sediment Transport. Advanced Series on Ocean Engineering, vol. 8, World Scientific, xviii + 369 p.

25. Fredsøe, J., Hansen, E.A., Mao, Ye and Sumer, B.M. (1988): Three-dimensional scour below pipelines. Trans. ASME, J. Offshore Mechanics and Arctic Engineering, vol. 110, 373-379.

26. Fredsøe, J., Sumer, B.M. and Arnskov, M. (1992): Time scale for wave/current scour below pipelines. International J. Offshore and Polar Engineering, vol. 2, No. 2, 13-17.

27. Gokce, T. and Gunbak, A.R. (1991): Self-burial and stimulated self burial of pipelines by waves. Proc. First International Offshore and Polar Engineering Conference, Edinburgh, U.K., vol. II, 308-314.

28. Gravesen, H. and Fredsøe, J. (1983): Modelling of liquefaction, scour and natural backfilling process in relation to marine pipelines. Offshore Oil and Gas Pipeline Technology, European Seminar, Feb. 2-3, 1983, Copenhagen.

29. Hansen, E.A. (1992): Scour below pipelines and cables: A simple model. Proc. 11th Offshore Mechanics and Arctic Engineering Conference, ASME, Calgary, Canada, vol. V-A, Pipeline Technology, 133-138.

30. Hansen, E.A., Fredsøe, J. and Mao, Ye (1986): Two dimensional scour below pipelines. Proc. 5th International Symp. on Offshore Mechanics and Arctic Engineering, ASME, Tokyo, Japan, vol. 3, 670-678.

2.8. REFERENCES

31. Hansen, E.A., Klomp, W.H.G., Smed, P.F., Chen, Z., Bijker, R. and Bryndum, M.B. (1995): Free span development and self-lowering of pipelines/cables. Proc. 14th Offshore Mechanics and Arctic Engineering Conference, ASME, Copenhagen, Denmark, vol. 5, Pipeline Technology, 409-417.

32. Hansen, E.A., Staub, C., Fredsøe, J. and Sumer, B.M. (1991): Time-development of scour induced free spans of pipelines. Proc. 10th Offshore Mechanics and Arctic Engineering Conference, ASME, Stavanger, Norway, vol. 5, Pipeline Technology, 25-31.

33. Henderson, F.M. (1966): Open Channel Flow. The Macmillan Company, 1966.

34. Herbich, J.B. (1981): Scour around pipelines and other objects. In: Offshore Pipeline Design Elements. Marcell Dekker, Inc. New York, NY., xvi + 233 p.

35. Herbich, J.B. (1985): Hydromechanics of submarine pipelines: Design problems. Can. J. Civil Engineering, vol. 12, No. 4, 863-887.

36. Herbich, J.B., Schiller, R.E., Jr., Watanabe, R.K. and Dunlap, W.A. (1984): Seafloor Scour. Design Guidelines for Ocean-Founded Structures, Marcell Dekker, Inc., New York, NY, xiv + 320 p.

37. Hjorth, P. (1975): Studies on the nature of local scour. Bull. Series A, No. 46, viii + 191 p., Department of Water Resources Engineering, Lund Institute of Technology/University of Lund, Lund, Sweden.

38. Hughes, S. A. (1993): Physical Models and Laboratory Techniques in Coastal Engineering. World Scientific.

39. Hulsbergen, C. H. (1984): Stimulated self-burial of submarine pipelines. Proc. 16th Offshore Technology Conference, Houston, Texas, May 7-9, 1984, paper No. OTC 4667, 171-177.

40. Hulsbergen, C. H. (1986): Spoilers for stimulated self-burial of submarine pipelines. Proc. 18th Offshore Technology Conference, Houston, Texas, May 5-8, 1986, Paper No. OTC 5339, 441-444.

41. Hulsbergen, C. H. and Bijker, R. (1989): Effect of spoilers on submarine pipeline stability. Proc. 21st Offshore Technology Conference, Houston, Texas, May 1-4, 1989, Paper No. OTC 6154, 337-350.

42. Ibrahim, A. and Nalluri, C. (1986): Scour prediction around marine pipelines. Proc. 5th International Symp. on Offshore Mechanics and Arctic Engineering, ASME, Tokyo, Japan, vol. 3, 679-684.

43. Jacobsen, V., Bryndum, M. and Fredsøe, J. (1984): Determination of flow kinematics close to marine pipelines and their use in stability calculations. Proc. 16th Annual Offshore Technology Conference, Paper No. OTC 4833, Conference Venue, vol. 3, 481-492.

44. Jensen, B.L. (1987): Large-Scale vortices in the wake of a cylinder placed near a wall. Proc. 2nd International Conference on Laser Anemometry, University of Strathclyde, Glasgow, Scotland, U.K., 21-23. September, 1987, 153-163.

45. Jensen, B.L., Sumer, B.M., Jensen, R. and Fredsøe, J. (1990): Flow around and forces on a pipeline near a scoured bed in steady current. Trans. ASME, J. Offshore Mechanics and Arctic Engineering, vol. 112, 206-213.

46. Jensen, H.R., Jensen, B.L., Sumer, B.M. and Fredsøe, J. (1989): Flow visualization and numerical simulation of the flow around marine pipelines on an erodible bed. Proc. 8th International Conference on Offshore Mechanics and Arctic Engineering, ASME, The Hague, The Netherlands, vol. V, 129-136.

47. Kjeldsen, S.P., Gjörsvik, O., Bringaker, K.G. and Jacobsen, J. (1973): Local scour near offshore pipelines. Second International Port and Ocean Engineering under Arctic Conditions, Conf. Reykjavik, 308-331.

48. Klomp, W.H.G. and Tonda, P.L. (1995): Pipeline cover stability. Proc. Fifth International Offshore and Polar Engineering Conference, The Hague, The Netherlands, 11-16. June, 1995, vol. II, 15-22.

49. Kozakiewicz, A., Fredsøe, J. and Sumer, B.M. (1995): Forces on pipelines in oblique attack: steady current and waves. Proc. 5th International Offshore and Polar Engineering Conference, ISOPE, The Hague, The Netherlands, Vol. 2, 174-183.

2.8. REFERENCES

50. Kristiansen, Ø. (1988): Current induced vibrations of pipelines on a sandy bottom. Thesis presented to the University of Trondheim, at Trondheim, Norway, in partial fulfillment of the requirements of the degree of Doctor of Philosophy.

51. Kristiansen, Ø. and Tørum, A. (1989): Interaction between current induced vibrations and scour of pipelines on a sandy bottom. Proc. 8th International Conference on Offshore Mechanics and Arctic Engineering, ASME, The Hague, The Netherlands, vol. V, 167-174.

52. Kroezen, M., Vellinga, P., Lindenberg, J. and Burger, A.M. (1982): Geotechnical and Hydraulic Aspects with Regard to Seabed and Slope Stability. Delft Hydraulics Laboratory, Publication No. 272, June 1982.

53. Leeuwenstein, W. (1985): Natural self-burial of submarine pipelines. MaTS- Stability of pipelines, scour and sedimentation. Coastal Engineering Group, Department of Civil Engineering, Delft University of Technology, Delft, The Netherlands.

54. Leeuwenstein, W. and Wind, H.G. (1984): The computation of bed shear in a numerical model. Proc. 19th International Conference on Coastal Engineering, Houston, TX, vol. 2, Chapter 114, 1685-1702.

55. Leeuwenstein, W., Bijker, E.A., Peerbolte, E.B. and Wind, H.G. (1985): The natural self-burial of submarine pipelines. Proc. 4th International Conf. on Behavior of Offshore Structures (BOSS), Elsevier Science Publishers, vol. 2, 717-728.

56. Li, F. and Cheng, L. (1999 a): Numerical model for local scour under offshore pipelines. J. Hydraulic Engineering, ASCE, vol. 125, No. 4, 400-406.

57. Li, F. and Cheng, L. (1999 b): Numerical simulation of pipeline local scour with lee-wake effects. International J. of Offshore and Polar Engineering, vol. 10, No. 3, 195-199.

58. Li, F. and Cheng, L. (2001): Prediction of lee-wake scouring of pipelines in currents. J. Waterway, Port, Coastal and Ocean Engineering, ASCE, vol. 127, no. 2, 106-112.

59. Li, F. and Cheng, L. (2002): Mathematical modelling of time-dependent scour below offshore pipelines. Personal communication.

60. Littlejohn, P.S.G. (1977): A study of scour around submarine pipelines. Report No. INT 113, Hydraulic Res. Station, Wallingford, England.

61. Lovera, F. and Kennedy, J.F. (1969): Friction factors for flat-bed floors in sand channels. J. Hydraulics Div., ASCE, vol. 95, HY4, 1227-1234.

62. Lucassen, R.J. (1984): Scour underneath submarine pipelines. Report No. PL-4 2A, Netherlands Marine Tech. Res., Netherlands Industrial Council for Oceanology, Delft University of Technology, Delft, the Netherlands, Sep. 1984. Student Thesis supervised by E.W. Bijker and W. Leeuwenstein.

63. Meneveau, C. and Katz, J. (2000): Scale-Invariance and turbulence models for Large-Eddy Simulation. In: Annu. Rev. Fluid Mech., vol. 32, 1-32.

64. Mao, Y. (1986): The interaction between a pipeline and an erodible bed. Series Paper 39, Tech. Univ. of Denmark, ISVA, in partial fulfillment of the requirement for the degree of Ph.D.

65. Mao, Y. (1988): Seabed scour under pipelines. Proc. 7th International Conference on Offshore Mechanics and Arctic Engineering Conference, ASME, Houston, TX, vol. V, 33-38.

66. Meyer-Peter, E. and Müller, R. (1948): Formulas for Bed-Load Transport. Rep. 2nd Meet. Int. Assoc. Hydraul. Struct. Res., Stockholm, 1948, 39-64.

67. Moncada-M., A.T. and Aguirre-Pe, J. (1999): Scour below pipeline in river crossings. J. Hydraulic Engineering, ASCE, vol. 125, No. 9, 953-958.

68. Müller, W. von (1929): Systeme von Doppelquellen in der ebenen Strömung. Zeitschrift f. angew. Math. und Mech., 9, Heft 3, 200-213.

69. Oumeraci, H. (1994): Scour in front of vertical breakwaters - Review of problems. Proceedings of International Workshop on Wave Barriers in Deep Waters. Port and Harbour Research Institute, Yokosuka, Japan, 281-307.

2.8. REFERENCES

70. Orgill, G., Barbas, S.T., Crossley, C.W., Carter, L.W. (1992): Current practice in determining allowable pipeline free spans. Proc. of the 11th Offshore Mechanics and Arctic Engineering Conf., June 7-11, 1992, Calgary, Canada, Pipeline Technology, vol. 5-A, 139-145.

71. Pluim-van der Velden, E.T.J.M. and Bijker, E.W. (1992): Local scour near submarine pipelines on a cohesive bottom. Behaviour of Offshore Structures (BOSS). Proc. Sixth International Conference, Supplement, 7-10. July, 1992, Imperial College of Science, Technology and Medicine, London. Bentham Press.

72. Rodi, W. (1980): Turbulence Models and Their Applications in Hydraulics. IAHR, Delft, The Netherlands, 104 p.

73. Roll, P. (1989): Selfburial of Marine Pipelines. Master's Thesis. Technical University of Denmark, ISVA, a study supervised by B.M. Sumer and J. Fredsøe.

74. Sarpkaya, T. and Isaacson, M. (1981): Mechanics of Wave Forces on Offshore Structures. Van Nostrand Reinhold Company. xvi + 651 p.

75. Seed, H.B. and Rahman, M.S. (1978): Wave-induced pore pressure in relation to ocean floor stability of cohesionless soil. Marine Geotechnology, 3, No. 2, 123-150.

76. Sidek, F.J. and Ibrahim, A.A. (1992): The armouring effects of shell fragments in seabeds beneath pipelines. Proc. 2nd International Offshore and Polar Engineering Conference, San Francisco, CA, vol. II, 92-100.

77. Smagorinski J. (1963): General circulation experiments with the primitive equations. I. The basic experiments. Mon. Weather Rev., vol. 91, No. 3, 99-164.

78. Stansby, P.K. and Starr, P. (1991): On a horizontal cylinder resting on a sand bed under waves and current, Int. J. Offshore Polar Eng. vol. 2, No. 4, pp. 262-266.

79. Staub, C. and Bijker, R. (1990): Dynamic numerical models for sand waves and pipeline self-burial. Proc. 22nd International Conference

on Coastal Engineering, Delft, The Netherlands, Chapter 190, vol. 3. 2508-2521.

80. Sumer, B.M. and Fredsøe, J. (1990): Scour below pipelines in waves. J. Waterway, Port, Coastal and Ocean Engineering, ASCE, vol. 116, No. 3, 307-323.

81. Sumer, B.M. and Fredsøe, J. (1991): Onset of scour below a pipeline exposed to waves. International J. Offshore and Polar Engineering, vol. 1, No. 3, 189-194.

82. Sumer, B.M. and Fredsøe, J. (1994): Self-burial of pipelines at span shoulders. International J. Offshore and Polar Engineering, vol. 4, No.1, 30-35.

83. Sumer, B.M. and Fredsøe, J. (1996): Scour around pipelines in combined waves and current. Proc. 7th International Conference on Offshore Mechanics and Arctic Engineering Conference, ASME, Florence, Italy, vol. V, Pipeline Technology, 595-602.

84. Sumer, B.M. and Fredsøe, J (1997): Hydrodynamics Around Cylindrical Structures. World Scientific, xviii + 530 p.

85. Sumer, B.M., Chua, L.H.C., Cheng, N.-S. and Fredsøe, J. (2002): Influence of turbulence on bedload sediment transport. Manuscript in preparation.

86. Sumer, B.M., Jensen, B.L. and Fredsøe, J. (1991): Effect of a plane boundary on oscillatory flow around a circular cylinder. J. Fluid Mechanics, vol. 225, 271-300.

87. Sumer, B.M., Jensen, R., Mao, Y. and Fredsøe, J. (1988 a): The effect of lee-wake on scour below pipelines in current. J. Waterway, Port, Coastal and Ocean Engineering, ASCE, vol. 114, No. 5, 599-614.

88. Sumer, B.M., Mao, Y. and Fredsøe, J. (1988 b): Interaction between vibrating pipe and erodible bed. J. Waterway, Port, Coastal and Ocean Engineering, ASCE, vol. 114, No. 1, 81-92.

89. Sumer, B.M., Pedersen, C., Yu, D. and Fredsøe, J. (1990): Bed shear-stress measurements in the vicinity of a pipeline in waves. Progress Report 71, Tech. Univ. of Denmark, ISVA, 61-72.

2.8. REFERENCES

90. Sumer, B.M., Truelsen, C., Sichmann, T. and Fredsøe, J. (2001 a): Onset of scour below pipelines and selfburial. Coastal Engineering, vol. 42, 4, 313-335.

91. Sumer, B.M., Whitehouse, R.J.S. and Tørum, A. (2001 b): Scour around coastal structures. A summary of recent research. Coastal Engineering.

92. Sutherland J. and Whitehouse R.J.S. (1998): Scale Effects in the Physical Modelling of Seabed Scour. HR Wallingford Report TR64.

93. Terzaghi, K. (1948): Theoretical Soil Mechanics. John Wiley Sons, Inc., New York.

94. van Beek, F.A. and Wind, H.G. (1990): Numerical modelling of erosion and sedimentation around pipelines. Coastal Engineering, vol. 14, No. 2, 107-128.

95. van der Meer, J.W. (1993): Conceptual Design of Rubble Mound Breakwaters. Delft Hydraulics, Publication no. 483, iv + 74 p.

96. Westerhortmann, J.H., Machemehl, J.L. and Jo, C.H. (1992): Effect of pipe spacing on marine pipeline scour. Proc. 2nd International Offshore and Polar Engineering Conference, San Francisco, CA, vol. II, 101-109.

97. Whitehouse R.J.S. (1998): Scour at Marine Structures, London: Thomas Telford. 216 pp.

98. Williamson, C.H.K. (1985): Sinusoidal flow relative to circular cylinders. J. Fluid Mechanics, vol. 155, 141-174.

Chapter 3

Scour around a single slender pile

In river hydraulics, a long tradition exists for studying scour around bridge piers, one of the most important causes of bridge failures. During the last 30 years, more than 1000 of about 600.000 bridges in the United States have failed, and 60% of those failures are due to scour, Briaud et al. (1999). More than 85.000 bridges in the U.S. are vulnerable scour (about 80.000 being scour-susceptible and about 7.000 scour-critical) (Lagasse, Thompson and Sabol, 1995). Two excellent accounts of scour at bridge piers have recently appeared in the literature. One is a compendium of papers presented in the ASCE Water Resources Engineering Conferences from 1991 to 1998 with 371 abstracts and 75 papers (Richardson and Lagasse, 1999). The other is a book by Melville and Coleman (2000) which, along with the recent knowledge, draws on the experiences on scour in New Zealand, and illustrates a great many examples of case studies.

Although piles are quite common in marine environments (in conjunction with pile-supported offshore/coastal structures (such as offshore platforms, jetties, offshore wind mills, etc.), pile works and subsea structures), scour around piles in the marine environment has not been studied as extensively as in the case of bridge piers, apparently due to the more complex flow and scour processes under waves and combined waves and current.

Basically, there are two kinds of flow regimes (and hence two kinds of scour processes) around a pile in a marine environment under waves: In the first one, the pile size is so small that the flow is separated, leading to the formation of separation vortices, as will be detailed in the next sections. This regime is termed the **slender-pile** regime. In the other, the body size would be so large that the flow is in the unseparated flow regime. Hence, the separation vortices (the mechanisms responsible for scour around slender bodies, as will be described later in the chapter) are not present. However, observations do show that scour occurs also in this regime. Clearly, the scour in this case must be related to mechanisms other than the previously mentioned vortex-flow processes. This regime is termed the **large-pile** regime.

The slender-pile regime prevails if the pile diameter, D, is small compared with the wave length, L, as will be detailed in Chapter 6. Otherwise, the presence of the pile will influence the waves, leading to the so-called diffraction effect. It is generally accepted that the diffraction effect becomes important when the ratio D/L becomes larger than $O(0.1)$ (Isaacson, 1979, see also, e.g., Sumer and Fredsøe, 1997). As shown by Isaacson (1979) (see also, e.g., Sumer and Fredsøe, 1997), this regime involves small Keulegan-Carpenter numbers $(KC < O(1))$.

The present chapter will concentrate on the slender-pile regime. The focus will be on flow and scour processes around a single slender pile exposed to currents, waves (regular/irregular), and combined waves and current. Scour around a group of slender piles will be studied in Chapter 4, while scour around large piles $(D/L > O(0.1)$ and $KC < O(1))$, the large-pile regime, will be analyzed in Chapter 6.

Finally, for a general discussion with regard to scale effects, the reader is referred to Section 2.5 in Chapter 2.

3.1 Flow around a slender pile

When a vertical circular pile is placed on a bed, the flow will undergo substantial changes (Fig. 3.1). First of all, a horseshoe vortex will be formed in front of the pile; secondly, a vortex flow pattern (in the form of vortex shedding in most of the practical cases) will be formed at the lee-side of the pile, and thirdly, the streamlines will contract at the side edges of the pile. In addition, there will be a downflow as a consequence of flow deceleration in front of the pile.

(There may also be the diffraction effect in the case of waves when the pile diameter is relatively large, $D/L > O(0.1)$, as mentioned in the preceding section. This will be studied in Chapter 6).

The overall effect of these changes is generally to increase the local sed-

3.1. FLOW AROUND A SLENDER PILE

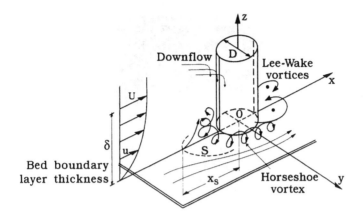

Figure 3.1: Definition sketch. S: Separation line associated with the bed boundary layer.

iment transport in the case of an erodible bed, resulting in a local scour around the pile, Figs. 3.20 and 3.21.

3.1.1 Horseshoe vortex in steady currents

The horseshoe vortex is caused by the rotation in the incoming flow; the boundary layer on the bed upstream of the pile undergoes a three-dimensional separation (along the dashed line, S, in Fig. 3.1) under the influence of the adverse pressure gradient induced by the presence of the structure itself. The separated boundary layer subsequently rolls up to form a spiral vortex around the structure, which then trails off downstream (Fig. 3.1).

Fig. 3.2 displays a hydrogen-bubble visualization of the horseshoe vortex formed in front of a 4 cm diameter model pile which is exposed to a steady current with a velocity of 10 cm/s in a flume. (The existence of the horseshoe vortex manifests itself in the picture by the absence of the bubbles in the immediate neighbourhood of the pile).

In the case of steady currents (wind flows, river flows, etc.), the horseshoe vortex has been investigated quite extensively (see Hjorth 1975; Baker, 1978, 1979, 1985, 1991; Niedoroda and Dalton, 1982; and Dargahi, 1989, among others). Various visualization techniques, such as the smoke technique in a wind tunnel (e.g. Schwind, 1962, and Baker, 1979) and the hydrogen bubble

152 CHAPTER 3. SCOUR AROUND A SINGLE SLENDER PILE

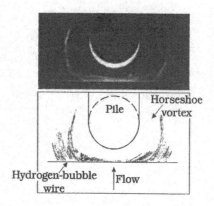

Figure 3.2: Horseshoe vortex visualized by hydrogen bubble technique.

technique in water (Dargahi, 1989), have been used to visualize the horseshoe vortex flow. Also, measurements of pressure and velocity beneath the horseshoe vortex have been carried out (Hjorth, 1975, Baker, 1979, Dargahi, 1989, Graf and Yulistiyanto, 1998 and Roulund, Sumer and Fredsøe, 2002). In Hjorth's and Baker's studies, the distribution of bed shear stress beneath the horseshoe vortex has been calculated from the measured velocity profiles. These latter studies have demonstrated that the bed shear stress can be amplified by a factor of 5-11 with respect to its undisturbed value, indicating the importance of the horseshoe vortex for scour processes, as will be seen later in the chapter.

Two "ingredients" necessary for the generation of the horseshoe vortex are: (1) An incoming boundary layer (with a thickness of δ, Fig. 3.1) must exist; and (2) The adverse pressure gradient induced by the pile must be sufficiently strong so that the boundary layer on the bed can separate to generate the horseshoe vortex (Fig. 3.1).

From dimensional grounds, the non-dimensional quantities describing the horseshoe vortex in the case of a steady current depend mainly on the following parameters (Baker, 1979)

$$\frac{\delta}{D}, \ Re_D \ (\text{or, alternatively, } Re_\delta) \text{ and Pile Geometry} \qquad (3.1)$$

in which δ/D = the ratio of the bed boundary-layer thickness to the pile

3.1. FLOW AROUND A SLENDER PILE

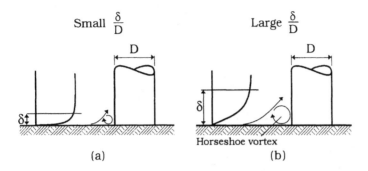

Figure 3.3: Effect of boundary layer thickness on horseshoe vortex.

diameter, Re_D = the pile Reynolds number,

$$Re_D = \frac{UD}{\nu}$$

and Re_δ = the bed-boundary-layer Reynolds number

$$Re_\delta = \frac{U\delta}{\nu}$$

in which U = the velocity at the outer edge of the bed boundary layer (the free-stream velocity), Fig. 3.1. (The Reynolds number is involved here, because it influences the separation of the bed boundary layer (i.e., the separation along S in Fig. 3.1), and therefore the horseshoe vortex, as will be detailed in the following paragraphs). Each parameter is now considered individually.

Effect of δ/D

Clearly, the separation of the bed boundary layer will be delayed if δ/D is small (i.e., a more uniform velocity distribution in the incoming boundary layer, Fig. 3.3 a), presumably leading to a smaller-size horseshoe vortex (cf. Figs. 3.3 a and b). For very small values of δ/D, the boundary layer may not even separate, and hence no horseshoe vortex will be formed.

Fig. 3.4 shows the results of various measurements of the distance x_s, the length characterizing the size of the horseshoe vortex (see Fig. 3.1 and also the definition sketch in Fig. 3.4) plotted versus δ/D (see also Baker, 1985). The pile is a circular cylinder.

154 CHAPTER 3. SCOUR AROUND A SINGLE SLENDER PILE

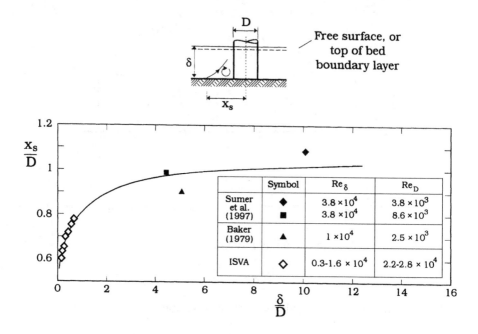

Figure 3.4: Separation distance as function of δ/D.

Fig. 3.4 clearly reveals the argument given in the preceding paragraph regarding the role of the boundary-layer-thickness-to-the-pile-diameter ratio: The smaller the value of δ/D, the smaller the size of the horseshoe vortex.

Effect of the Reynolds number

Similar to the effect of δ/D, the boundary-layer separation will be delayed if the Reynolds number is small (i.e., a larger viscosity) in the case of a separating *laminar* boundary layer. This is because the boundary layer will "face" more resistance to separation when the Reynolds number is small; therefore the separation will be delayed, leading to a smaller-size horseshoe vortex. For very small values of the Reynolds number, the boundary layer may not even separate at all; therefore, the horseshoe vortex may not come into existence for such small values of the Reynolds number.

Baker (1979) measured the variation of the distance of the horseshoe vortex from the pile centre, x_v, in the case of a separating *laminar bound-*

3.1. FLOW AROUND A SLENDER PILE

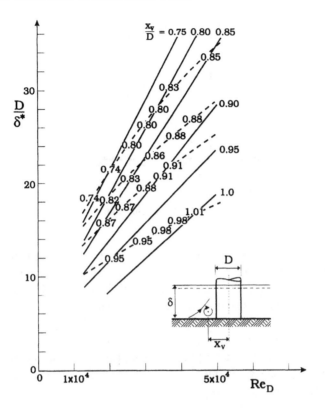

Figure 3.5: Separation distance as function of Re_D and D/δ^*. Baker (1979).

ary layer. Similar to x_s (Fig. 3.4), the quantity x_v is a representative length scale of the horseshoe vortex. The results of Baker's measurements are reproduced in Fig. 3.5 where δ^* is the displacement thickness of the undisturbed boundary layer, defined by (Fig. 3.6)

$$U\delta^* = \int_0^\infty (U - u)dy$$

Fig. 3.5 clearly reveals the argument given in the preceding paragraph: For a given boundary-layer-thickness-to-pile-diameter ratio δ^*/D it is seen that the smaller the Reynolds number, the smaller the value of x_v/D, and therefore the smaller the horseshoe vortex.

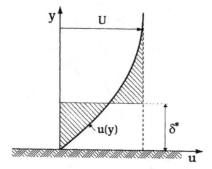

Figure 3.6: Definition sketch.

It may be noted, however, that, for the case of a separating *turbulent boundary layer*, the role of the Reynolds number may be reversed, i.e., the size of the horseshoe vortex may be decreased with increasing Reynolds number. This is due to the increased momentum exchange between the layers of fluid in the *turbulent* boundary layer (and therefore due to the delay in the boundary layer separation) with increasing Reynolds number. We shall return to this issue later in the chapter in conjunction with the mathematical modelling of flow and scour processes at piles.

Effect of pile geometry. Cross-sectional shape

The pile geometry obviously influences the adverse pressure gradient caused by the presence of the pile. While a "streamlined" cross-section induces a small adverse pressure gradient, a pile with a square cross section (with 90^0 orientation) will generate a large adverse pressure gradient. Clearly, the horseshoe vortex in the former case will be relatively small. Sumer, Christiansen and Fredsøe (1997) report the values of the separation distance measured for three different pile geometries, namely for a circular pile, for a square pile with 90^0 orientation, and for a square pile with 45^0 orientation, see Fig. 3.7. The figure clearly shows that the pile geometry is an influencing factor in the development and formation of the horseshoe vortex - the more streamlined the cross-sectional shape of the pile, the smaller the horseshoe vortex.

3.1. FLOW AROUND A SLENDER PILE

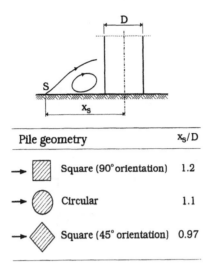

Figure 3.7: $Re_D = 3.8 \times 10^4$. Square (90^0) and Circular: $\delta/D = 10$. Square (45^0): $\delta/D = 7$.

Effect of pile geometry. Pile height

The pile height in the preceding paragraphs is considered to be infinitely large. Obviously, the adverse pressure gradient generated by the presence of the pile, and thus the resulting horseshoe vortex, will be influenced by the pile height in the case of finite-height piles; the smaller the pile height, the smaller the adverse pressure gradient, and therefore the smaller the size of the horseshoe vortex. The data displayed in Fig. 3.8 clearly reveals this.

When the cross-flow dimension of a rectangular cylinder, L, is large with respect to its height, H, the horseshoe vortex branches into smaller vortices in the cross-flow direction. Chou and Chao's (2000) measurements have shown that a horseshoe vortex first evolves into a wavy structure in the cross-flow direction, and for aspect ratios L/D equal to or larger than 10, the wavy horseshoe vortex branches itself into smaller regular vortices. The number of branched vortices increases as the aspect ratio increases.

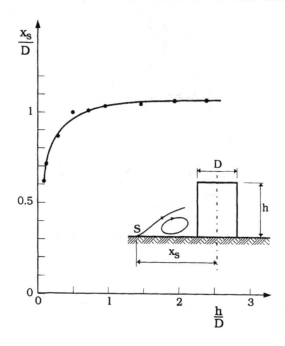

Figure 3.8: Effect of pile height. Circular pile. $Re_{\delta^*} = 9.5 \times 10^3$, $\delta^*/D = 0.066$. Baker (1985).

The following paragraphs will describe two more features of the horseshoe vortex, namely (1) the transition to turbulence in the horseshoe vortex; and (2) the bed shear stress underneath the horseshoe vortex, an important aspect with regard to scour.

Laminar horseshoe vortex versus turbulent horseshoe vortex

From the above considerations it may be expected that, for very small values of δ/D and Re_D, the horseshoe vortex will be in the laminar regime. Baker (1991) reports that the laminar horseshoe-vortex systems are subject to oscillations before they become fully turbulent. Baker's (1991) experiments indicate that the so-called primary oscillations (the oscillations of the separated flow systems) first emerge when

$$Re_D(\delta^*/D)^{1/2} = 800 \qquad (3.2)$$

3.1. FLOW AROUND A SLENDER PILE

while the so-called secondary oscillations (those of the vortex core) first emerge when

$$Re_{\delta^*} = 150 \tag{3.3}$$

in which Re_{δ^*} = the Reynolds number based on the boundary-layer thickness δ^* (see Fig. 3.6 for the definition of δ^*),

$$Re_{\delta^*} = \frac{\delta^* U}{\nu}$$

(It may be noted that, regarding the data depicted in Figs. 3.4 and 3.5, the parameters Re_D and δ^*/D are such that the regime of the horseshoe vortex is well beyond the critical values given in the preceding equations).

For horseshoe-vortex flows encountered in practice, the boundary-layer-thickness-to-the-pile-diameter ratio and the Reynolds number are such that the horseshoe vortex flow is normally in the turbulent regime.

Bed shear stress underneath a horseshoe vortex

Fig. 3.9 displays the mean bed shear stress measured along the principal axis x normalized by the undisturbed mean bed shear stress, $\overline{\tau}/\overline{\tau}_\infty$ (Fig. 3.1). (The overbar in the latter quantity and throughout the chapter denotes time averaging in the case of steady current and ensemble averaging in the case of waves). Fig. 3.9 clearly shows that the bed shear stress underneath the horseshoe vortex just in front of the pile can be as much as a factor of 5 larger (or more) than the undisturbed bed shear stress. (Note that, although the data corresponding to the cross symbols in the figure (the smallest boundary-layer-thickness-to-the-pipe-diameter ratio) seem to indicate that the size of the horseshoe vortex is smaller than the other two, confirming the arguments in the preceding paragraphs, it is not particularly easy to see the influence of Re_D and δ/D on the bed shear stress itself from Fig. 3.9 because of the somewhat small size of the data).

Fig. 3.10, on the other hand, displays $\overline{\tau}/\overline{\tau}_\infty$ (the amplification in the bed shear stress, Section 1.2) over the entire bed area in the neighbourhood of the pile (Hjorth, 1975). The figure shows that the amplification can be as much as 11 at the midway between the front and the side edges of the pile (A in Fig. 3.10). The combined action of the horseshoe vortex and the contraction of the flow near the side edges of the pile apparently amplifies the bed shear stress tremendously.

CHAPTER 3. SCOUR AROUND A SINGLE SLENDER PILE

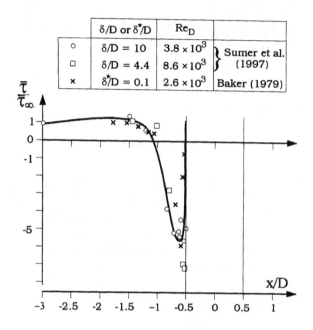

Figure 3.9: Bed shear stress at the horseshoe-vortex side of pile. $x = 0$ coincides with the pile axis.

Obviously, the above mentioned increase in the bed shear stress would induce a tremendous amount of sediment transport in front of and at the side edges of the pile, presumably leading to the formation of a scour hole around the pile in a fairly short period of time, when the bed is erodible. We shall return to this point, namely the scour process around a pile and its time scale, later in the chapter.

3.1.2 Horseshoe vortex in waves

In the case of waves, an additional parameter, the Keulegan-Carpenter number, KC, emerges (in addition to the parameters given in Eq. 3.1). The KC number is defined by

$$KC = \frac{U_m T_w}{D} \tag{3.4}$$

3.1. FLOW AROUND A SLENDER PILE

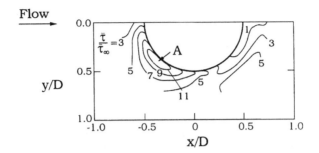

Figure 3.10: Bed shear stress. $D = 7.5$ cm, $V = 30$ cm/s, $\delta = 20$ cm/s, $\delta/D = 2.7$, $Re_D = 2.3 \times 10^4$. Hjorth (1975).

in which U_m = the maximum value of the undisturbed orbital velocity at the bed, T_w = the wave period. If the orbital velocity is assumed to vary sinusoidally, namely

$$U = U_m \sin(\omega t) \quad (3.5)$$

then KC will be

$$KC = \frac{2\pi a}{D} \quad (3.6)$$

in which a = the amplitude of the undisturbed orbital motion of the water particles at the bed, $a = U_m T_w/(2\pi)$, and ω = the angular frequency, $\omega = 2\pi/T_w$.

As seen, the KC number is proportional to $2a/D$ in which $2a$ = the stroke of the orbital motion of the water particles at the bed.

Small KC numbers therefore mean that the orbital motion of water particles is small relative to the width of the pile. When KC is very small, the horseshoe vortex may not even be formed. This is because the stroke of the motion is not large enough for the incoming bed boundary layer to separate (along S, Fig. 3.1).

For very large KC numbers, on the other hand, the stroke of the motion is so large that the flow for each half period resembles that in a steady current. Therefore, for such large KC numbers, the horseshoe vortex may be expected to behave in much the same way as in the case of the steady current.

Sumer et al. (1997) studied the variations of the various flow characteristics with respect to KC. The following paragraphs will summarize the results of Sumer et al.'s (1997) work.

162 CHAPTER 3. SCOUR AROUND A SINGLE SLENDER PILE

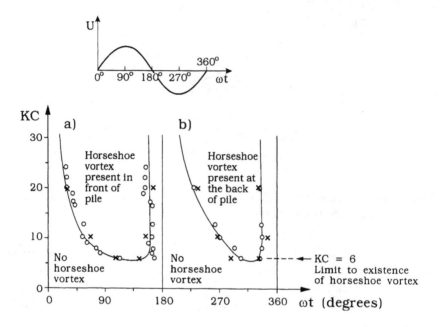

Figure 3.11: Horseshoe vortex in phase space. o: flow visualization. ×: bed-shear-stress measurements. Sumer et al. (1997).

The existence of a horseshoe vortex

Fig. 3.11 displays the results of Sumer et al.'s (1997) study regarding the existence of a horseshoe vortex in waves. In the figure, $\omega t = 0^0$ corresponds to the zero upcrossing in the orbital velocity at the bed ($\omega t = 0^0 - 180^0$ being the crest half period, and $\omega t = 180^0 - 360^0$ the trough half period). The asymmetry observed in Fig. 3.11 between the two half periods is due to the asymmetry in the waves.

Fig. 3.11 indicates that no horseshoe vortex is formed for $KC < 6$.

For the Re_D experienced in the tests presented in Fig. 3.11 ($Re_D = O(10^3)$), the flow over the pile surface separates at $KC = 1$ (Sarpkaya, 1986, see also, e.g., Sumer and Fredsøe, 1997, Fig. 3.15), whereas the experimental results in Fig. 3.11 show that the flow in front of the pile separates at $KC = 6$, a much higher KC than that needed for the flow separation at the pile surface. The question is then: what mechanism suppresses the boundary

3.1. FLOW AROUND A SLENDER PILE

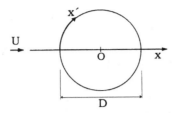

Figure 3.12: Definition sketch.

layer separation in front of the pile for KC below 6 ? This is related to the adverse pressure gradient. Using the potential-flow theory, the pressure gradient in front of the pile (along the x-axis) can be written as (Sumer et al., 1997)

$$\frac{\partial p}{\partial x} = \frac{1}{2}\frac{\rho U^2}{D}\left[1 - \frac{1}{4(x/D)^2}\right]\left[\frac{1}{(x/D)^3}\right] \quad (3.7)$$

Likewise, the pressure gradient over the surface of the pile, again using the potential-flow theory, can be written as

$$\frac{\partial p}{\partial x'} = -8\frac{\rho U^2}{D}\sin(2x'/D)\cos(2x'/D) \quad (3.8)$$

in which x' is the distance along the pile surface measured from the stagnation point (Fig. 3.12). From Eqs. 3.7 and 3.8, it can be found that the maximum value of the adverse pressure gradient in front of the pile (Eq. 3.7) is a factor 5 smaller than that over the surface of the pile (Eq. 3.8). This explains why the separation is "delayed" in front of the pile until after KC reaches the value of 6, a considerably larger value than that necessary for the flow separation over the pile surface, $KC = 1$.

Sumer et al. (1997) have also investigated the influence of the cross-sectional shape of the pile and that of a superimposed current on the existence of a horseshoe vortex.

The results show that the horseshoe vortex exists for KC larger than 4 for the case of a square pile with 90^0 orientation. This is obviously related to the adverse pressure gradient generated in front of the pile. This adverse pressure gradient is larger in the case of the square-section pile than that in the case of the circular pile. Hence, the horseshoe vortex comes into existence at a smaller KC in the case of the square-section pile.

Regarding the influence of a superimposed current, the results show that the horseshoe vortex exists for smaller and smaller KC with increasing current velocity. This is again related to the adverse pressure gradient; the adverse pressure gradient becomes larger and larger with increasing current velocity.

It should be noted that, in Sumer et al.'s (1997) work, the wave boundary layer was in the laminar regime, which usually is the case when one is working with real waves in the laboratory (in a wave flume with a smooth bottom). No study is yet available, investigating the influence of a *turbulent* incoming wave boundary layer. Likewise, the influence of large Re_D on the horseshoe vortex in waves is also unknown.

Lifespan of a horseshoe vortex

The horseshoe vortex first emerges some time after the flow reverses (Fig. 3.13, Frame 2), it "lives" for a certain period of time during the half cycle (Fig. 3.13, Frames 2-4), and it finally disappears (Fig. 3.13, Frame 5). The disappearance of the horseshoe vortex occurs when the flow reverses again; the horseshoe vortex is literally destructed with the flow reversal (Fig. 3.13, Frame 5), and, obviously, a new horseshoe vortex begins to emerge in the next half cycle of the motion, at the other side of the pile.

As has already been seen in Fig. 3.11, the horseshoe vortex is maintained over a larger and larger span of ωt, as KC is increased. In other words, the life span of a horseshoe vortex increases with increasing KC (Fig. 3.11).

Fig. 3.11 shows that, for example, for $KC = 10$, the horseshoe vortex first emerges at about $\omega t = 50^0$, and it disappears at about $\omega t = 160^0$, its life span being 110^0, while, for $KC = 25$, the corresponding figures are 23^0 and 160^0, respectively, the life span of the horseshoe vortex in this latter case being 137^0. Obviously, the increase in the life span of the horseshoe vortex with increasing KC is linked to the fact that the adverse pressure gradient, necessary for the formation of the horseshoe vortex, will be maintained over a longer and longer period of the half cycle, as KC is increased. This is simply because the stroke of the motion gets larger and larger with increasing KC.

Sumer et al. (1997) have also investigated the influence of the cross-sectional shape on the life span of the horseshoe vortex. They found that the life span of the horseshoe vortex is larger in the case of a square pile than in the case of a circular pile. The difference is particularly pronounced for smaller KC numbers.

3.1. FLOW AROUND A SLENDER PILE

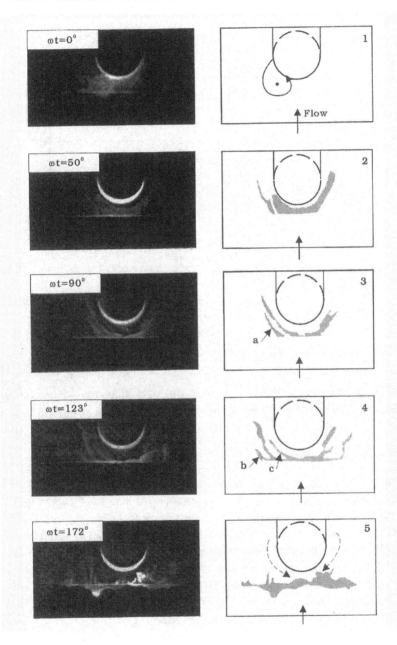

Figure 3.13: Time development of horseshoe vortex. $KC = 10.3$. Sumer et al. (1997).

166   CHAPTER 3. SCOUR AROUND A SINGLE SLENDER PILE

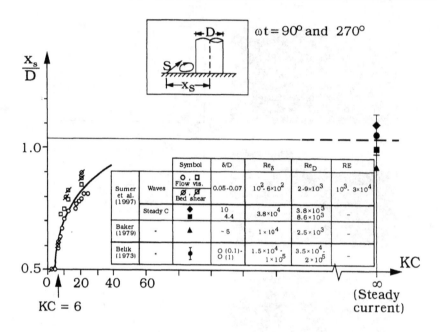

Figure 3.14: Separation distance. Circular pile. Sumer et al. (1997).

Likewise, the life span of a horseshoe vortex is increased with increasing current in a wave/current flow environment (Sumer et al., 1997).

Separation position

Fig. 3.14 shows the separation distance (Fig. 3.1) normalized by the pile diameter, plotted against the KC number. x_s is a function of the phase ωt, and the data given in Fig. 3.14 corresponds to $\omega t = 90^0$ and 270^0 (the circles and squares, respectively). The difference between the two half cycles is attributed to the asymmetry between the two half periods of the waves. As seen, x_s first experiences a steep increase with KC, and then it tends to approach its asymptotic value (the steady current value) as $KC \to \infty$.

The scatter in the data for $KC = \infty$ (the steady-current case) in Fig. 3.14 may be due to the Reynolds number effect.

Sumer et al. (1997) found that the influence of the cross-sectional shape on the separation distance is important. The trend is much the same as in

3.1. FLOW AROUND A SLENDER PILE

the case of the steady current (Fig. 3.7). The influence of a superimposed current was also found important; the separation distance increases quite markedly with increasing value of the current velocity.

Bed shear stress beneath the horseshoe vortex

Fig. 3.15 shows the variation of the bed shear stress along the x−axis (Fig. 3.1) at the phase values $\omega t = 90^0$ and 270^0 (cf. Fig. 3.9).

Fig. 3.16, on the other hand, displays the amplification in the bed shear stress underneath the horseshoe vortex (at a streamwise distance of $0.1D$ from the upstream edge of the pile) as a function of KC. The figure shows that the bed shear stress increases with increasing KC. This is a direct result of the increased presence of the horseshoe vortex with increasing KC (Figs. 3.11 and 3.14). The scatter in the data for $KC = \infty$ (the steady current) in Fig. 3.16 is due to the Reynolds number effect. As seen from Fig. 3.4, the larger the Reynolds number, the larger the separation distance, therefore the larger the bed shear stress under the horseshoe vortex. The steady-current data in Fig. 3.16 appear to reveal this.

Fig. 3.17 depicts the root-mean-square (r.m.s.) value of the fluctuating component of the bed shear stress under the horseshoe vortex (at the same location), plotted against KC. A non-zero r.m.s. value means that the horseshoe vortex is not in the laminar regime. From the figure it is inferred that the transition to turbulence in the horseshoe vortex begins to occur somewhere between $KC = 10$ and 20 (recall that the incoming wave boundary layer for these KCs, and indeed for all the other KCs in Sumer et al.'s (1997) experiments is in the laminar regime).

Fig. 3.17 implies that the transition to turbulence in the horseshoe vortex flow is determined not only by δ/D and Re_D, as discussed in the previous section, but also by KC. Clearly, the regime of the incoming boundary layer on this transition must also be an influencing factor. However, no detailed study is available yet, investigating the transition to turbulence in the horseshoe vortex as a function of the governing parameters, namely δ/D, Re_D and KC as well as the turbulence in the incoming boundary layer.

Finally, it may be noted that the transition displayed in Fig. 3.17 agrees satisfactorily with the transition criteria given by Baker (1991) (Eqs. 3.2 and 3.3) (see Sumer et al., 1997 for the details).

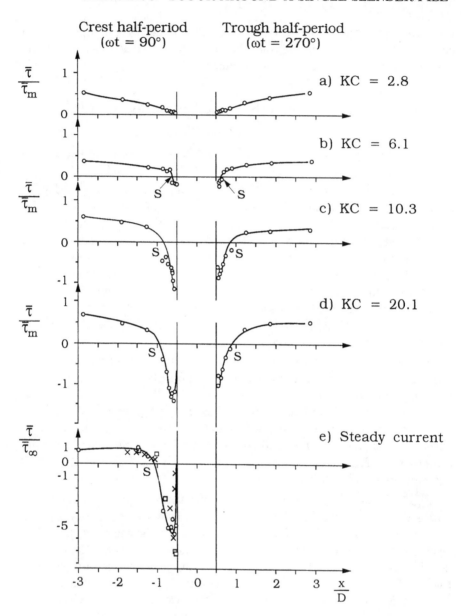

Figure 3.15: Bed shear stress at horseshoe-vortex side of pile. ×: Baker (1979). From Sumer et al. (1997).

3.1. FLOW AROUND A SLENDER PILE

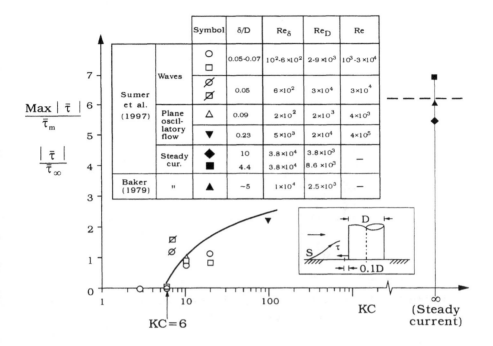

Figure 3.16: Amplification in horseshoe-vortex bed sear stress. Filled symbols: Turbulent bed boundary layer. Sumer et al. (1997).

3.1.3 Lee-wake vortex flow

The lee-wake vortices are caused by the rotation in the boundary layer over the surface of the pile; the shear layers emanating from the side edges of the pile roll up to form these vortices in the lee wake of the pile (Fig. 3.1).

In the case of a steady current, the lee-wake flow is described mainly by

$$Re_D, \text{ Pile Geometry} \tag{3.9}$$

The steady-current, lee-wake flow is well understood today. A detailed account of mechanisms of vortex shedding, vortex shedding frequency, effect of Re_D, effect of the roughness of pile surface, effect of cross-sectional shape, effect of incoming turbulence, effect of shear in the incoming turbulence can be found in, e.g., Sumer and Fredsøe (1997).

The non-dimensional quantities describing the lee-wake vortex flow in

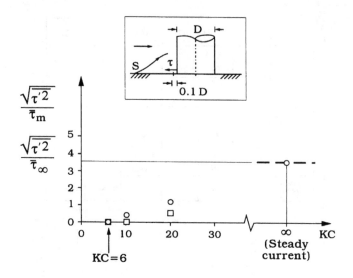

Figure 3.17: R.m.s. value of fluctuating bed shear stress under horseshoe vortex. o: $\omega t = 90^0$. □: $\omega t = 270^0$. Sumer et al. (1997).

waves, on the other hand, depend mainly on

$$KC, \ Re_D, \ \text{Pile Geometry} \qquad (3.10)$$

In the case of a rough pile, the relative roughness, k_s/D, emerges as an additional parameter in Eqs. 3.9 and 3.10. Here, k_s = the roughness of the pile surface. In contrast to the steady-current case, the lee-wake vortex flow is essential for the scour characteristics in the case of waves, as will be described in the following paragraphs.

An extensive volume of knowledge has accumulated over the past two decades on the vortex flow behind a *free* cylinder subjected to an oscillatory flow (e.g., Sarpkaya and Isaacson, 1981; Bearman, Graham, Naylor and Obasaju, 1981; and Williamson, 1985; see also Sumer and Fredsøe, 1997 for a detailed review of the subject), and the complex behaviour of vortex motions in various regimes is well understood. These studies have demonstrated that the vortex flow is primarily governed by KC.

In the case of a vertical cylinder confined with the seabed (the pile problem), the subject has been investigated in the study by Sumer et al. (1997)

3.1. FLOW AROUND A SLENDER PILE

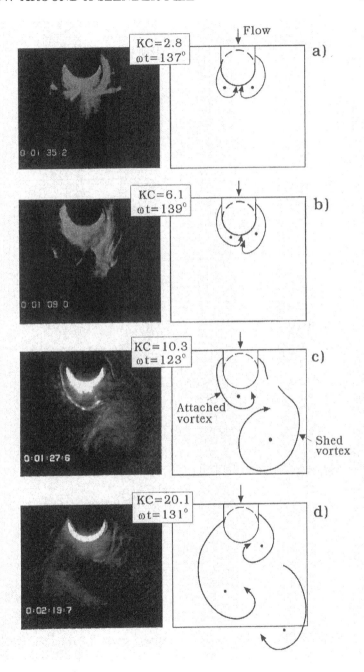

Figure 3.18: Near-bed wake vortices. Sumer et al. (1997).

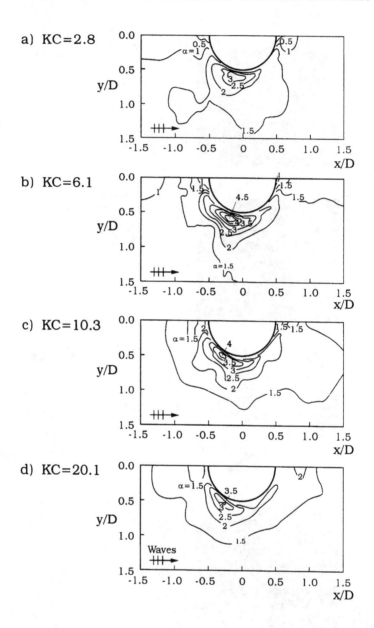

Figure 3.19: Amplification of bed shear stress. Sumer et al. (1997).

3.1. FLOW AROUND A SLENDER PILE

Figure 3.20: Scour hole in the case of clear-water scour. Eadie and Herbich (1986).

referred to in the preceding paragraphs in conjunction with the horseshoe vortex. Sumer et al. (1997) identified the following vortex-flow regimes :

1. When $2.8 \leq KC < 4$, separation behind the pile occurs in the form of a pair of symmetric vortices (Fig. 3.18 a). These vortices are washed around the pile when the flow reverses.

2. When $4 < KC < 6$, the symmetry breaks up, but the vortices are still attached, i.e., no shedding yet exists (Fig. 3.18 b).

3. When $6 < KC < 17$, vortex shedding occurs, with one vortex shed in each half period (Fig. 3.18 c). This regime corresponds to the single-pair regime of Williamson's (1985) two-dimensional sinusoidal plane oscillatory flow regimes.

4. When $17 < KC < 23$, again, vortex shedding occurs, but now two vortices are shed in each half cycle, presumably leading to a larger length of the lee-wake behind the pile for each half period (Fig. 3.18 d). This regime corresponds to the double-pair regime of Williamson's (1985) two-dimensional sinusoidal plane oscillatory flow regimes.

Apparently, the pile-flow regimes are generally in accord with the regimes experienced in the case of a two-dimensional free cylinder exposed to a sinusoidal flow. This is attributed to the extremely small wave-boundary layer thickness ($\delta/D = O(0.05)$ in Sumer et al.'s, 1997, tests); the vortex-flow regimes are therefore practically uninfluenced by the three-dimensionality of the flow. Although no test was made with KC below 2.8 in Sumer et al.'s study, the preceding results imply that the flow is unseparated for KC numbers smaller than $O(1)$, similar to the free-cylinder case (see, e.g., Sumer and Fredsøe, 1997, Figs. 3.15 and 3.16).

3.1.4 Contraction of streamlines

This effect is described by the contour plots of the bed shear stress, Fig. 3.19. The figure shows that there is a concentration of bed shear stress at or near the side edges of the pile. As seen, the amplification in the bed shear stress with respect to its undisturbed value is $\alpha = O(4)$ in the case of waves, while it is $\alpha = O(10)$ in the case of a steady current (Fig. 3.10). The tremendous increase in the amplification in the case of the steady current (Fig. 3.10) is linked to the very strong presence of the horseshoe vortex, as discussed earlier in conjunction with Fig. 3.10.

3.2 Scour around a slender pile

3.2.1 Scour around a slender pile in steady currents

In steady currents, the key element in the scour process is the horseshoe vortex. Combined with the effect of the contraction of streamlines at the side edges of the pile, this vortex can erode a significant amount of sediment away from the neighbourhood of the pile (see the amplification in the bed shear stress underneath the horseshoe vortex and in the bed area where the streamlines are contracted in Figs. 3.9 and 3.10), leading to a truncated-cone-shaped scour hole around the pile (Figs. 3.20, 3.21 and 3.22). Observations show that while the upstream slope of the scour hole is more or less equal to the angle of internal friction, the downstream slope is relatively less steep (Fig. 3.23). This is mainly due to the effect of gravity.

Scour around a pile in steady currents has been investigated quite extensively (particularly in the context of scour at bridge piers), Hjorth (1975), Melville (1975), Breusers, Nicollet and Shen (1977), Ettema (1976), Melville and Raudkivi (1977), Raudkivi and Ettema (1977), Ettema (1980), Imberger, Alach and Schepis (1982), Raudkivi and Ettema (1983), Chiew (1984), Raud-

3.2. SCOUR AROUND A SLENDER PILE 175

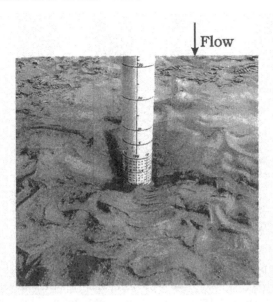

Figure 3.21: Scour hole in the case of live-bed scour. Roulund et al. (2002).

kivi (1986), Chiew and Melville (1987), Melville and Sutherland (1988), Melville and Dongol (1992), Melville and Raudkivi (1996), Ettema, Melville and Barkdoll (1998) and Melville and Chiew (1999) among others. Reviews of the subject can be found in the books by Breusers and Raudkivi (1991), Hoffmans and Verheij (1997), Whitehouse (1998), Raudkivi (1998) and Melville and Coleman (2000).

Scour depth

The previously mentioned research has shown that the scour depth is influenced by various factors such as the Shields parameter; the sediment gradation; the boundary-layer-depth-to-pile-size ratio; the sediment-size-to-pile-size ratio; the shape factor; and the alignment factor. Each parameter is now considered individually.

Shields parameter. Fig. 3.24 illustrates schematically the variation of the scour depth with θ (see, e.g., Melville and Coleman, 2000). The scour depth increases with increasing θ for the clear-water scour ($\theta < \theta_{cr}$); the effect

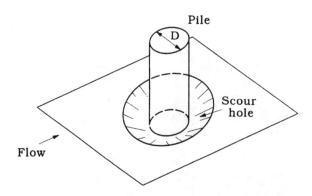

Figure 3.22: Scour hole around a vertical pile.

is much the same as in the case of the pipeline scour (Chapter 2, Fig. 2.26). For the live-bed scour, for θ values just beyond the critical value, the scour depth decreases slightly after experiencing a maximum. This is because of the commencement of the sediment transport over the entire bed, and, as a result, backfilling of the scour hole (albeit slightly) by the inflow of the sediment from upstream. However, as the Shields parameter is increased further, the scour depth reaches a second peak, the live-bed peak. This second peak occurs at about the transition-to-the-flat-bed sediment transport, i.e., at the commencement of the sheet-flow regime ($\theta = O(0.5)$, Sumer et al., 1996).

Fig. 3.25, which is reproduced from Melville and Sutherland (1988), illustrates the variation of the normalized scour depth S/D with the velocity ratio U/U_{cr} for different values of the geometric standard deviation, σ_g, in which U = the mean approach flow velocity and U_{cr} = the mean approach flow velocity corresponding to the initiation of motion at the bed (in the undisturbed case). (Note that the parameters U/U_{cr} or θ/θ_{cr} are used alternatively in the literature (cf. Imberger et al., 1982); U/U_{cr} is adopted in favour of θ/θ_{cr} in the plot in Fig. 3.25). Basically, the data displayed in Fig. 3.25 reveal the variation depicted in Fig. 3.24.

Sediment gradation. Work by Ettema (1976) showed that scour depths are reduced tremendously as the geometric standard deviation, $\sigma_g = d_{84}/d_{50}$ (representing the sediment gradation) increases. Ettema's tests were all clear-water tests. Live-bed tests by Baker (1986) (reproduced here in Fig. 3.25 from Melville and Sutherland, 1988) showed, however, that the reductions

3.2. SCOUR AROUND A SLENDER PILE

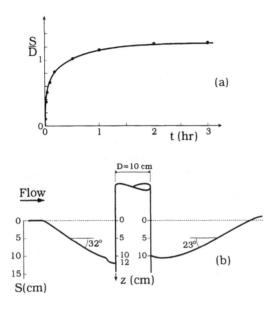

Figure 3.23: Scour measured in the laboratory. Equilibrium profile (b). $D = 10$ cm, h (flow depth) $= 40$ cm, δ (boundary-layer thickness) $= 20$ cm, V (mean-flow velocity) $= 46$ cm/s, $d_{50} = 0.26$ mm. Angle of internal friction $= 30^0$.

are not radically large. (It may be noticed that the first peak in Fig. 3.25, the threshold peak (see Fig. 3.24), is progressively shifted to the right. This is because of the increase in the critical velocity with increasing sediment gradation).

Fig. 3.25 reveals that the larger the sediment gradation, the smaller the scour depth, apparently due to the armouring effect, as discussed in connection with the pipeline scour (Chapter 2, Fig. 2.35). As seen from Fig. 3.25, the effect is most pronounced for clear-water scour. For example, for $U/U_{cr} = 1.5$, the scour depth S/D is about $S/D = 2$ for $\sigma_g = 1.3$, while it is only about $S/D = 0.3$ when σ_g is increased from 1.3 to 5.2, a tremendous reduction in the scour depth.

Boundary-layer-depth-to-pile-size ratio. The boundary-layer depth/flow depth is also an influencing factor. The data collected by Melville and Sutherland (1988) is reproduced in Fig. 3.26. In the figure, S is the scour

CHAPTER 3. SCOUR AROUND A SINGLE SLENDER PILE

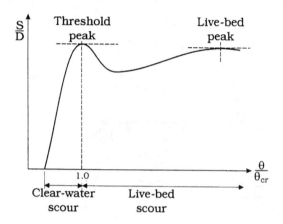

Figure 3.24: Scour depth against Shields parameter for uniform sediment. Melville and Coleman (2000).

depth and S_0 is the scour depth experienced for large values of δ/D ($\delta/D \gtrsim 4$) in which δ = the flow depth in a river flow (the boundary-layer thickness) and D = the pile diameter. Fig. 3.26 indicates that the scour depth increases with increasing δ/D. This is due to the influence of δ/D on the horseshoe vortex; the larger the value of δ/D, the larger the size of the horseshoe vortex (Fig. 3.3), as described earlier, and therefore the larger the scour depth.

Pile-size-to-sediment-size ratio. The data compiled by Melville and Sutherland (1988) is reproduced in Fig. 3.27. (The data originally stems from the works of Ettema, 1980, and Chiew, 1984). In the figure, S_0 is the scour depth experienced for large values of D/d_{50} ($D/d_{50} \gtrsim 50$). The data shows that sediment that is large relative to the pile size limits the scour depth. In the extreme case where the sediment size is comparable to the pile size, the near-pile flow (including the horseshoe vortex) will be "destructed" by the presence of the sediment, and therefore the scour depth, as a result, will be reduced. As seen from the figure, the effect of the sediment size disappears when $D/d_{50} \gtrsim 50$.

3.2. SCOUR AROUND A SLENDER PILE

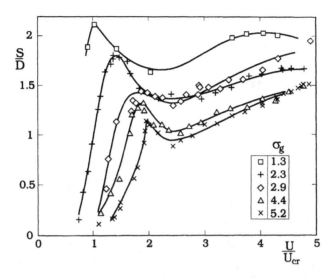

Figure 3.25: Effect of sediment gradation. Data by Baker (1986). $d_{50} = 0.6$ mm. From Melville and Surtherland (2000).

Shape in plan	Length/Width	I	II	III	IV
Circular	1.0	1.0	1.0	1.0	1.0
Lenticular	2.0	-	0.97	-	-
"	3.0	-	0.76	-	-
"	4.0	0.67	-	0.73	-
"	7.0	0.41	-	-	-
Parabolic nose	-	-	-	-	0.56
Triangular nose, 60°	-	-	-	-	0.75
Triangular nose, 90°	-	-	-	-	1.25
Elliptic	2.0	-	0.91	-	-
"	3.0	-	0.83	-	-
Ogival	4.0	0.86	-	0.92	-
Joukowski	4.0	-	-	0.86	-
"	4.1	0.76	-	-	-
Rectangular	2.0	-	1.11	-	-
"	4.0	1.4	-	1.11	-
"	6.0	-	1.11	-	-

Table 3.1. Shape factor K_s, compiled by Melville and Sutherland (1988).

Shape factor. The cross-sectional shape of the pile is also important. Sumer, Christiansen and Fredsøe (1993) found that the scour depth normal-

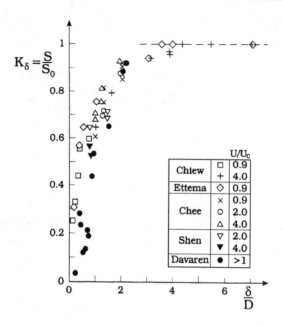

Figure 3.26: Effect of boundary-layer thickness. Data compiled by Melville and Sutherland (1988).

ized with the pile size, S/D, is $S/D = 1.3$ for a circular pile, while it is $S/D = 2.0$ for a square section pile (90^0 orientation). This is due partly to the size of the horseshoe vortex (the larger the horseshoe vortex, the larger the scour depth, cf. Fig. 3.7), and partly to the plan-view extent of the separation zone around the pile itself (the larger this area, obviously the larger the scour depth). Note that the scour depth in the latter study for a square cross-section with 45^0 orientation was found to be slightly smaller than that for the 90^0-orientation square section pile.

Melville and Sutherland (1988) compiled the data regarding the shape effect. This data is reproduced in Table 3.1. Here, the shape factor is defined by $K_s = S/S_o$ in which $S_o =$ the scour depth in the case of the circular pile. In the table, I : Tison (1940), II: Laursen and Toch (1956), III: Chabert and Engeldinger (1956) and IV: Venkartadri et al. (1965).

Another shape factor is the pile height. When the pile height is not infinitely large, the horseshoe vortex will be influenced (Fig. 3.8); its size

3.2. SCOUR AROUND A SLENDER PILE

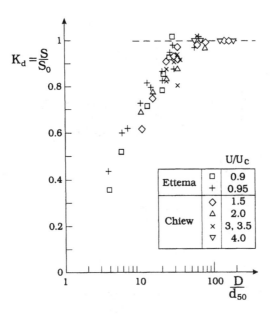

Figure 3.27: Effect of sediment size. Data compiled by Melville and Sutherland (1988).

will be reduced, and, as a result, the scour depth will be affected. Fig. 3.28 reveals this; the smaller the pile height, the smaller the size of the horseshoe vortex (Fig. 3.8), and therefore the smaller the scour depth. The data in Fig. 3.28 can be expressed by the following empirical relation

$$\frac{S}{S_0} = 1 - \exp(-0.55\frac{h}{D}) \qquad (3.11)$$

in which h = the pile height, and S_0 = the scour depth for infinitely tall piles. The data, although limited, seems to indicate that the effect of the finite height of the pile practically disappears when $h/D \gtrsim O(5)$.

Example 8 *Scour depth around a stone*

A pile with a height equal to its diameter may, to a first approximation, be considered to represent a stone. From Fig. 3.28, the scour depth in

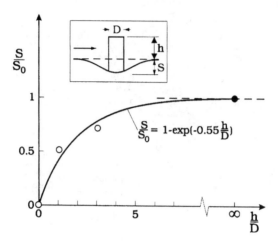

Figure 3.28: Effect of pile height. Pile circular. Empty circles: DHI/Snamprogetti (1992), $\theta = \theta_{cr}$. By courtesy of DHI and Snamprogetti.

this case will be $S/D \simeq 0.65$ (where S_0 is taken as $1.3D$, see Eq. 3.12). The same exercise achieved with a sphere, on the other hand, gives a scour depth of $S/D \simeq 0.15 - 0.45$ (in the case of live bed, $\theta > \theta_{cr}$) (Fig. 3.30). Although these results are at best suggestive, the scour depth around a stone may therefore be taken as $S/D = O(0.5)$. Furthermore, drawing an analogy between the self-burial of a pipeline (Section 2.4.2) and that of a stone, the self-burial depth of a stone in steady currents may be taken as $e/D = O(0.5)$.

Alignment factor. Alignment factors taken from Laursen (1958) are shown in Fig. 3.29. Here $K_\alpha = S/S_0$ in which S_0 = the scour depth for the value of the angle of attack, $\alpha = 0$. The figure indicates that the angle of attack for cross-sectional shapes other than the circular one may be quite significant. The increase in the scour depth is linked to the adverse pressure gradient created by the pile; the larger the angle of attack, the larger the adverse pressure gradient, the larger the size of the horseshoe vortex, and therefore the larger the scour depth. See also the discussion in Ettema, Mostafa, Melville and Yassin (1998).

Of the above factors, the boundary-layer-depth-to-pile-size ratio, the shape factor and the alignment factor are related to the flow. The analysis in Section 3.1.1 implies that the Reynolds number, Re_D, also is another factor

3.2. SCOUR AROUND A SLENDER PILE

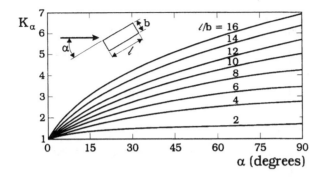

Figure 3.29: Effect of alignment. Laursen (1958).

influencing the flow, and therefore the scour depth. However, no data is available to reveal the influence of the Reynolds number on the scour depth.

The reported range of the scour depth in the literature for circular piles, S/D, is $O(1-2.5)$ for the *live-bed* scour, the actual value depending on the previously mentioned parameters (Breussers et al., 1977, Melville and Sutherland, 1988; Raudkivi, 1998; Melville and Coleman, 2000, among others).

From the data reported in Breussers et al. (1977), the following statistics can be worked out for the scour depth, as was done by Sumer et al. (1992 a):

$$\text{Mean value of } \frac{S}{D} = 1.3, \text{ with } \sigma_{S/D} = 0.7 \quad (3.12)$$

(This is for the live-bed scour, $\theta > \theta_{cr}$). One can take $S/D = 1.3 + \sigma_{S/D} = 1.3 + 0.7 = 2$, or $S/D = 1.3 + 2\sigma_{S/D} = 1.3 + 2 \times 0.7 = 2.7$ as the maximum scour depth for the live-bed scour for design purposes.

Melville and Sutherland (1988) adopt a value of $S/D = 2.4$ for this purpose for the live-bed scour where the sediment gradation also is accounted for. The latter authors present a design method for the estimation of the scour depth according to which the maximum scour depth, namely $S/D = 2.4$, is reduced using multiplying factors:

$$\frac{S}{D} = K_I K_\delta K_d K_s K_\alpha \quad (3.13)$$

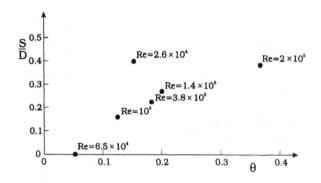

Figure 3.30: Scour depth below a sphere. Data from Technical University of Denmark, ISVA.

in which

$$K_I = 2.4, \text{ if } \frac{U - (U_a - U_{cr})}{U_{cr}} > 1 \qquad (3.14)$$

and

$$K_I = 2.4 \left| \frac{U - (U_a - U_{cr})}{U_{cr}} \right|, \text{ if } \frac{U - (U_a - U_{cr})}{U_{cr}} < 1 \qquad (3.15)$$

Here, U_a = the mean approach flow velocity at the armour peak, $U_a = 0.8 U_{ca}$, and U_{ca} = the mean approach flow velocity beyond which armouring of channel bed is impossible. Each sediment has a unique value of U_a and U_{cr}. If the sediment is a uniform sediment, or behaving as a uniform sediment, then $U_a = U_{cr}$. Otherwise, U_a needs to be calculated as outlined in Melville and Sutherland (1988, under Limiting Armour Conditions). The remaining factors in Eq. 3.13, namely K_δ is found from Fig. 3.26, K_d from Fig. 3.27, K_s from Table 3.1, and K_α from Fig. 3.29.

Scour depth in cohesive sediment

Scour at piles founded in cohesive sediment such as clay is another topic of interest. Here, the key issue is that clays scour much more slowly (0.1-100 cm/hr, Briaud et al., 1999) than sand, and therefore the equilibrium scour may not be reached during, say, a 100-year flood. Hence, the knowledge of the time development of scour becomes important to avoid overly conservative (and therefore expensive) estimate of the scour depth by considering the

3.2. SCOUR AROUND A SLENDER PILE

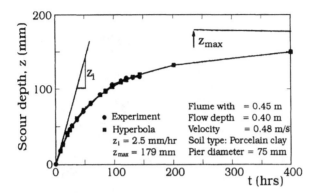

Figure 3.31: Example of flume test results. Cohesive sediment. Biraud et al. (1999).

maximum scour depth. To this end, Briaud et al. (1999) have developed a method, called SRICOS (Scour Rate In COhesive Soils) to predict the scour depth versus time curve around a cylindrical bridge pier standing in the way of a constant velocity flow and founded in a uniform cohesive soil. The following paragraphs will describe this work.

Briaud et al. (1999) made 42 experiments with model piers with the diameter ranging from $D = 2.5 - 7.6$ cm in a small flume (46 cm in width), and $D = 7.6 - 22.9$ cm in a large flume (153 cm in width). Four different soils were used: three clays and one sand. The water depth varied in the range 16-40 cm in the small flume and 25-40 cm in the large flume, and the velocity in the range 20-83 cm/s.

An example of the scour depth versus time curve is depicted in Fig. 3.31. As seen, the scouring process is extremely slow; even with $t = 400$ hr, the scour depth has not reached its equilibrium depth (cf. Fig. 3.23 and 3.31). Biraud et al. (1999) approximated the variation in Fig. 3.31 by a hyperbola:

$$z(t) = (\frac{1}{\dot{z}_i} + \frac{t}{z_{max}})^{-1} t \tag{3.16}$$

in which z = the scour depth; z_{max} = the maximum scour depth (the ordinate of the asymptote), $\dot{z}_i$ = the initial slope of the z versus time curve, i.e., the initial scour rate, and overdot denotes the time derivative. Biraud et al. (1999) plotted the z_{max} values obtained in this way against the pier Reynolds

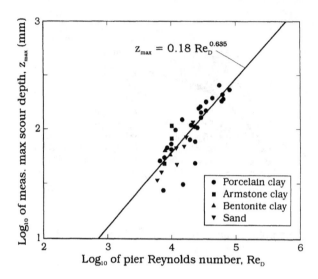

Figure 3.32: Measured scour depth versus Re_D. Biraud et al. (1999).

number (Fig. 3.32), and expressed the data in this figure by the following relationship:

$$z_{max} \text{ (mm)} = 0.18 Re_D^{0.635} \qquad (3.17)$$

They note that they plotted z_{max} against various parameters, and that it was found that the most well-behaved relationship was obtained when z_{max} was plotted against Re_D. It is interesting to notice that the sand results and the clay results in Fig. 3.32 practically collapse on the same curve, meaning that the maximum scour depth is the same irrespective of the category of the bed sediment (sand or clay). Although the time scale of the scour processes differs enormously for the two kinds of sediment, the equilibrium scour depth does not. This may be attributed to the fact that the plan-view extent of the flow effects, namely the horseshoe vortex, the contraction of streamlines and the lee-wake flow, obviously is the same for both sediment beds; thus, the end effect will be the same when the process attains its equilibrium. Biraud et al. (1999) report that an existing, widely used empirical equation obtained for sand fits data on clay quite well. This is an important result.

The procedure how to find the scour depth corresponding to a certain duration of flood flow is given as follows (Biraud et al. 1999):

3.2. SCOUR AROUND A SLENDER PILE

1. Collect sediment samples at the site where the pile is to be built. Biraud et al. (1999) propose 76.2-mm-diameter Shelby tube samples for this.

2. Perform tests on the samples to obtain the curve linking the erosion rate $\dot{z}$ and the bed shear stress. One such example is shown in Fig. 3.33. Biraud et al. (1999) suggest what they call Erosion Function Apparatus (EFA) "standard" tests for the latter, as described in their paper. However, these tests can also be achieved in an ordinary flume facility, or in a rotating "Couette flow" facility, provided that the bed shear stress can be measured accurately. (The bed shear stress measurements in an ordinary flume can be made, using a flush-mounted hot film in a corresponding smooth, rigid bed test, in the same fashion as described in Sumer, Chua, Chen and Fredsøe, 2002).

3. Determine the maximum bed shear stress τ_{max} which will exist on the bed around the pier at the beginning of the scour process (when the bed is plane). For this, consult the diagrams similar to that given in Fig. 3.10, or use the relationship given by Biraud et al. (1999):

$$\tau_{max} = 0.094 \rho V^2 \left(\frac{1}{\log_{10} Re_D} - \frac{1}{10} \right) \tag{3.18}$$

4. Obtain the initial scour rate $\dot{z}_i$. This is related to τ_{max}, i.e., the initial scour will occur at the location where $\tau = \tau_{max}$. To find the initial scour rate $\dot{z}_i$, go to the $\dot{z}$ versus τ diagram established in Step 2, and pick up the $\dot{z}$ value corresponding to $\tau = \tau_{max}$. This is $\dot{z}_i$.

5. Calculate the scour depth from Eq. 3.16 for the duration of flood t. In this latter equation, z_{max} is to be calculated from Eq. 3.17.

Example 9 *A numerical example for the prediction of scour depth in cohesive sediment (Biraud et al., 1999)*

A bridge pier is 2 m in diameter in a river that will experience a flood velocity of 2 m/s for a duration of 4 days. Samples of the clay recovered at the site lead to the $\dot{z}$ versus τ diagram depicted in Fig. 3.33. The maximum shear stress around the pier before scour begins is from Eq. 3.18

$$\tau_{max} = 0.094 \times 1000 \times 2^2 \left(\frac{1}{\log \frac{2 \times 2}{10^{-6}}} - \frac{1}{10} \right) = 19.4 \text{ N/m}^2$$

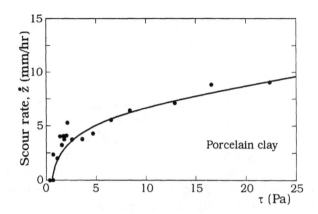

Figure 3.33: Example of scour rate versus shear stress curve. Clay. Biraud et al. (1999).

For this τ_{max}, the initial rate of scour $\dot{z}_i$ is found from Fig. 3.33

$$\dot{z}_i = 8.5 \text{ mm/h}$$

The maximum scour depth z_{max} from Eq. 3.17

$$z_{max} \text{ (mm)} = 0.18(\frac{2 \times 2}{10^{-6}})^{0.635} = 2803 \text{ mm}$$

Then, the scour depth after 4 days is from Eq. 3.16

$$z(t = 4 \text{ days}) = (\frac{1}{8.5} + \frac{4 \times 24}{2803})^{-1}(4 \times 24) = 632 \text{ mm}$$

As seen, the scour depth, 63.2 cm, is only 22.5% of the maximum scour depth, 2.8 m.

3.2.2 Scour around a slender pile in waves

In the case of waves, two major changes occur with regard to the near-bed flow. First, the horseshoe vortex undergoes substantial changes, as seen in Section 3.1.2. Second, the lee-wake flow is no longer a "passive" flow feature; rather, the lee-wake vortices act as a convection mechanism to transport the eroded sediment away from the pile during each half period of the motion

3.2. SCOUR AROUND A SLENDER PILE

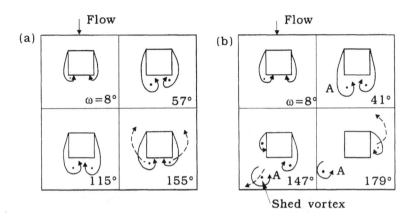

Figure 3.34: Sequence of flow picture around a square pile. (a): $KC = 10$. (b): $KC = 13$. Sumer et al. (1993).

(e.g., Vortex A in Fig. 3.34). From Section 3.1, these two processes are primarily governed by KC

$$KC = \frac{U_m T_w}{D} \qquad (3.19)$$

Therefore, the scour process itself is expected to be governed by the same parameter.

Sumer et al. (1992 a) and Kobayashi and Oda (1994) demonstrated that the Keulegan-Carpenter number is the main parameter governing the scour process (and thereby the scour depth) on a live bed ($\theta > \theta_{cr}$).

Sumer et al. (1993) further investigated the influence of pile geometry on scour, employing a circular pile and a square pile (the latter with two different orientations, namely 90^0 and 45^0). One of the key findings of Sumer et al.'s works (1992 a and 1993) is that the onset of scour coincides with the onset of vortex shedding: This occurs when KC reaches 3 for 45^0-orientation square piles, 6 for circular piles, and 11 for 90^0-orientation square piles.

The scour process was observed to occur in the following way: The flow sweeps the sediment into the core of the *shed* vortex (such as Vortex A in Fig. 3.34 b). Next, the vortex carries the sediment away from the pile while it is convected downstream, thus causing a net scour around the pile. In the absence of shedding, the vortices that form behind the pile (Fig. 3.34 a) sweep the sediment grains into their core regions, but do not carry them

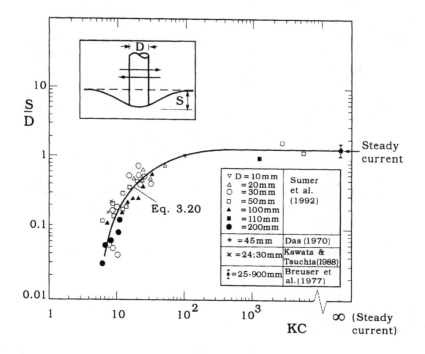

Figure 3.35: Equilibrium scour depth. Circular pile. Live bed ($\theta > \theta_{cr}$). Sumer et al. (1992 a).

away from the pile, because they themselves are not removed from the pile. Therefore, the end result is that the mean scour around the pile is nil. Sumer et al.'s works (1992 a and 1993) therefore linked the "onset" of net scour to the lee-wake vortex flow behind the pile.

Scour depth

As the KC number is increased, the previously mentioned flow processes, namely the lee-wake vortex flow and the horseshoe vortex, are influenced (Sections 3.1.2 and 3.1.3). As for the former, the length of the lee-wake becomes larger and larger with increasing KC, meaning that larger and larger lengths of the bed area will be exposed to the shed vortices, and therefore their scour action. Regarding the horseshoe vortex, its life span and its size increase with increasing KC (Figs. 3.11 and 3.14), as discussed in Section

3.2. SCOUR AROUND A SLENDER PILE

3.1.2. These two effects result in an increase in the scour depth. Therefore, the scour depth will increase with increasing KC. Sumer et al.'s works (1992 a and 1993) revealed this; see Fig. 3.35 where the data in the case of circular piles are plotted. This result was later confirmed by Kobayashi and Oda (1994).

Fig. 3.35 further shows that the scour depth approaches a constant value, as $KC \to \infty$. This is because (1) the contribution of the lee-wake vortices to the equilibrium scour depth should approach a constant value, considering the finite lifetime of vortices; and (2) the contribution of the horseshoe vortex to the equilibrium scour depth should also approach a constant value since the KC dependence of the horseshoe vortex disappears for large values of the KC number.

Since the constant value attained by the scour depth is the same as that obtained for steady currents (namely $S/D \to 1.3$ as $KC \to \infty$, see Fig. 3.35), it can be concluded that the contribution to the equilibrium scour depth for large KC numbers (such as $KC > O(100)$, Fig. 3.35) is predominately coming from the horseshoe vortex.

For small values of KC (such as $KC < O(10)$), on the other hand, first of all, the onset of net scour is directly related to the lee-wake vortex flow (in the form of vortex shedding) as discussed in the preceding paragraphs, and secondly, the size and life span of the horseshoe vortex are rather small (Figs. 3.11 and 3.14), and, for these reasons, the scour process should be predominantly governed by the lee-wake flow for such small KC numbers.

Sumer et al. (1992 a) give the following empirical expression for the data in Fig. 3.35:

$$\frac{S}{D} = 1.3 \left\{ 1 - \exp\left[-0.03(KC - 6)\right] \right\}; KC \geq 6 \qquad (3.20)$$

It may be noted that the standard-deviation sign in Fig. 3.35 indicates the standard deviation of the data for the steady current case. The scatter with regard to this latter case may be attributed to the effects of various factors outlined in the previous section, such as the Shields parameter; the sediment gradation; the boundary-layer-depth-to-pile-size ratio; the sediment-size-to-pile-size ratio and the Reynolds number. Since the focus in the work of Sumer et al. (1992 a) was the variations with respect to the Keulegan-Carpenter number, no attempt was made to resolve the results in terms of the aforementioned factors. However, for design purposes, the above empirical

192 CHAPTER 3. SCOUR AROUND A SINGLE SLENDER PILE

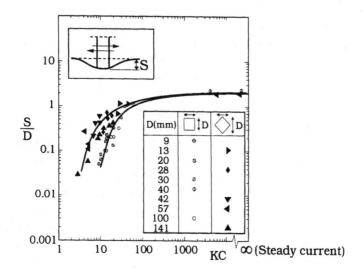

Figure 3.36: Equilibrium scour depth. Square pile. Live bed ($\theta > \theta_{cr}$). Sumer et al. (1993).

expression may be written in the following form

$$\frac{S}{S_c} = 1 - \exp\left[-0.03(KC - 6)\right]; KC \geq 6 \qquad (3.21)$$

which makes it possible for the influence of the aforementioned factors to be accommodated in the formulation through S_c, the scour depth experienced in the current-alone case (e.g., Eq. 3.13). Note that Eqs. 3.20 and 3.21 are valid for live-bed conditions.

Influence of cross-sectional shape. The influence of cross-sectional shape on the scour depth has been investigated by Sumer et al. (1993). The results of their experiments are shown in Figs. 3.36 and 3.37. The figures reveal a significant effect, particularly for small values of KC. The latter is related to the onset of vortex shedding; the scour depth is largest for the cross section where the critical value of KC (for the onset of vortex shedding) is smallest, namely the 45^0 square pile. It is smallest for the cross section where the critical value of KC is largest, namely 90^0 square pile. As seen from the figures, the difference between the three cases becomes relatively smaller, as

3.2. SCOUR AROUND A SLENDER PILE

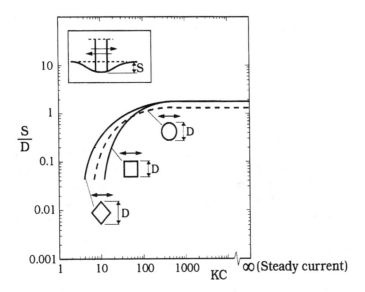

Figure 3.37: Equilibrium scour depth. Effect of cross-sectional shape. Live bed ($\theta > \theta_{cr}$). Sumer et al. (1993).

KC is increased. This effect may be attributed to the decreasing importance of the vortex shedding with increasing KC, considering the finite lifetime of the vortices.

Empirical expressions are given for the scour depth (Sumer et al., 1993); for 90^0-orientation square piles:

$$\frac{S}{D} = 2\left\{1 - \exp\left[-0.015(KC - 11)\right]\right\}; KC \geq 11 \qquad (3.22)$$

and, for 45^0-orientation square piles:

$$\frac{S}{D} = 2\left\{1 - \exp\left[-0.019(KC - 3)\right]\right\}; KC \geq 3 \qquad (3.23)$$

or, in the format of Eq. 3.21, for 90^0-orientation square piles:

$$\frac{S}{S_c} = 1 - \exp\left[-0.015(KC - 11)\right]; KC \geq 11 \qquad (3.24)$$

and, for 45^0-orientation square piles:

$$\frac{S}{S_c} = 1 - \exp\left[-0.019(KC - 3)\right]; KC \geq 3 \qquad (3.25)$$

in which S_c = the scour depth experienced in the current-alone case. These equations are valid for scour on a live bed ($\theta > \theta_{cr}$).

Scour depth in irregular waves. Sumer and Fredsøe (2001) have studied experimentally the influence of irregular waves on scour. A measured in-situ water elevation spectrum for the North Sea storm conditions was used as the control spectrum to produce the wave-generator signal. (This spectrum is well described by the JONSWAP wave spectrum).

The KC number in the case of irregular waves may be defined in several ways, such as $KC = U_m T_z/D$; $U_m T_s/D$; $U_m T_p/D$; $U_s T_z/D$; $U_s T_s/D$; $U_s T_p/D$. Here U_m is defined by

$$U_m = \sqrt{2}\sigma_U \qquad (3.26)$$

in which σ_U = the r.m.s. value of the orbital velocity U at the bed

$$\sigma_U^2 = \int_0^\infty S(f) df \qquad (3.27)$$

in which $S(f)$ = the power spectrum of U, and f = the frequency. The quantity U_s is

$$U_s = 2\sigma_U \qquad (3.28)$$

and may be interpreted as the "significant" velocity amplitude, analogous to the half of the significant wave height. The periods, T_z; T_s; and T_p, on the other hand, are the mean zero-upcrossing period, the significant wave period, and the peak period ($= 1/f_p$), respectively (see, for example Sumer and Fredsøe, 1997, p. Chapter 7).

The authors have measured the scour depth, $(S/D)_{irregular}$, and compared it with that calculated from Eq. 3.20, where KC was calculated in six different ways as described in the preceding paragraph. Apparently, the definition $KC = U_m T_p/D$ has given the best representation. It may be noted that the Keulegan-Carpenter number defined in this way reduces to the ordinary KC number in the case of regular waves, because $\sqrt{2}\sigma_U \rightarrow U_m$, and $T_p \rightarrow T_w$ in this latter case.

To conclude, the scour depth in the case of irregular waves can be predicted using the empirical expression given in Eq. 3.20 provided that KC is calculated by $KC = U_m T_p/D$ in which $U_m = \sqrt{2}\sigma_U$.

3.2. SCOUR AROUND A SLENDER PILE

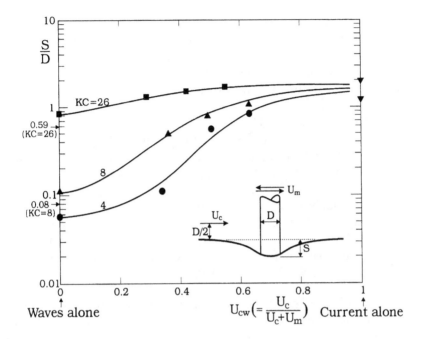

Figure 3.38: Equilibrium scour depth. Co-directional waves and current. Arrows on vertical axis: Waves-alone by Eq. 3.20. Live bed ($\theta > \theta_{cr}$). Sumer and Fredsøe (2001).

Scour depth in combined waves and current. This has been investigated by Herbich and his co-workers (Wang and Herbich, 1983; Herbich et al., 1984; and Eadie and Herbich, 1986). The ranges of the parameters in these works are $2 < KC < 25$ and $0.35 < U_{cw} < 0.7$ in which U_c = the current velocity and U_{cw}

$$U_{cw} = \frac{U_c}{U_c + U_m} \tag{3.29}$$

The results of these latter works indicate that the scour depth in the case of the combined flow is not radically different from that experienced in steady currents, apparently due to the mainly current-dominated flow environment.

Subsequently, Sumer and Fredsøe (2001) made a systematic study of the influence of a superimposed current on the scour depth in combined waves and current.

Fig. 3.38 displays the data obtained by Sumer and Fredsøe (2001) for the case when the waves were propagating in the same direction as the current. Here U_c = the undisturbed current velocity measured at the distance $D/2$ from the bed, representing the near-bed current velocity.

The following conclusions can be deduced from Fig. 3.38.

1. The data reveals that S/D goes to the expression given in Eq. 3.20 (the arrows on the vertical axis), as the parameter $U_{cw} \to 0$ (the waves-alone case), while it approaches the steady-current values reported in the literature, as $U_{cw} \to 1$ (the current-alone case; e.g., Eq. 3.12).

2. The data implies that, for small KC numbers, even a slight current superimposed on waves would cause the scour depth to increase significantly. This is due to the presence of a strong horseshoe vortex (both in time and space) in front of the pile even in the case of a weak current (Sumer et al., 1997, Figs. 13 and 16).

3. It is seen from Fig. 3.38 that the scour depth is apparently dominated by the current component of the combined flow when $U_{cw} \geqslant 0.7$, since the scour depth approaches the values obtained in the case of the current alone for $U_{cw} \geqslant 0.7$. This result is attributed to the constant presence of the lee wake at the downstream side of the pile, and its complete disappearance at the upstream side of the pile for such large values of U_{cw}. Hence, the flow picture in this case will look like that of the case of the current alone.

Fig. 3.39 compares Sumer and Fredsøe's (2001) results with the those of Wang and Herbich (1983), and Eadie and Herbich (1986, Figs. 9-13).

Fig. 3.39 shows that the results of Eadie and Herbich's experiments (1986) are consistent with Sumer and Fredsøe's results. Fig. 3.39 also shows that there is a general, qualitative agreement between Sumer and Fredsøe's results and the results of Wang and Herbich (1983), although the scour depths measured in the latter study seem to be generally larger. Nevertheless, the Wang and Herbich data apparently indicate the same kind of KC variation as in the case of the Sumer and Fredsøe study.

Fig. 3.40 depicts the entire data obtained in Sumer and Fredsøe's (2001) experiments including the case where the waves were propagating perpendicular to the current, plotted in the same format. Fig. 3.40 shows that

3.2. SCOUR AROUND A SLENDER PILE

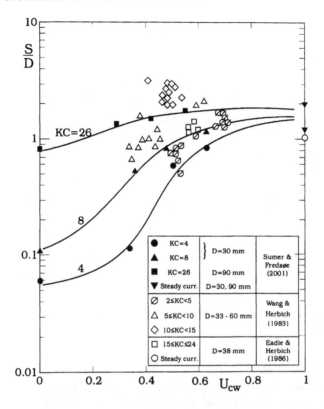

Figure 3.39: Equilibrium scour depth. Co-directional waves and current. Comparison. Sumer and Fredsøe (2001).

the influence of the current on the results in the case of the waves propagating perpendicular to the current is as important as in the case of the co-directional waves and current.

The behaviour of the scour process at the two ends of the range $0 \leq U_{cw} \leq 1$ in Fig. 3.40 is as follows. The near-bed video recording made in the study showed that the vortex shedding (the key element in the scour process in the case of small KC numbers) is not influenced very much by the presence of the current for the wave-dominated regime (i.e., at the lower end of the range $0 \leq U_{cw} \leq 1$). (Sumer and Fredsøe report that the near-bed vortex shedding in the direction of wave propagation was clearly identified

CHAPTER 3. SCOUR AROUND A SINGLE SLENDER PILE

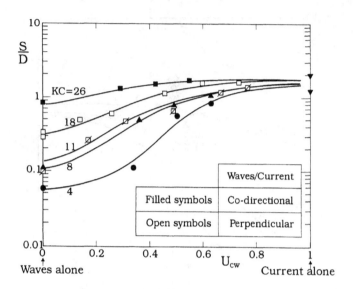

Figure 3.40: Equilibrium scour depth. Waves and current. Live bed ($\theta > \theta_{cr}$). Sumer and Fredsøe (2001).

from their video recordings for $KC = 8$ even for values of U_{cw} as large as 0.5). Therefore, the direction of the current will not be a significant factor for the scour depth at this end of the range $0 \leqslant U_{cw} \leqslant 1$. The direction of the current at the "current" end, i.e., $U_{cw} = 1$, will not be a significant factor either, simply because the horseshoe vortex, the major element for the scour in this current-dominated regime, will be practically uninfluenced by the presence of the waves.

Based on Sumer and Fredsøe's data given in the preceding figures, the following empirical expression can be given for the scour depth:

$$\frac{S}{D} = \frac{S_c}{D}[1 - \exp\{-A(KC - B)\}]; \quad KC \gtrsim 4 \qquad (3.30)$$

in which S_c = the scour depth in the case of the steady-current alone (Eq. 3.12 or Eq. 3.13), and the quantities A and B are given as follows:

$$A = 0.03 + \frac{3}{4}U_{cw}^{2.6} \qquad (3.31)$$

$$B = 6\exp(-4.7U_{cw}) \qquad (3.32)$$

3.2. SCOUR AROUND A SLENDER PILE

Caution must be exercised, however, when the previous equations are implemented for KC numbers larger than the upper boundary of the tested range of KC, namely for $KC \gtrsim 30$. Also, note that the preceding formulation is valid for the live-bed scour regime.

Example 10 *A numerical example for the prediction of scour depth*

Consider a 30 cm diameter pile. Predict the scour depth when this pile is exposed to waves with a period of $T_w = 10$ s and a wave height of $H = 2$ m. The water depth is $h = 10$ m. The sand size is 0.2 mm.

1. Calculate the quantity L_0 (the deep-water wave length)

$$L_0 = \frac{gT_w^2}{2\pi} = \frac{9.81 \times 10^2}{2\pi} = 156 \text{ m}$$

2. Calculate the parameter h/L_0

$$\frac{h}{L_0} = \frac{10}{156} = 0.064$$

3. From the wave tables for sinusoidal waves

$$\sinh(kh) = 0.733 \text{ for } \frac{h}{L_0} = 0.064$$

4. Calculate the amplitude of the orbital motion of water particles at the seabed, assuming that the small-amplitude sinusoidal wave theory is applicable (Appendix A)

$$a = \frac{H}{2}\frac{\cosh(k(z+h))}{\sinh(kh)} = \frac{H}{2}\frac{\cosh(k(-h+h))}{\sinh(kh)} = \frac{H}{2}\frac{1}{\sin(kh)} = \frac{2}{2}\frac{1}{0.733} =$$
$$= 1.36 \text{ m}$$

in which $z =$ the vertical distance measured from the mean water level which is put equal to $z = -h$ (the seabed). The maximum value of the velocity at the bed is (Appendix A)

$$U_m = \frac{\pi H}{T_w}\frac{\cosh(k(z+h))}{\sinh(kh)} = \frac{\pi \times 2}{10}\frac{1}{0.733} = 0.86 \text{ m/s}$$

CHAPTER 3. SCOUR AROUND A SINGLE SLENDER PILE

5. Check if the sinusoidal theory is applicable

$$U \text{ (the Ursell parameter)} = \frac{HL^2}{h^3} < 15$$

in which L, the wave length, is $L = L_0 \tanh(kh)$. (Otherwise, use the cnoidal theory). From the wave tables

$$\tanh(kh) = 0.591 \text{ for } \frac{h}{L_0} = 0.064$$

Therefore

$$L = 156 \times 0.591 = 92 \text{ m}$$

and then

$$U = \frac{HL^2}{h^3} = \frac{2 \times 92^2}{10^3} = 17$$

which is only slightly larger than 15. Therefore we may assume that the sinusoidal theory still is applicable.

6. Calculate the Keulegan-Carpenter number at the seabed

$$KC = \frac{2\pi a}{D} = \frac{2 \times \pi \times 1.36}{0.3} = 28.5$$

7. Predict the scour depth from Eq. 3.21

$$\frac{S}{S_c} = 1 - \exp\left[-0.03(KC - 6)\right] = 1 - \exp\left[-0.03 \times (28.5 - 6)\right] = 0.49$$

(3.33)

in which S_c, the scour depth in the case of the steady-current alone, can be found from Eq. 3.12 or Eq. 3.13.

8. From Eq. 3.12, one can take $S_c/D = 1.3 + \sigma_{S/D} = 1.3 + 0.7 = 2$, or $S_c/D = 1.3 + 2\sigma_{S/D} = 1.3 + 2 \times 0.7 = 2.7$ as the maximum scour depth for the live-bed scour for design purposes. Taking the latter, $S_c/D = 2.7$, the scour depth will be

$$S = 0.49 S_c = 0.49 D \frac{S_c}{D} = 0.49 \times 0.3 \times 2.7 = 0.4 \text{ m}$$

3.2. SCOUR AROUND A SLENDER PILE

9. Calculate the Shields parameter

$$\theta = \frac{U_{fm}^2}{g(s-1)d} = \frac{\frac{f_w}{2}U_m^2}{g(s-1)d} = \frac{\frac{0.004}{2} \times 0.86^2}{9.81 \times (2.65-1) \times 0.0002} = 0.46$$

where f_w is calculated from

$$f_w = 0.035 RE^{-0.16}$$

as 0.004 (Fredsøe and Deigaard, 1992, p. 29), assuming that the bed is acting as a smooth wall ($dU_{fm}/\nu \lesssim 10$). Here, $RE = aU_m/\nu$, the wave-boundary-layer Reynolds number. As seen, the Shields parameter, $\theta = 0.46$, is larger than the critical value, $\theta_{cr} \simeq O(0.05)$, i.e., the bed is live, therefore the equation used to calculate the scour depth (Eq. 3.20) is valid.

Example 11 *A numerical example for the prediction of scour depth. Combined wave and current*

Now, consider that the pile is exposed to a combined wave and current climate. The waves are the same as in the previous example while the current velocity at the distance $D/2$ from the bed $U_c = 0.6$ m/s. Calculate the scour depth.

1. Calculate U_{cw}

$$U_{cw} = \frac{U_c}{U_c + U_m} = \frac{0.6}{0.6 + 0.86} = 0.41$$

2. Calculate the scour depth from Eqs. 3.30-3.32:

$$A = 0.03 + \frac{3}{4}U_{cw}^{2.6} = 0.03 + \frac{3}{4} \times 0.41^{2.6} = 0.104 \quad (3.34)$$

$$B = 6\exp(-4.7U_{cw}) = 6 \times \exp(-4.7 \times 0.41) = 0.87 \quad (3.35)$$

$$\frac{S}{D} = \frac{S_c}{D}[1 - \exp\{-A(KC - B)\}] \quad (3.36)$$

$$= \frac{S_c}{D}[1 - \exp\{-0.104 \times (28.5 - 0.87)\}] = \frac{S_c}{D} \times 0.94 \quad (3.37)$$

in which S_c, the scour depth in the case of the steady-current alone, can be found from Eq. 3.12 or Eq. 3.13

3. Taking the latter, $S_c/D = 2.7$, from the previous example, the scour depth will be

$$S = \frac{S_c}{D} \times 0.94 \times D = 2.7 \times 0.94 \times 0.3 = 0.76 \text{ m}$$

Influence of non-linear waves. Research was also carried out to study the influence of the wave non-linearity on local scour around a single pile (Carreiras, Larroude, Seabra-Santos and Mory, 2000). The characteristics of non-linear waves evolve in the nearshore domain and hence the scour processes may depend on the position of the piles, i.e., on the local characteristics of the waves.

For linear waves, the shape and characteristics of the wave are kept constant all over the domain, and the quantities entering the KC number do not depend on the pile position. According to the linear theory, the KC number is equivalently defined as a function of the stroke of the horizontal excursion, $2a$, relative to the diameter of the cylinder, D

This form accounts for the physical processes producing local scour in terms of boundary layer separation and vortex shedding (Sumer and Fredsøe, 1997). Non-linear waves behave in a different way. Initial sinusoidal variations evolve into asymmetric profiles as the waves approach the coastline. The scour induced by one of these waves around a pile may be different depending on the exact location of the pile, i.e., depending on the local characteristics of the wave.

A set of 19 experiments with an isolated pile exposed to non-linear wave propagation was conducted for different wave conditions and for different pile diameters. The wave period and the pile diameter varied in the ranges 1.0-2.6 s and 10-30 mm, respectively , and the mean water depth was 0.10 or 0.15 m. The experiments were conducted in a wave flume of width 0.3 m and length 7.5 m. The piles were fixed in a flat horizontal sand pit or in a sand pit forming a 1/20 sloping beach. The sand grain size was $d_{50} = 0.27$ mm. The scour depths were measured around the piles, and bed form changes were measured over an area containing the piles, enabling to investigate both local and global scour processes. The Shields parameter varied in the range 0.04-0.11.

The variation of the local scour depth with the KC number was found to agree with the form of Eq. 3.20, provided that the local characteristics of

3.2. SCOUR AROUND A SLENDER PILE

waves were used to estimate KC. The best agreement between Eq. 3.20 and the experimental data was obtained when the Keulegan-Carpenter number was determined using Eq. 3.6, i.e.,

$$KC = \frac{2\pi a}{D} \qquad (3.38)$$

and the stroke of the motion close to the bottom, $2a$, was computed by integrating the measured local velocity. Also, the coefficient $m = 0.06$ in

$$\frac{S}{D} = 1.3\{1 - \exp[-m(KC - 6)]\}$$

fitted the data better instead of 0.03 (see Eq. 3.20) for 19 tests over the range in KC number from 11-23.

The local definition of the KC number was found to give good results for scour prediction for piles placed at different positions in the flume, taking advantage of the variation in wave non-linearity with distance along the flume. The time scaling also must take account of the wave non-linearity.

Scour depth in breaking waves. The effect of the breaking waves on scour was studied experimentally (Carreiras et al. , 2000) for different forcing conditions. In each set of 5 experiments (in which the flow conditions were invariable) the pile was placed at a different position related to the point of wave breaking on a 1:20 sloping beach. The mean water depth in the horizontal bottom was 0.17 m.

The pile was moved in steps of 0.15 m; 2 tests had piles onshore of breaking, 2 offshore and one was coincident with the point of wave breaking (Fig. 3.41). The bed morphodynamic changes were measured and compared with those that occurred for the same flow conditions without a pile. There was not a significant undertow velocity in these experiments.

Visualizations showed that vortices produced around the pile were initially dominant in the development of the scour, and a rapid increase of the scour depth was observed. However, the high turbulence produced by the dissipation of the wave energy in the surf zone induced an intense sediment transport.

Depending on the position of the pile relative to the breaking point, the ripple formation and dynamics (when the pile is offshore relative to the point of wave breaking), and the formation of the bar (when the pile is onshore), were found to have a main influence on the evolution of scour. In fact,

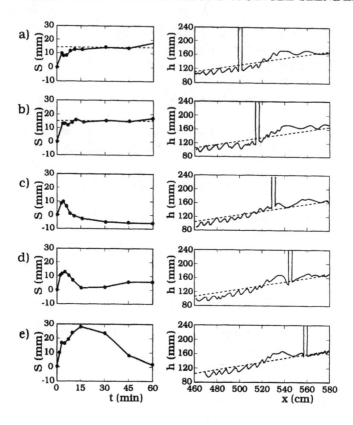

Figure 3.41: Scour-depth time development and equilibrium bed profile. Carreiras et al. (2000).

the final bed changes resulted from the superposition of the large scale bed evolution due to the wave breaking and the small scale bed scour.

In the experiments in which the pile is *offshore* relative to the point of wave breaking (Fig. 3.41 a and b) the equilibrium scour depth seems to be well represented by Eq. 3.20 with coefficient $m = 0.06$, provided, once more, that the KC number is calculated with Eq. 3.38. In the other three experiments, where the pile location was *onshore* relative to the point of wave breaking or *coincident* to it (Fig. 3.41 c-d), the erosion is highly influenced by the formation of the bar, a few minutes after the beginning of the test.

Even for large KC number conditions the scour patterns are completely

3.2. SCOUR AROUND A SLENDER PILE

modified when a single pile is located in the surf zone of breaking waves. When the pile is located at the breaking point or onshore, global large-scale bed changes, namely the formation of the bar, are superposed to the local scour processes.

The influence of breaking waves on scour has also been investigated by Bijker and de Bruyn (1988). Their key result is that while the scour depth decreases when a non-breaking wave is superimposed on a steady current (in agreement with Fig. 3.38), it increases when a breaking wave is superimposed on the same current. Their experiment showed that the scour depth can be 1.46 times larger in the case of the combined waves and current (with breaking waves) than in the case of the current alone. Considering that the combined-wave-and-current scour is always smaller than the current-alone scour (when the waves do not break) (Fig. 3.38), this 1.46 factor increase in the scour depth is apparently linked with the breaking waves. The authors related this to the observed increase in the orbital velocity under breaking waves.

Scour depth in cohesive sediment. To the authors' knowledge, no study is yet available, investigating the scour depth in the case of cohesive-sediment bed. (Refer to the brief account given in Section 3.2.1 of scour in cohesive sediment in the case of steady currents).

3.2.3 Scour around a cone-shaped object

Sumer, Fredsøe, Christiansen and Hansen (1994) measured the bottom shear stress for a cone-shaped structure to observe the influence of the side slope on the shear stress amplification. Fredsøe and Sumer (1997) in a follow-up study reported the results of similar bottom-shear stress measurements and scour tests made with cone-shaped structures.

Figs. 3.42-3.44 present their bottom-shear-stress results. Their results regarding the scour tests will be presented in Chapter 6.

It is clear from Figs. 3.42-3.44 that the overall effect of side slopes in front of the cone-shaped structure is to decrease the bottom shear stress, apparently due to the large decrease in the adverse pressure gradient. It is interesting to note that the horseshoe vortex still exists in the case of the cone (Fig. 3.42), but its strength is greatly reduced. Similarly, the bottom shear stress decreases significantly at the side edges of the cone, and this is due partly to the rather small contraction of streamlines, and partly to the rather weak horseshoe vortex. Likewise, the bottom shear stress also

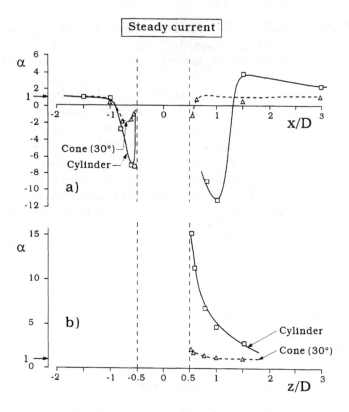

Figure 3.42: Bed-shear-stress amplification. Sumer et al. (1994).

experiences similar large decreases in the lee-wake area, and this is linked to the weak occurrence of vortex shedding in the case of the cone.

To the authors' knowledge, no data is available on scour at a "slender" cone-shaped structure in steady currents and in waves. Some data has been reported in Fredsøe and Sumer (1997). However, this was for a large cone. We shall return to this issue in Chapter 6.

3.3 Time scale

Time scale of scour (see Section 1.3 in Chapter 1) has been studied by Sumer, Christiansen and Fredsøe (1992 b) and Sumer et al. (1993). On dimensional

3.3. TIME SCALE

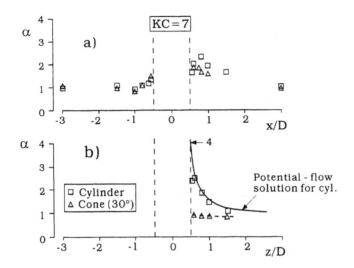

Figure 3.43: Bed-shear-stress amplification. Sumer et al. (1994).

grounds, the time scale can be written in the following non-dimensional form. For steady currents:

$$T^* = T^*(\frac{\delta}{D}, \theta) \tag{3.39}$$

and, for waves:

$$T^* = T^*(KC, \theta) \tag{3.40}$$

in which T^* is the normalized time scale defined in the same way as in Eq. 2.44, and δ = the flow depth (the boundary layer thickness). These relations have been tested against the data obtained for circular piles in Sumer et al. (1992 b), and further in Sumer et al. (1993) (for waves), which also gives data for other pile geometries, namely the square piles with the 90^0- and 45^0-orientations. Fig. 3.45 displays the results for the case of the waves.

The data in Fig. 3.45 shows that T^* decreases with increasing θ, consistent with the results obtained in the case of pipelines (Section 2.3.4), while it shows that T^* increases with increasing KC. The latter behaviour is attributed to the fact that the volume of sediment to be eroded increases with increasing KC.

Fig. 3.45 (the two bottom panels) illustrates the cross-section influence on the time scale. From the data, the time scale for the 45^0-arrangement square

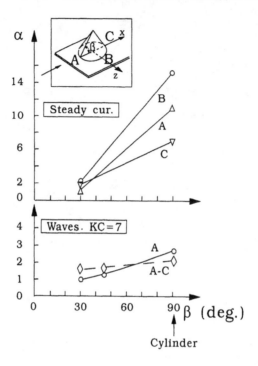

Figure 3.44: Effect of side slope on bed-shear-stress amplification. Sumer et al. (1994).

pile is not radically different from that for circular piles, although T^* for the former case appears to be slightly smaller than that for the latter situation. Regarding the 90^0-arrangement square piles, however, the figure shows that the time scale in this case is somewhat smaller than that for the circular piles. The figure also shows that this discrepancy becomes more pronounced with decreasing θ. Sumer et al. (1993) linked this behaviour to the strength of the shed vortices. They showed that the vortices are stronger in the case of the 90^0- orientation square pile than in the case of the circular pile and the 45^0- orientation square pile. (They showed that the vortex strength in the latter two cases is apparently much the same). Now, the stronger the vortices, the faster the scour process. Therefore, the time scale in the case of the 90^0 orientation square pile should be smaller than in the other two cases, as exhibited in Fig. 3.45.

3.3. TIME SCALE

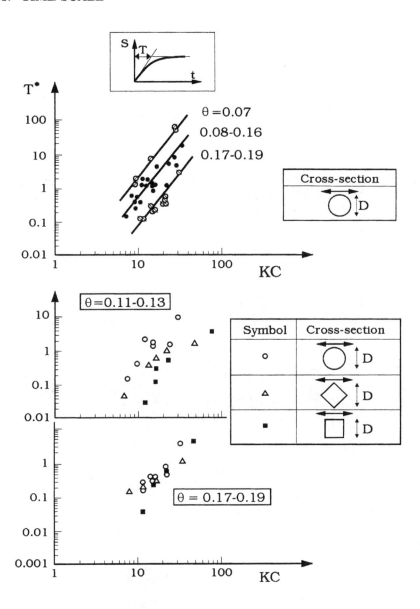

Figure 3.45: Time scale of scour around pile. Live bed ($\theta > \theta_{cr}$). Sumer et al. (1993).

210 CHAPTER 3. SCOUR AROUND A SINGLE SLENDER PILE

Empirical expressions have been given in Sumer et al. (1992 b), relating the normalized time scale, T^*, to KC and θ. These are, for steady currents:

$$T^* = \frac{1}{2000}\frac{\delta}{D}\theta^{-2.2} \quad (3.41)$$

in which δ = the flow depth (the boundary layer thickness) and, for waves:

$$T^* = 10^{-6}(\frac{KC}{\theta})^3 \quad (3.42)$$

A detailed study of the temporal development of clear-water scour at piles in steady currents (in conjunction with bridge piers) has been made by Melville and Chiew (1999). Data is presented and used to quantify the influence of flow duration on the depth of local scour.

Example 12 *A numerical example for the prediction of time scale in steady currents*

Given that $D = 30$ cm, $d_{50} = 0.5$ mm, what is the time scale of the scour process at a pile which is exposed to a mean current of $V = 0.6$ m/s. The water depth is $h = 10$ m.

1. Calculate the undisturbed bed friction velocity from

$$U_f = \frac{V}{2.5\left[\ln(\frac{30h}{k_s}) - 1\right]}$$

in which $k_s = 2.5 d_{50}$:

$$U_f = \frac{0.6}{2.5 \times \left[\ln(\frac{30 \times 10}{2.5 \times (0.5 \times 10^{-3})}) - 1\right]} = 0.021 \text{ m/s}$$

2. Calculate the undisturbed Shields parameter:

$$\theta = \frac{U_f^2}{g(s-1)d_{50}} = \frac{0.021^2}{9.81 \times (2.65 - 1) \times (0.5 \times 10^{-3})} = 0.055$$

3. Calculate T^* from Eq. 3.41:

$$T^* = \frac{1}{2000}\frac{\delta}{D}\theta^{-2.2} = \frac{1}{2000} \times \frac{10}{0.3} \times (0.055)^{-2.2} = 9.8$$

3.3. TIME SCALE

4. Find T from Eq. 2.44:

$$T = \frac{D^2}{(g(s-1)d_{50}^3)^{1/2}} T^* =$$

$$= \frac{0.30^2}{(9.81 \times (2.65-1) \times (0.5 \times 10^{-3})^3)^{1/2}} \times 9.8 = 19608 \text{ s} \simeq 5 \text{ h}$$

Example 13 *A numerical example for the prediction of time scale in waves*

Calculate the time scale of scour when the pile in the previous example is exposed to waves with a period of $T_w = 10$ s and a wave height of $H = 2$ m. The water depth is $h = 10$ m.

1. From Example 10 in Section 3.2.2, the amplitude a and the maximum value of the velocity U_m of the orbital motion of water particles at the seabed and KC are

$$a = 1.36 \text{ m}, \quad U_m = \frac{2\pi a}{T_w} = 0.86 \text{ m/s}; \quad KC = \frac{2\pi a}{D} = \frac{2 \times \pi \times 1.36}{0.3} = 28.5$$

2. Find the friction factor for the wave boundary layer, assuming that the bed is a rough boundary (from, for example, Fredsøe and Deigaard, 1992, p. 25):

$$f_w = 0.04 \left(\frac{a}{k_s}\right)^{-\frac{1}{4}}, \quad \frac{a}{k_s} > 50$$

$$f_w = 0.007 \text{ for } \frac{a}{k_s} = \frac{a}{2.5 d_{50}} = \frac{1.36}{2.5 \times (0.5 \times 10^{-3})} = 1088$$

3. Calculate the maximum bed friction velocity for the wave boundary layer:

$$U_{fm} = \sqrt{\frac{f_w}{2}} U_m = \sqrt{\frac{0.007}{2}} 0.86 = 0.051 \text{ m/s}$$

4. Check if the bed is acting as a rough wall: $dU_f/\nu = 0.05 \times 5.1/0.01 = 26$, larger than 10; therefore, the calculation of the friction velocity assuming that the bed is a rough wall may be justified.

5. Calculate the undisturbed Shields parameter:

$$\theta = \frac{U_{fm}^2}{g(s-1)d_{50}} = \frac{0.051^2}{9.81 \times (2.65-1) \times (0.5 \times 10^{-3})} = 0.32$$

Figure 3.46: Scour protection (riprap). Melville and Coleman (2000).

6. Calculate T^* from Eq. 3.42

$$T^* = 10^{-6}(\frac{KC}{\theta})^3 = 10^{-6}(\frac{28.5}{0.32})^3 = 0.71$$

7. Find T from Eq. 2.44:

$$T = \frac{D^2}{(g(s-1)d_{50}^3)^{1/2}}T^* = \frac{0.30^2}{(9.81 \times (2.65-1) \times (0.5 \times 10^{-3})^3)^{1/2}} \times 0.71$$
$$= 1420 \text{ s} \simeq \frac{1}{2} \text{ hr}$$

3.4 Scour protection

Photographs in Figs. 3.46 and 3.47 give two examples of scour protection at bridge piers.

Methods to protect piles against the scour are basically the same as those used for pipelines: The bed area around the pile is covered by a protection layer, either in the form of a stone protection layer, or in the form of a protective mattress (Section 2.6). The key issue here is to determine the plan-view extent and thickness of the protection layer. Although, given the

3.4. SCOUR PROTECTION

Figure 3.47: Scour protection. Sakura Bridge (Kino River). Tsujimoto (1987).

angle of internal friction of the seabed sediment and the scour depth, the plan-view extent of the scour hole can be determined, and, hence, the plan-view extent of the protection layer can be designed accordingly, this would be only a first and probably a conservative estimate.

Breusers and Raudkivi (1991, p. 91) and Melville and Coleman (2000, p. 348) give detailed accounts of protection work in the case of bridge piers (the steady-current case). Fig. 3.48 outlines the scour protection measures for circular and elongated bridge-piers from the latter reference. The protection-material (riprap) size diagram is based on the method by Parola (1993, 1995). As indicated in Fig. 3.48, Melville and Coleman (2000) recommend that the ideal level for placement of the stone layer is with the surface of the layer at about the level of the lowest bed form expected in the vicinity of the bridge. This will minimize the possibility of bed-form destabilization of the protection stones, Melville and Coleman (2000, Section 9.5.1). See also the discussion in item 4 under Failure of Scour Protection in the following paragraphs.

No systematic study is available, linking the dimensions of the protection layer to the bulk flow parameters in wave-dominated flows. The stone size may be determined from the Shields criterion for the threshold of the motion at the surface of the stone layer. The plan-view extent of the protection

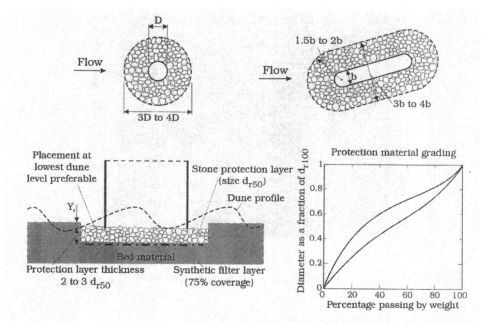

Figure 3.48: Recommendations for the use of stone protection as a pier scour countermeasure. Melville and Coleman (2000).

layer may be found from the information on the scour depth and the internal friction of the seabed sediment, as mentioned previously. The stone layer may be placed in the form of a berm around the pile (as in Figs. 3.46 and 3.47). However, for site-specific situations, the common practice is usually to make physical model studies of the scour, and its related protection work for the final design. Fig. 3.49 presents the recommended layout of scour protection (as a result of a physical-model study) for the piers of the Western Bridge across the western channel of the Great Belt in Denmark (Hebsgaard, Ennemark, Spangenberg, Fredsøe and Gravesen, 1994; see also Hebsgaard, 1998).

Failure of scour protection. A stone-protection (riprap) layer may be subject to failure. This is caused by different *mechanisms*:

1. The stones may be moved by the action of the flow; in this case, the critical value of the Shields parameter for threshold at the surface of the

3.4. SCOUR PROTECTION

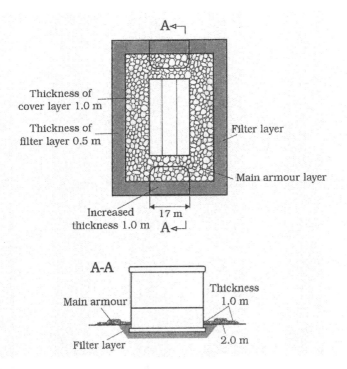

Figure 3.49: Scour protection for piers of Western Bridge, Great Belt, Denmark. Hebsgaard et al. (1994).

protection layer, θ_{cr}, is exceeded, $\theta > \theta_{cr}$. Clearly, the stone size should be selected so that the condition $\theta < \theta_{cr}$ is met for the stability of the protection layer. The Shields parameter for threshold is given in Fig. 1.2. In the calculation of the Shields parameter, θ, on the other hand, the large amplification in the bed shear stress due to the presence of the pile (Figs. 3.10 and 3.19) must be taken into consideration. The amplification in the bed shear stress may also be due to the placement of the riprap at an elevation higher than the undisturbed bed, it may be due to the additional turbulence generated by the pile itself, etc., and obviously, these additional effects should also be taken into consideration.

2. The underlying finer bed material is winnowed ("sucked") from between the riprap stones (Fig. 3.50). Experiments of Chiew (1995) show that

216   CHAPTER 3. SCOUR AROUND A SINGLE SLENDER PILE

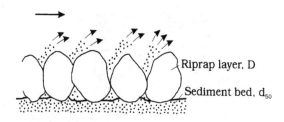

Figure 3.50: Definition sketch for suction removal of sediment from between armour blocks.

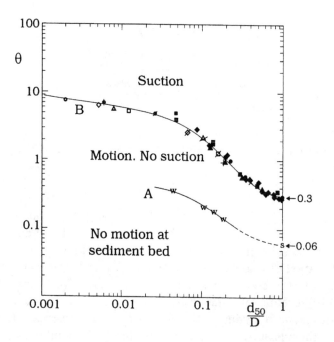

Figure 3.51: Sediment will be removed from between armour blocks (see the previous figure) when (θ and d_{50}/D) falls in "Suction" region. Sumer et al. (2001).

3.4. SCOUR PROTECTION

a thick protection layer can prevent winnowing in the absence of a filter layer.

Sumer, Cokgor and Fredsøe (2001) give a design diagram (Fig. 3.51) for the following two critical conditions for *steady currents*: (1) Curve A in Fig. 3.51 represents the critical condition for the incipient motion at the sediment base (data taken from Worman, 1992, and Sumer, 1986); and (2) Curve B represents the critical condition for the suction removal of the bed sediment from between riprap stones. Curve B can be represented by the following empirical expression:

$$\theta_{su} = 0.3 + 3(\frac{d_{50}}{D})^{-0.15} \exp(-7.5\frac{d_{50}}{D}) \qquad (3.43)$$

In Fig. 3.51 and in the preceding equation, d_{50} is the size of the finer bed material, D is the size of stones/armour blocks and θ is the Shields parameter defined by

$$\theta = \frac{U_f^2}{g(s-1)d_{50}} \qquad (3.44)$$

in which U_f is the friction velocity at the riprap bed. The stone size must be designed such that the Shields parameter calculated from the preceding equation is smaller than θ_{su} given in Eq. 3.43, meaning that no winnowing will occur.

Sumer et al.'s (2001) study shows that the vortices which form in the holes between the armour blocks are the key elements in the process. The sediment swept into these vortices is entrained into the main body of the flow by these vortices.

Sumer et al. (2001) also give the time scale of the suction process and the downward displacement of stones (the general lowering of the riprap layer) when the suction removal of sediment takes place.

3. Observations, both in steady currents (Belcher, Xu and Hunt, 1990) and in oscillatory flows (waves) (Fredsøe, Sumer, Laursen and Pedersen, 1993) show that the bed shear stress at the junction between the riprap and the bed is generally subject to an increase. In the case of the waves, the abrupt change in the roughness between the riprap and the bed may even induce a steady streaming (Fredsøe et al., 1993). Also, there will be an increase in the bed shear stress in the vicinity of each individual stone, in much the same way as in the case of a single pile. These effects may result in local scour at the edge of the riprap, termed the edge scour. When local scour holes form, the riprap stones at the edge of the protection layer may

slump down into these local scour holes, affecting the stability of the riprap layer. Chiew (1995) has investigated the "edge" failure in steady currents. However, with a proper design of the protection layer, slumping down of the stones into the scour hole presumably form a protective slope, which may be considered as a positive effect for riprap protection.

4. Under live-bed conditions, destabilization by bed-form progression may become important. This is dependent on the destabilizing effect of bed-form troughs as they pass the protection layer. Lauchlan and Melville (2001) studied this effect. Their study showed that the deeper the placement level of riprap stones within the sediment bed, the less exposed the riprap was to destabilizing of the bed forms and the better the protection against the scour. The latter authors proposed a pier riprap size-prediction equation which included a parameter to account for placement level.

5. The protection material may sink into the seabed when no filter material is used. The sinking of the protection material into the seabed may be caused by different mechanisms: (1) It may be due to scour below the individual stones, see Example 8 in Section 3.2.1; (2) It may be due to the momentary liquefaction (which may or may not be helped by the seepage flow). (Note that the momentary liquefaction itself is caused by the upward directed pore pressure gradient experienced during the passage of the wave trough); or (3) It may be caused by the liquefaction due to the buildup of pore pressure in the seabed soil. This latter buildup of pore pressure may be induced by waves, or it may be induced by the cyclic loading on the soil due to earthquakes or cyclic motions executed by the pile itself under waves. The latter two effects (Items 2 and 3) will be studied in detail in Chapter 10.

Finally, it proves necessary to monitor the scour development (if any) and the scour protection after the protection work is applied in the field. This obviously enables the operator to detect any scour at the bed developing at / around the protection layer, to locate if there is any local failure of the protection work itself, and to take remedial measures if and when necessary.

3.5 Mathematical modelling

As has been seen in the preceding sections, the horseshoe vortex is an essential part of the 3-D flow around a slender pile/vertical cylinder (Fig. 3.1). In recent years, the numerical treatment of the horseshoe vortex has been made by several researchers.

3.5. MATHEMATICAL MODELLING

Briley and McDonald (1981) made Navier-Stokes (N.-S.) computations of a laminar, steady horseshoe vortex at the junction between an elliptic strut and a flat plate.

Using a 3-D incompressible Navier-Stokes (N.-S.) code, Kwak, Rogers, Kaul and Chang (1986) computed the laminar, steady junction flow.

Deng and Piquet (1992) studied the 3-D turbulent flow about an airfoil/flat plate junction, where the main features of the horseshoe vortex are captured by the study. An iterative, fully decoupled technique was applied to the Reynolds-averaged N.-S. equations in this study.

A comprehensive review of the work until early nineties is given by Deng and Piquet (1992).

3-D numerical calculations of the flow around a vertical wall-mounted cylinder have also been carried out with the purpose of studying the scour. A partial list of these studies is given in Table 3.2. The following paragraphs will give a brief account of these studies.

Kobayashi's (1992) study. In connection with oscillatory flows, Kobayashi (1992) made an attempt to simulate the 3-D flow around a circular pile in the case of an oscillatory flow, using the so-called vortex-segment model.

This model is basically an extension of the familiar 2-D discrete-vortex model (see Sumer and Fredsøe, 1997, for a review of discrete vortex models) to the 3-D case. In Kobayashi's study, the KC number was $KC = 5$.

No horseshoe vortex and no shedding were obtained, since the KC number was below the critical values for these flow features to exist.

However, the model did produce the unsteady behaviour of the lee-wake vortices, and also, it facilitated a good comparison with the measured velocity field.

Olsen and Melaaen's (1993) and Olsen and Kjellesvig's (1998) studies. Olsen and Melaaen (1993) made an attempt to compute the scour around a circular cylinder in a steady current.

The flow was calculated, using the steady-state N.-S. equations on a 3-D non-orthogonal grid. The Reynolds stress terms were solved, using the $k - \varepsilon$ turbulence model (see Eqs. 2.86-2.93).

Sediment transport equations (Eqs. 2.85 and 2.87 with an equation for bed concentration, similar to Eq. 2.97-2.98) were incorporated into the model for scour calculations. Only the clear-water scour case was simulated in the calculations. Comparison was made with scour patterns measured in an experiment, and good agreement was obtained.

The study appears to produce the horseshoe vortex in front of the cylinder (Olsen and Melaaen, 1993, Fig. 4). As the flow model was a steady model, no shedding was obtained.

Author	Model	Remarks
Kobayashi (1992)	Discrete vortex model; Laminar	Oscillatory flow; Rigid, plane bed
Olsen & Melaaen (1993)	$k - \varepsilon$	Steady cur.; Bed morphology
Olsen & Kjellesvig (1998)	"	"
Richardson & Panchang (1998)	Prandtl's mixing length; RNG	Rigid, plane bed; Rigid, scoured bed
Tseng, Yen & Song (2000)	LES	Rigid, plane bed Also square pile
Yuhi, Ishida & Umeda (2000)	Laminar	Rigid, plane bed; Rigid, scoured bed; Oscillatory flow
Roulund, Sumer & Fredsøe (2001)	$k - \omega$	Rigid, plane bed; Bed morphology; Steady cur.; Oscillatory flow

Table 3.2. A partial list of studies on mathematical modelling of flow and scour around a pile.

Olsen and Kjellesvig (1998) continued the research reported in Olsen and Melaaen (1993). They essentially used the same formulation as in Olsen and Mallaen (1993). The calculation was carried out until the scour process reached the state of equilibrium. The scour depth obtained in the calculations agreed well with the scour depth obtained from empirical formulae.

Richardson and Panchang's (1998) study. Richardson and Panchang (1998) made a 3-D calculation of the flow around a vertical circular cylinder. The calculations were made for three cases: one with a plane bed, the other with an "intermediate" scour hole and the third one with an equilibrium scour hole. The latter scour holes were the "copies" of the scour holes measured in Melville and Raudkivi's (1977) clear-water experiment. In the calculations, a computational fluid dynamics code, FLOW-3D, was used, which

3.5. MATHEMATICAL MODELLING

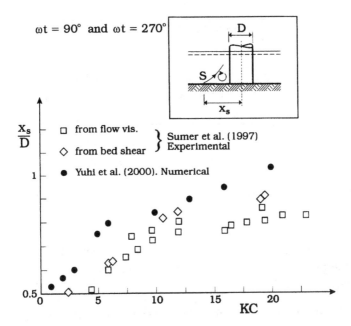

Figure 3.52: Separation distance. Circular pile. $Re_D = 2\text{-}9\times 10^3$, $\delta/D = 0.05\text{-}0.07$ in Sumer et al. (1997) and $Re_D = 2\times 10^3$, $\delta/D = 0.1$ in Yuhi et al. (2000).

solves the 3-D transient N.-S. equations by the volume-of-fluid method. The model supports turbulent closure through a number of advanced schemes, including Prandtl's mixing-length theory, the eddy viscosity model, the two-equation $k - \varepsilon$ model, and the renormalized group (RNG) theory. The latter model is basically an extension of the $k - \varepsilon$ model; however, it requires the less reliance on empirical constants and provides a better solution in areas affected by high shear (Richardson and Panchang, 1998). In Richardson and Panchang's calculations, two different turbulence closure schemes were used, namely Prandtl's mixing-length theory and the RNG model. However, the authors report that the results of simulations are largely similar. The calculations were carried out for the same test conditions as in the experiments of Melville and Raudkivi (1977). The horseshoe vortex was resolved in the calculations, and the results appear to be in good agreement with the experiments. Richardson and Panchang (1998) supplemented their results with Lagrangian particle tracking. The focus in Richardson and Panchang's

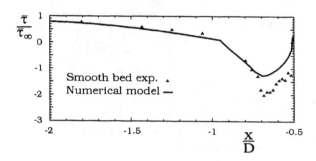

Figure 3.53: Bed shear stress at horseshoe-vortex side of pile. Roulund et al. (2002).

(1998) study was the flow in front of the cylinder. Except some results regarding the particle-tracking simulations, no results were given with regard to the lee-wake flow.

Tseng, Yen and Song's (2000) study. Tseng, Yen and Song (2000) made computations of 3-D flow around a vertical, wall-mounted cylinder (exposed to a steady current), using the Large Eddy Simulation (LES), which is described in the previous chapter (Eqs. 2.105-2.112). In the computations, both the horseshoe vortex and the vortex shedding have been resolved. The model results have been validated against the results of Dargahi's (1989) experiments. Subsequently, simulation runs have been undertaken to study various properties of the 3-D flow around both circular and square cylinders. The computations have been carried out for the case of a rigid, plane bottom. However, the implications of the results with regard to scouring have been discussed.

Yuhi, Ishida and Umeda's (2000) study. Yuhi, Ishida and Umeda (2000) have achieved a 3-D numerical solution of the N.-S. equations (Eqs. 2.83 and 2.84 with the last term on the right hand side of the equation omitted) for both steady currents and oscillatory flows. These numerical calculations have been undertaken for two kinds of beds: for a rigid, plane bed and for a rigid, truncated-cone-shaped bed (simulating a scoured bed) in the case of the steady current; and for a rigid, plane bed in the case of the oscillatory flow. The calculations are for laminar-regime flows. The paper by Yuhi et al. (2000) mainly concentrates on the horseshoe vortex and the

3.5. MATHEMATICAL MODELLING

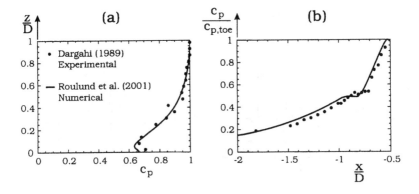

Figure 3.54: (a): Pressure at the upstream edge of pile. (b): Pressure at horseshoe-vortex side of pile. Roulund et al. (2002).

effect of the contraction of streamlines while the results concerning the wake flow have been reported to be presented in a follow-up publication by the same authors. Yuhi et al.'s (2000) steady-flow calculations show that the maximum amplification in the bed shear stress beneath the horseshoe vortex at the streamwise principal axis (cf. Fig. 3.9) is about 10 for the flat-bed situation, while it increases to as much as 20 in the case of the scoured-bed case with $S/D = 0.3$, and then it drops back to the value of 10 for the case of the scoured bed with $S/D = 1.2$. Their oscillatory-flow calculations, on the other hand, show that the corresponding amplification factor is rather low, $O(0-1.5)$, for the tested range of the Keulegan-Carpenter number, namely $1 \leq KC \leq 20$, depending on KC, in agreement with the results obtained from the experiments of Sumer et al. (1997) (Fig. 3.15).

Fig. 3.52 compares the separation distance (a quantity characterizing the size of the horseshoe vortex) obtained by Yuhi et al. (2000) with the experimental results of Sumer et al. (1997). The numerical results are in fairly good agreement with the results obtained from the bed-shear-stress measurements (diamonds), while the results from the flow visualization measurements (squares) seem to underpredict the separation distance, as anticipated. Note that the experiments (Sumer et al., 1997) were conducted with real waves (with some asymmetry), as pointed out in conjunction with Fig. 3.14, whereas the numerical calculations were obtained for a sinusoidally varying oscillatory flow.

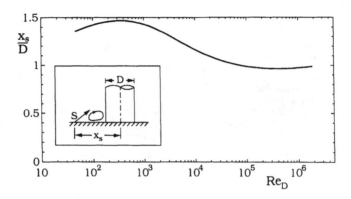

Figure 3.55: Separation distance. Smooth bed. Numerical model. Roulund et al. (2002).

Roulund, Sumer and Fredsøe's (2002) study. These authors have simulated the flow around a pile, using a 3-D code, EllipSys3D, a three-dimensional general-purpose flow solver developed at Risø National Laboratory, Denmark and Department of Energy Engineering at the Technical University of Denmark (Sørensen, 1995, and Michelsen, 1992). The code basically solves the 3-D N.-S. equations. In the calculations of Roulund et al., the $k - \omega$ turbulence model (Wilcox, 1993, and Menter, 1992) was used for closure. The latter closure model was adopted because of its better performance for flows with strong pressure gradients. In the $k - \omega$ turbulence model, k stands for the turbulence energy (in the same way as in the $k - \varepsilon$ turbulence model (Chapter 2, Section 2.7.2)) and ω stands for the specific dissipation of turbulent kinetic energy defined by

$$\omega = \frac{\varepsilon}{k\beta^*} \qquad (3.45)$$

in which ε = the dissipation of turbulent kinetic energy and β^* is one of the model closure constants, taken to be 0.09. The quantity ω satisfies an equation similar to k. Detailed information about $k - \omega$ turbulence model can be found in Wilcox (1993).

The numerical results in Roulund et al.'s (1993) work were validated against experiments. Fig. 3.53 compares the bed shear stress obtained by the model with the experiments where the bed shear stress was measured

3.5. MATHEMATICAL MODELLING

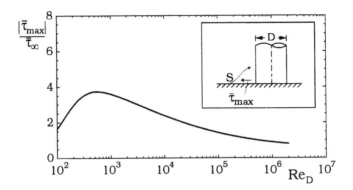

Figure 3.56: Bed shear stress. Smooth bed. Numerical model. Roulund et al. (2002).

with a hot-film probe (cf. Fig. 3.9). Fig. 3.54, on the other hand, compares the pressure obtained from the model with the corresponding experimental results from Dargahi (1989). (c_p in the latter figure is the pressure coefficient, defined in the same way as in Fig. 2.1). More validation results are given in Roulund et al. (2002).

The model, tested and validated, was used to study the flow process (the horseshoe vortex and the vortex shedding) around the pile. The influence of the boundary layer thickness, the influence of the Reynolds number, the influence of the bed roughness and the influence of the pile inclination on the horseshoe vortex were investigated. Figs. 3.55 and 3.56 present the results obtained by Roulund et al. regarding the influence of the Reynolds number in the case of a circular pile in steady current. The boundary layer thickness in the calculation was kept constant, at $\delta/D = 8$, while the Reynolds number was changed. Fig. 3.55 indicates that, for small Reynolds numbers, x_s, the length characterizing the size of the horseshoe vortex, increases with increasing Re_D while, for large Reynolds numbers, the opposite holds true. This is precisely what is expected from the physical arguments given in Section 3.1.1 under the heading Effect of the Reynolds Number. Furthermore, Fig. 3.57 illustrates the influence of the inclination of the pile on the amplification of the bed shear stress, obtained by the model.

The model is mainly solved for steady flows, i.e., $\partial/\partial t = 0$. However, the influence of the transient calculations on the end results has also been

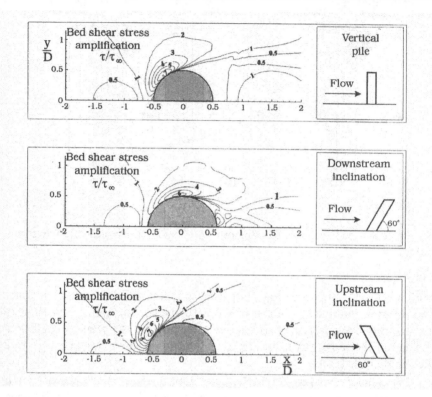

Figure 3.57: Bed shear stress. Effect of pile inclination. Smooth bed. Numerical model. Roulund et al. (2002).

discussed. No significant difference between the steady-flow results and the results obtained from time averaging of the transient calculations has been observed.

A morphological model was developed with a two-dimensional bedload sediment transport description. This is basically the 2-D version of the equation of the continuity of sediment (Eq. 2.100) in Chapter 2. A two-dimensional bedload description, similar to that of Kovacs and Parker (1994), was implemented. Furthermore, a description of surface-layer sand slides for bed slopes exceeding the angle of repose was derived and implemented. The morphological model was applied to calculation of scour around a vertical circular pile. The morphological model captured all the observed bed fea-

3.5. MATHEMATICAL MODELLING

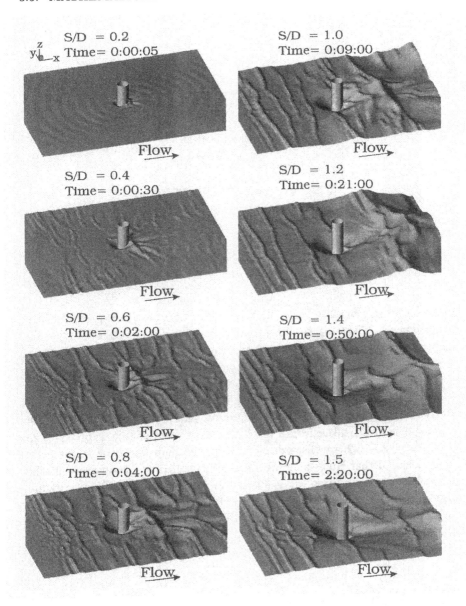

Figure 3.58: Time development of scour hole. Numerical model. Roulund et al. (2002).

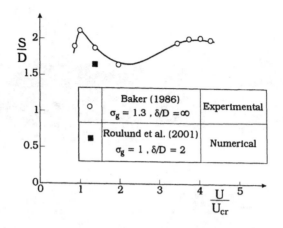

Figure 3.59: Scour depth. Comparison between numerical model and experiment. Roulund et al. (2002).

tures, i.e. the scour hole and the formation of a downstream dune at the initial stage. Migrating ripples on the bed were also resolved. Fig. 3.58 illustrates how the scour hole develops around a circular pile subjected to a steady current. In this simulation, the boundary layer thickness was 20 cm, the mean flow velocity was 46 cm/s, the pile diameter was 10 cm, the sand size was 0.26 mm. The maximum equilibrium scour depth obtained from this simulation is plotted in Fig. 3.59 (the square symbol). The curve in the same figure is taken from Fig. 3.25. The rest of the data in Fig. 3.25 for $\sigma_g = 2.3 - 5.2$ are discarded to keep the figure relatively simple. (It may be noted that Roulund et al.'s morphological model cannot handle the nonuniform sediment size ($\sigma_g \neq 1$)). It is seen from Fig. 3.59 that the numerical result apparently agrees quite well with the experimental data. The small difference (slightly larger than 10%) between the numerical result and the experiments may be attributed to the difference in the boundary-layer thickness (cf. Fig. 3.26).

3.6 References

1. Baker, C.J. (1978): Vortex Flow Around the Bases of Obstacles. Ph. D. Thesis, University of Cambridge, Cambridge, U.K.

3.6. REFERENCES

2. Baker, C.J. (1979): The laminar horseshoe vortex. J. Fluid Mechanics, vol. 95, part 2, 347-367.

3. Baker, C.J. (1985): The position of points of maximum and minimum shear stress upstream of cylinders mounted normal to flat plates. J. Wind Eng. and Industrial Aerodyn., vol. 18, 263-274.

4. Baker, C.J. (1991): The oscillation of horseshoe vortex systems. Trans. ASME I: J. Fluids Engineering, vol. 113, 489-495.

5. Baker R. A. (1986): Local Scour at Bridge Piers in Non-Uniform Sediment. Thesis presented to the University of Auckland, New Zealand, in partial fulfillment of the requirements for the degree of Master of Philosophy.

6. Bearman, P.W., Graham, J.M.R., Naylor, P. and Obasaju, E.D. (1981): The role of vortices in oscillatory flow about bluff cylinders. Proc. International Symposium on Hydrodynamics in Ocean Engineering, Trondheim, Norway, vol. 1, 621-643.

7. Belcher, S.E., Xu, D.P. and Hunt, J.C.R. (1990): The response of a turbulent boundary layer to arbitrarily distributed two-dimensional roughness changes. Quart. J. Royal Met. Soc., vol. 116, 611-635.

8. Belik, L. (1973): The secondary flow about circular cylinders mounted normal to a flat plate. Aeronautical Quarterly, February, 47-54.

9. Bijker, E.W. and de Bruyn, C.A. (1988): Erosion around a pile due to current and breaking waves. Proc. 21st Coastal Engineering Conference, Costa del Sol, Malaga, Spain, vol. 2, 1368-1381.

10. Breusers, H.N.C., Nicollet, G. and Shen, H.W. (1977): Local scour around cylindrical piles, J. Hyd. Res., vol. 15, No. 3, 211-252.

11. Breusers, H.N.C. and Raudkivi, A.J. (1991): Scouring. A.A. Balkema, Rotterdam, viii + 143 p.

12. Briaud, J.-L., Ting, F.C.K., Chen, H.C., Gudavalli, R., Perugu, S. and Wei, G. (1999): SRICOS: Prediction of scour rate in cohesive soils at bridge piers. J. Geotechnical and Geoenvironmental Engineering, ASCE, vol. 125, No. 4, 237-246.

13. Briley, W.R. and McDonald, H. (1981): Computation of three-dimensional horseshoe vortex flow using the Navier-Stokes equations. Proc. 7th International Conference on Numerical Methods in Fluid Dynamics, Stanford, CA, W.C. Reynolds and R.W. MacCormack (eds.), Lecture Notes in Physics, vol. 141, Springer, 91-98.

14. Carreiras, J., Larroudé, Ph.,Santos, F.J. Seabra and Mory, M. (2000): Wave scour around piles. In: Proc. 27th Coastal Engineering Conference, Sydney, Australia, vol. 2, 1860-1870.

15. Chabert, J. and Engeldinger, P. (1956): Etude des affouillements autour des piles des ponts. Laboratoire National d'Hydraulique, Chatou, France (in French).

16. Chiew, Y.M. (1984): Local Scour at Bridge Piers. Report No. 355, Ph. D. Thesis. University of Auckland, Department of Civil Engineering.

17. Chiew, Y.M. and Melville, B.W. (1987): Local scour around bridge piers, J. Hyd. Res., vol. 25, No. 1, 15-26.

18. Chiew, Y.M. (1995): Mechanics of riprap failure at bridge piers. J. Hydraulic Engineering, ASCE, vol. 121, No. 9, 635-643.

19. Chou, J. H. and Chao, S.Y. (2000): Branching of a horseshoe vortex around surface-mounted rectangular cylinders. Experiments in Fluids, vol. 28, 394-402.

20. Dargahi, B. (1989). The turbulent flow field around a circular cylinder. Experiments in Fluids, vol. 8, 1-12.

21. Das, M.M. (1970): A literature review on bed-load transport due to wave action and localized scour in non-cohesive sediments. Final Report HEL 21-6: Literature Review on Erosion and Deposition of Sediment near Structures in the Ocean, H.A. Einstein and R.L. Wiegel, eds., Hydraulic Engineering Laboratory, College of Engineering, University of California, Berkeley, CA.

22. Deng, G.B. and Piquet, J. (1992): Navier-Stokes computations of horseshoe vortex flows. International J. for Numerical Methods in Fluids, vol. 15, 99-124.

3.6. REFERENCES

23. DHI/SNAMPROGETTI (1992): SISS Project. Sea bottom instability around small structures. Erodible Bed Laboratory Tests (Phase 1). Final Report, Text and Drawings. DHI (Danish Hydraulic Institute) and Snamprogetti, Contract INGE91/SP/03060, June 1992.

24. Eadie, R.W. IV, and Herbich, J.B. (1986): Scour about a single, cylindrical pile due to combined random waves and current. Proc. 20th Coastal Engineering Conference, Taipei, Taiwan, ASCE, New York, N.Y., vol. 3, 1858-1870.

25. Ettema, R. (1976): Influence of bed material gradation on local scour. Report No. 124, University of Auckland, Department of Civil Engineering.

26. Ettema, R. (1980): Scour at Bridge Piers. Report No. 216, Ph.D. Thesis, University of Auckland, Department of Civil Engineering.

27. Ettema, R., Melville, B.W. and Barkdoll, B. (1998): Scale effects in pier-scour experiments. J. Hydraulic Engineering, ASCE, vol. 124, No. 6, 639-643.

28. Ettema, R., Mostafa, E.A., Melville, B.W. and Yassin, A.A. (1998): Local scour at skewed piers. J. Hydraulic Engineering, ASCE, vol. 124, No. 7, 756-759.

29. Fredsøe, J. and Deigaard, R. (1992): Mechanics of Coastal Sediment Transport. Advanced Series on Ocean Engineering, vol. 8, World Scientific, xviii + 369 p.

30. Fredsøe, J. and Sumer, B.M. (1997): Scour at the round head of a rubble-mound breakwater. Coastal Engineering, vol. 29, pp. 231-262.

31. Fredsøe, J., Sumer, B.M., Laursen T. and Pedersen, C. (1993): Experimental investigation of wave boundary layers with a sudden change in roughness. J. Fluid Mech., vol. 252, 117-145.

32. Graf, W.H. and Yulistiyanto, B. (1998): Experiments on flow around a cylinder; the velocity and vorticity field. J. Hyd. Res., vol. 36, No. 4, 637-653.

33. Hebsgaard, M. (1998): Scour protection of bridge piers. In: East Bridge. Ed. N.J. Gimsing. The Storebælt Publications. A/S Storebæltsforbindelsen. Copenhagen.

34. Hebsgaard, M., Ennemark, F., Spangenberg, S., Fredsøe, J. and Gravesen, H. (1994): Scour model tests with bridge piers. International Assoc. of Navigation Congresses, PIANC, Bulletin No. 82, 84-92.

35. Herbich, J.B., Schiller, R.E., Jr., Watanabe, R.K. and Dunlap, W.A. (1984): Seafloor Scour. Design Guidelines for Ocean-Founded Structures, Marcell Dekker, Inc., New York, NY, xiv + 320 p.

36. Hjorth, P. (1975): Studies on the nature of local scour. Bull. Series A, No. 46, viii + 191 p., Department of Water Resources Engineering, Lund Institute of Technology/University of Lund, Lund, Sweden.

37. Hoffmans, G.J.C.M. and Verheij (1997): Scour Manual. A.A. Balkema/Rotterdam.

38. Isaacson, M. (1979): Wave-induced forces in the diffraction regime. In: Mechanics of Wave-Induced Forces on Cylinders, T.L. Shaw (ed.), Pitman Advanced Publishing Program, 68-89.

39. Imberger, J., Alach, D. and Schepis, J. (1982): Scour behind circular cylinders in deep water. Proc. 18th International Conference on Coastal Engineering, ASCE, vol. 2, 1522-1554.

40. Kawata, Y. and Tsuchiya, Y. (1988): Local scour around cylindrical piles due to waves and currents combined. Proc. 21st International Conference on Coastal Engineering, Costal del Sol, Malaga, ASCE, vol. 2, 1310-1322.

41. Kobayashi, T. (1992): Three-dimensional analysis of the flow around a vertical cylinder on a scoured seabed. Proc. 23rd International Conference on Coastal Engineering, Venice, Italy, vol. 3, Chapter 99, 3482-3495.

42. Kobayashi, T. and Oda, K. (1994): Experimental study on developing process of local scour around a vertical cylinder. Proc. 24th International Conference on Coastal Engineering, Kobe, Japan, vol. 2, Chapter 93, 1284-1297.

3.6. REFERENCES

43. Kovacs, A. and Parker, G. (1994): A new vectorial bedload formulation and its application to the time evolution of straight river channels. J. Fluid Mechanics, vol. 267, 153-183.

44. Kwak, D., Rogers, S.E., Kaul, U.K. and Chang, J.L.C. (1986): A numerical study of incompressible juncture flows. NASA Technical Memorandum 88319, Ames Research Center, Moffet Field, CA 94035.

45. Lagasse, P.F., Thompson, P.L. and Sabol, S.A. (1995): Guarding against scour. Civil Engineering, June issue, 56-59.

46. Lauchlan, C.S. and Melville, B.W. (2001): Riprap protection at bridge piers. J. Hydraulic Engineering, ASCE, vol.127, No. 5, 412-418.

47. Laursen, E.M. (1958): Scour at Bridge Crossings. Bulletin No. 8, Iowa Highway Research Board, Ames, Iowa.

48. Laursen, E.M. and Toch, A. (1956): Scour around bridge piers and abutments. Bulletin No. 4, Iowa Highway Research Board, Ames, Iowa.

49. Melville, B.W. (1975): Local Scour at Bridge Sites. Report No. 117, Ph.D. Thesis, University of Auckland, Department of Civil Engineering.

50. Melville, B.W. and Chiew, Y.M. (1999): Time scale for local scour at bridge piers. J. Hydraulic Engineering, ASCE, vol. 125, No. 1, 59-65.

51. Melville, B.W. and Coleman, S.E. (2000): Bridge Scour. Water Resources Publications, LLC, CO, USA, xxii + 550 p.

52. Melville, B.W. and Dongol, D.M.S. (1992): Bridge pier scour with debris accumulation. J. Hydraulic Engineering, ASCE, vol.118, No. 9, 1306-1310.

53. Melville, B.W. and Raudkivi, A.J. (1977): Flow characteristics in local scour at bridge piers. J. Hyd. Res., vol. 15, 373-380.

54. Melville, B.W. and Raudkivi, A.J. (1996): Effects of foundation geometry on bridge pier scour. J. Hydraulic Engineering, ASCE, vol. 122, No. 4, 203-209.

55. Melville, B.W. and Sutherland, A.J. (1988): Design methods for local scour at bridge piers. J. Hydraulic Engineering, ASCE, vol. 114, No. 10, 1210-1226.

56. Menter, F.R. (1992): Improved Two-Equation $k - \omega$ Turbulence Models for Aerodynamic Flows. NASA Technical Memorandum 103975, NASA, Ames Research Center, California.

57. Michelsen, J.A. (1992): Basis3D - A Platform for Development of Multiblock PDE Solvers. Technical Report. Department of Fluid Mechanics, Technical University of Denmark, AFM 92-05, ISSN 0590-8809.

58. Niedoroda, A.W. and Dalton, C. (1982): A review of the fluid mechanics of ocean scour, Ocean Engineering, vol. 9, No. 2, 159-170.

59. Olsen, N.R.B. and Melaaen, M.C. (1993): Three-dimensional calculation of scour around cylinders. J. Hydraulic Engineering, ASCE, vol. 119, No. 9, 1048-1054.

60. Olsen, N.R.B. and Kjellesvig, H.M. (1998): Three-dimensional numerical flow modelling for estimation of maximum local scour depth. J. Hyd. Res., vol. 36, No. 4, 579-590.

61. Parola, A.C. (1993): Stability of riprap at bridge piers. J. Hydraulic Engineering, ASCE, vol. 119, No. 10, 1080-1093.

62. Parola, A.C. (1995): Boundary stress and stability of riprap at bridge piers. In: River, Coastal and Shoreline Protection: Erosion Control Using Riprap and Armourstone, ed. C.R. Thorne et al., John Wiley & Sons, Inc., New York, U.S.A.

63. Raudkivi, A.J. (1986): Functional trends of scour at bridge piers. J. Hydraulic Engineering, ASCE, vol. 112, No. 1, 1-13.

64. Raudkivi, A.J. (1998): Loose Boundary Hydraulics. A.A. Balkema / Rotterdam.

65. Raudkivi, A.J. and Ettema, R. (1977): Effect of sediment gradation on clear water scour. J. Hydraulic Engineering, ASCE, vol. 103, No. 10, 1209-1213.

3.6. REFERENCES

66. Raudkivi, A.J. and Ettema, R. (1983): Clear water scour at cylindrical piers. J. Hydraulic Engineering, ASCE, vol. 109, No. 3, 338-350.

67. Richardson, E.V. and Lagasse, P.F. (1999): Stream Stability and Scour at Highway Bridges. Compendium of Papers ASCE Water Resources Engineering Conferences 1991 to 1998. Richardson and Lagasse (eds.). ASCE, 1801 Alexander Bell Drive, Reston, Virginia 20191-4400. xxxi + 1040 p.

68. Richardson J.E. and Panchang V.G. (1998): Three-dimensional simulation of scour-inducing flow at bridge piers, J. Hydraulic Engineering, ASCE, vol.124, No. 5, 530-540.

69. Roulund, A. (2000): Three-Dimensional Numerical Modelling of Flow Around a Bottom-Mounted Pile and its Application to Scour. Technical University of Denmark, Department of Hydrodynamics and Water Resources, Ph. D. Thesis supervised by B.M. Sumer and J. Fredsøe.

70. Roulund, A., Sumer, B.M. and Fredsøe, J. (2002): Numerical and experimental study of flow and scour around a pile. Manuscript in preparation.

71. Sarpkaya, T. (1986): Forces on a circular cylinder in viscous oscillatory flow at low Keulegan-Carpenter numbers. J. Fluid Mechanics, vol. 165, 61-71.

72. Sarpkaya, T. and Isaacson, M. (1981): Mechanics of Wave Forces on Offshore Structures. Van Nostrand Reinhold Company, xvi + 651 p.

73. Schlichting, H. (1979): Boundary-Layer Theory. 7th ed., McGraw-Hill Book Company, xxii + 817.

74. Schwind, R. (1962): The three-dimensional boundary layer near a strut. Gas Turbine Lab. Rep., Massachusetts Institute of Technology.

75. Sumer, B.M. (1986): Recent Developments on the Mechanics of sediment suspension. General-Lecture Paper, in the book Euromech 192: Transport of suspended solids in open channels (eds. Bechteler and Vollmers), A.A. Balkema Publishers, Rotterdam.

76. Sumer, B.M. and Fredsøe, J. (1997): Hydrodynamics Around Cylindrical Structures. World Scientific, xviii + 530 p.

77. Sumer, B.M. and Fredsøe, J. (2001): Scour around a pile in combined waves and current. J. Hydraulic Engineering, ASCE, vol. 127, No. 5, 403-411.

78. Sumer, B.M., Christiansen, N. and Fredsøe, J. (1992 b): Time scale of scour around a vertical pile. Proc. 2nd International Offshore and Polar Engineering Conference, San Francisco, CA, vol. III, 308-315.

79. Sumer, B.M., Christiansen, N. and Fredsøe, J. (1993): Influence of cross section on wave scour around piles. J. Waterway, Port, Coastal and Ocean Engineering, ASCE, vol. 119, No. 5, 477-495.

80. Sumer, B.M., Christiansen, N. and Fredsøe, J. (1997): Horseshoe vortex and vortex shedding around a vertical wall-mounted cylinder exposed to waves. J. Fluid Mechanics, vol. 332, 41-70.

81. Sumer, B.M., Chua, L.H.C., Cheng N.-S. and Fredsøe, J. (2002): The influence of turbulence on bedload sediment transport. Manuscript in preparation.

82. Sumer, B.M., Cokgor, S. and Fredsøe, J. (2001): Suction of sediment from between armour blocks. J. Hydraulic Engineering, ASCE, vol. 127, No. 4, 293-306.

83. Sumer, B.M., Fredsøe, J. and Christiansen, N. (1992 a): Scour around a vertical pile in waves. J. Waterway, Port, Coastal and Ocean Engineering, ASCE, vol.. 117, No. 1, 15-31.

84. Sumer, B.M., Fredsøe, J., Christiansen, N. and Hansen, S.B. (1994): Bed shear stress and scour around coastal structures. Proc. 24th International Coastal Engineering Conference, ASCE, Kobe, Japan, vol. 2, 1595-1609.

85. Sumer, B.M., Kozakiewicz, A., Fredsøe, J. and Deigaard, R. (1996): Velocity and concentration profiles in the sheet flow layer of movable bed. J. Hydraulic Engineering, ASCE, vol. 122, No. 10, 549-558.

3.6. REFERENCES

86. Sutherland, J. and Whitehouse, R.J.S. (1998): Scale effects in the physical modelling of seabed scour, HR Wallingford Report TR64, September 1998.

87. Sørensen, N.N. (1995): General Purpose Flow Solver Applied to Flow over Hills. Ph.D. Thesis, Risø National Laboratory, Roskilde, Denmark, Risø-R-827 (EN).

88. Tison, L.J. (1940): Erosion autour des piles de ponts en riviere. Annales des Travaux Publics de Belgique, vol. 41, No. 6, 813-817 (in French).

89. Tseng, M.-H., Yen, C.-L. and Song, C.C.S. (2000): Computation of three-dimensional flow around square and circular piers. International J. for Numerical Methods in Fluids, vol. 34, 207-227.

90. Tsujimoto, T. (1987): Local scour around bridge piers in rivers and its protection works. Memoirs of the Faculty of Technology, Kanazawa University, vol. 20, No. 1, 11-21.

91. Venkatadri, C., Rao, G.M., Hussain, S.T. and Asthana, K.C. (1965): Scour around bridge piers and abutments. Irrig. Power, Jan., 35-42.

92. Yuhi, M., Ishida, H. and Umeda, S. (2000): A numerical study of three-dimensional flow fields around a vertical cylinder mounted on a bed. In: Proceedings of Coastal Structures' 99, Santander, Spain, ed. I. Losada, A.A. Balkema/Rotterdam., vol. 2, 783-792.

93. Wang, R.-K. and Herbich, J.B. (1983). Combined current and wave-produced scour around a single pile. Texas Eng. Expt. Station, Dept. Civ. Eng., Texas Univ. System, CEO Report No. 269.

94. Whitehouse, R. (1998): Scour at Marine Structures. Thomas Telford, xix + 198 p.

95. Wilcox, D.C. (1993): Turbulence Modelling for CFD. DCW Industries, Inc., La Canada, California, USA, xx + 460 p.

96. Williamson, C.H.K. (1985): Sinusoidal flow relative to circular cylinders. J. Fluid Mechanics, vol. 155, 141-174.

97. Worman, A. (1992): Incipient motion during static armoring. J. Hydraulic Engineering, ASCE, vol. 118, No. 3, 496- 501.

Chapter 4

Scour around a group of slender piles

Pile groups are widely used in practice to support marine structures. Because of the interference of the flows around individual pile members, the resulting scour picture in the case of pile groups may differ significantly from that occurring around a single pile.

In this chapter, we shall first focus on scour around pile groups in steady currents, then we shall concentrate our attention on scour around pile groups in waves. Subsequently, we shall give a brief account of global and local scour at pile groups.

4.1 Pile group in steady currents

As seen in the previous chapter (Section 3.2.1), the horseshoe vortex is one of the key elements in the scour process around a pile in steady currents. In the case of a pile group, the lee-wake vortex flow may also be important; for instance, in the case of a two-pile, tandem arrangement (Fig. 4.1 b), the lee-wake vortex of the upstream pile is certainly an important element in the scour process around the downstream pile. Obviously, both the horseshoe-vortex flow and the lee-wake vortex flow will undergo substantial changes in the case of a pile group due to the interference effect, presumably leading to substantial changes in the scour process at the pile group.

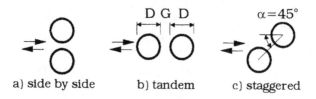

Figure 4.1: Definition sketch for two-pile groups.

4.1.1 Two-pile group

Flow

Basically, there are three kinds of two-pile configurations: the side-by-side arrangement, the tandem arrangement, and the staggered arrangement (Fig. 4.1). Clearly the interference effect for both the lee-wake vortex flow and the horseshoe vortex flow will strongly depend on the pile-group configuration.

First we consider the **interference effect for the lee-wake vortex flow.**

This subject has been investigated quite extensively in the case of 2-D flow around multiple cylinders (in view of its practical applications in the area of offshore engineering, and nuclear engineering, as regards the forces on and vibrations of such structures). A comprehensive review of the subject can be found in Zdravkovich (1987).

Fig. 4.2 illustrates various flow regimes experienced by a two-cylinder system. Fig. 4.3 gives a detailed picture of the flow regimes in the two fundamental cases, the side-by-side arrangement and the tandem arrangement.

As seen from Fig. 4.3, in the case of the side-by-side arrangement, the two cylinders act as a single body, and hence the flow is in the single-vortex-street regime when $G/D < 0.25$. Fig. 4.3 further shows that the flow goes through the biased-gap-flow regime and the coupled-vortex-street regime, as the gap-to-diameter ratio G/D increases. It is only after G/D reaches the value of about 3 that the vortex streets behind the cylinders become "uncoupled".

In the case of the tandem arrangement, on the other hand, the two cylinders act as a single body, and therefore there is only one vortex street when $G/D < 0.15$ (Fig. 4.3). From figure, G/D needs to be increased to as much as 3, for a two-vortex-street regime to emerge. For $G/D > 3$, the

4.1. PILE GROUP IN STEADY CURRENTS

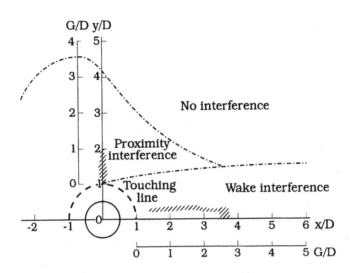

Figure 4.2: Definition of regions of flow interference for two-cylinder arrangements. Hatched area: bistable flow regions. Zdravkovich (1987).

vortex street behind the downstream cylinder is called binary vortex street by Zdravkovich on the ground that each vortex in this case consists of two vortices (one is formed behind the upstream cylinder and the other behind the downstream cylinder).

Next we consider the **interference effect for the horseshoe-vortex flow**. Apparently, no study is available investigating this important interference effect. However, the following assessment can be made from the information available on single piles (Section 3.1.1).

1. In the case of the side-by-side arrangement of the two-pile system (Fig. 4.1 a), no interference effect for the horseshoe vortices can be expected for the values of the pile spacing $G/D > O(2)$, considering that the width of the legs of the individual horseshoe vortices, normalized by the pile diameter, is like $O(1)$ (Fig. 3.2)

2. Again, in the case of the side-by-side arrangement, the two piles are expected to act like a single pile, therefore one single horseshoe vortex is expected to form (with a size much larger than the single-pile case, however), when G/D is very small, like $G/D < O(0.1)$.

242   CHAPTER 4. SCOUR AROUND A GROUP OF SLENDER PILES

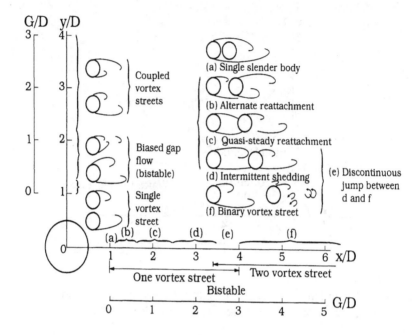

Figure 4.3: Interference flow regimes for side-by-side and tandem arrangements. Zdravkovich (1987).

3. For the values of the pile spacing between these two limits, namely $O(0.1) < G/D < O(2)$, the interference effect should be expected.

4. In the case of the tandem arrangement (Fig. 4.1 b), the horseshoe vortex in front of the rear pile will obviously be destructed for small values of the pile spacing. It may be inferred from the studies regarding the lee-wake vortex-flow interference (Fig. 4.3) that the downstream-pile horseshoe vortex is expected to exist only when $G/D > O(3)$. However, the size of this horseshoe vortex is expected to be relatively smaller than that of the single-pile case due to "shielding".

5. The horseshoe vortex at the front pile in the case of the tandem arrangement may be influenced slightly by the presence of the downstream pile. This is because of the slightly larger "blockage" caused by the downstream pile (the lee-wake has now a larger width because of the

4.1. PILE GROUP IN STEADY CURRENTS

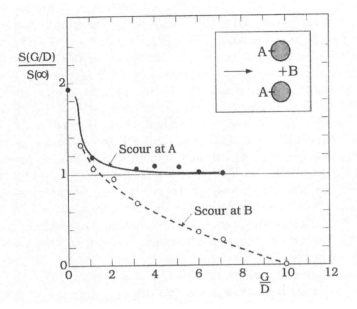

Figure 4.4: Scour depth for side-by-side, two-pile group. Hannah (1978). See also Breusers and Raudkivi (1991).

increased size of the lee-wake vortices). This induces a relatively larger adverse pressure gradient, presumably leading to a relatively larger horseshoe vortex. From Fig. 4.3, it may be expected that this occurs for $G/D < O(3)$.

Scour for the two-pile, side-by-side arrangement

Scour at two-pile groups in steady current and with clear-water conditions was studied by Hannah (1978). Breusers and Raudkivi (1991) summarize the results of this latter work.

The sediment used in the experiments was $d_{50} = 0.75$ mm, the pile diameter $D = 3.3$ cm, the flow velocity $V = 28.5$ cm, and the flow depth $h = 14$ cm. The Shields parameter was $\theta \simeq 0.03$, a value slightly smaller than the critical value, indicating clear-water conditions. The results reported in the study were for runs for 7 hours (Note that tests in the study showed that scour depths were 80 % of equilibrium scour depths after 7 hours).

Fig. 4.4 presents the scour data related to the side-by-side arrangement. $S(\infty)$ in the figure is the scour depth measured for the single-pile situation under exactly the same flow conditions. The figure indicates that the scour at the individual piles (scour at A) is increased considerably for very small pile spacings (a factor of 2 increase as $G/D \to 0$). This may be due partly to the increased size of the horseshoe vortex for such small pile spacings (as pointed out in item (2) in the preceding subsection), and partly to the very strong gap flow between the two neighbouring piles. However, the figure indicates that this scour (scour at A) falls quite rapidly with increasing pile spacing, and reaches practically its single-pile value for gap ratios $G/D > O(2)$. This is because, for such large values of the pile spacing, the interference effect for the horseshoe vortex flow would be not very significant, as remarked in the preceding paragraphs (item (1) in the preceding subsection). However, note that, even for such large pile spacings, the scour is apparently larger than the single-pile value (about 5 % larger). This indicates that a slight pile interference can exist even for pile spacings of as large as $G/D = 7$.

Fig. 4.4 further indicates that the scour depth at the midway between the piles (scour at B) becomes smaller and smaller with increasing pile spacing for $G/D > O(2)$. Two completely independent scour holes are obtained only after G/D is increased to 10.

Scour for the two-pile, tandem arrangement

Fig. 4.5 displays the data obtained in the case of the tandem arrangement. The figure shows that the scour at the front pile first increases with increasing G/D, and reaches a maximum value (a value which is about 30 % higher than the single-pile value) at $G/D \simeq 2$, and then it decreases, and eventually reaches its single-pile value for $G/D > 10$. This increase in the scour depth may be due partly to the anticipated increase in the horseshoe-vortex flow at the front pile for pile spacing $G/D < O(3)$, as pointed out in the preceding paragraphs.

Fig. 4.5 further shows that the scour depth at the rear pile always is smaller than the single-pile case. Furthermore, the scour apparently begins to decrease when G/D reaches the value of $O(3)$. This coincides with the pile spacing, $O(3)$, at which the downstream horseshoe vortex begins to form (see item 4 above). The reason why the scour depth at the rear pile is smaller than the single-pile value can be explained by the fact that the horseshoe vortex at the rear pile has a relatively smaller size than that of a single pile

4.1. PILE GROUP IN STEADY CURRENTS

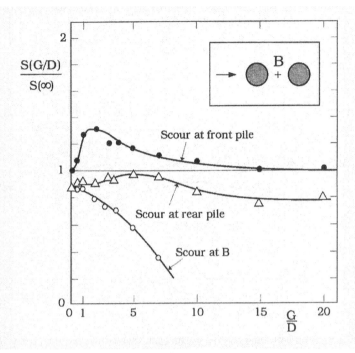

Figure 4.5: Scour depth for tandem, two-pile group. Hannah (1978). See also Breusers and Raudkivi (1991).

(item 4 above).

Similar to the side-by-side arrangement, scour at the midway between the piles gradually decreases with increasing G/D, and tends to zero when G/D becomes about 10.

Effect of angle of attack

Diagrams similar to those presented in the previous two figures are given also for the staggered-pile case with $\alpha = 45^0$ in Breusers and Raudkivi (1991).

Fig. 4.6 illustrates how the angle of attack affects the scour depth for a pile group with $G/D = 5$. Generally, the scour depth first increases, and subsequently it decreases with increasing α.

For the front pile, the increase can be explained by the slight increase of the size of the horseshoe vortex, apparently due to the increased projection

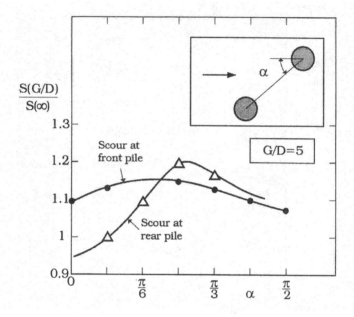

Figure 4.6: The effect of angle of attack on scour depth for two-pile group. Hannah (1978). See also Breusers and Raudkivi (1991).

area of the pile group (hence due to the increased adverse pressure gradient). However, for larger values of α, each pile will begin to have a separate projection area rather than a joint one, and hence the horseshoe vortex will begin to transform to that of a single-pile, and consequently, the scour depth will begin to decrease, as revealed by Fig. 4.6.

For the rear pile, on the other hand, the scour depth begins to increase with increasing α. This is because the shielding effect for the rear pile will begin to diminish, as the angle of attack is increased. This will obviously result in larger and larger scour depths with increasing α. However, this will continue until a certain value of α ($\sim 45^0$) is reached. For further increase of α, the piles will begin to act as a single pile, and the scour depth will have to converge to this single-pile value.

4.2. PILE GROUP OF IN WAVES

4.1.2 Three-pile group

Pile configuration and measurement location	$\frac{G}{D}$	$\frac{S(G/D)}{S(\infty)}$
a) Flow → Measurement location	0.75	1.04
	2.0	1.0
	5.0	1.17
b) →	0.75	1.03
	2.0	1.0
	5.0	1.22
c) →	0.75	1.0
	1.6	1.0
	3.3	0.92
d) →	0.75	1.12
	1.6	1.08
	3.3	1.03

Table 4.1. Scour depth for three-pile group. Gormsen and Larsen (1984).

Some limited experiments were carried out by Gormsen and Larsen (1984) for two kinds of three-pile groups (Table 4.1). The test conditions were: $D = 7.5$ cm, $V = 56$ cm/s, h (the flow depth) $= 22.5$ cm, $d_{50} = 0.55$ mm, and the corresponding value of the Shields parameter, $\theta = 0.1$. The results are summarized in Table 4.1. Clearly, the data size for each case is too small for the results to be conclusive. However, the general trend is that, for the tested gap-to-diameter ratios, the scour depth increases (albeit slightly, by 5-15 %) with respect to that experienced in the case of the single pile, with the exception that the scour depth remains unchanged, or even decreases at the downstream pile for the three-pile group displayed in (c) in Table 4.1. This latter behaviour is linked to the shielding effect.

248 CHAPTER 4. SCOUR AROUND A GROUP OF SLENDER PILES

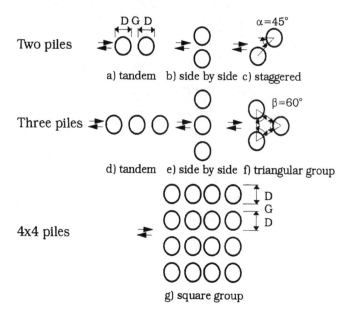

Figure 4.7: Pile groups tested in the study of Sumer and Fredsøe (1998).

4.2 Pile group of in waves

While a substantial amount of knowledge has accumulated about wave scour around single piles over the past decade or so (Wang and Herbich (1983), Herbich et al. (1984), Eadie and Herbich (1986), Sumer, Fredsøe and Christiansen (1992), Sumer, Christiansen and Fredsøe (1993), and Kobayashi and Oda (1994), see Chapter 3), comparatively little is known about wave scour at pile groups.

Chow and Herbich's (1978) study is apparently the first one to deal with the wave scour at pile groups. Chow and Herbich (1978) (see also Herbich et al. (1984, p. 148)) studied the scour around six-, four-, and three-legged pile structures. One of the key parameters influencing the scour turned out to be the pile spacing. Chow and Herbich studied this influence for relatively large spacing, namely for $G/D \geq 3$. The pile groups such as two-pile tandem arrangement, two-pile side-by-side arrangement, and their three-pile counterparts (the basic "building blocks" of various pile groups encountered in prac-

4.2. PILE GROUP OF IN WAVES

tice) were not included in the study. Furthermore, at the time of the conduct of Chow and Herbich's study, the knowledge of the hydrodynamic processes was quite sparse, and, particularly, the role of the Keulegan-Carpenter number, KC was unknown. Therefore, Chow and Herbich's analysis was mainly based on empirical information.

Subsequently, Sumer and Fredsøe (1998) investigated in a systematic manner the wave scour around pile groups. The studied pile-group configurations are displayed in Fig. 4.7. Sumer and Fredsøe's work is actually an extension of their previous work on wave scour around single piles (Sumer et al., 1992 and 1993) to the pile groups displayed in Fig. 4.7. The study basically focuses on the variations with the pile spacing, covering the entire range, from G/D equal to zero to ∞ (the single-pile case). The study also investigates the variations with respect to the KC number.

In the following paragraphs, scour around pile groups in waves will be described by reference to a number of canonical examples, based on Sumer and Fredsøe's (1989) study.

4.2.1 Two-pile group

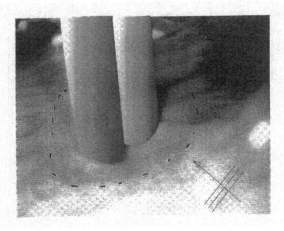

Figure 4.8: Scour hole at the equilibrium stage for two-pile side-by-side arrangement. $KC = 6$, and $G/D = 0.4$. Sumer and Fredsøe (1998).

CHAPTER 4. SCOUR AROUND A GROUP OF SLENDER PILES

For the case of a single pile exposed to waves, the live-bed ($\theta > \theta_{cr}$) scour depth mainly depends on the Keulegan-Carpenter number, KC, as described in the previous chapter (Section 3.2.2, Eq. 3.20):

$$\frac{S}{D} = f(KC) \qquad (4.1)$$

In the case of a pile group (Fig. 4.7), there will be additional parameter, namely the nondimensional pile spacing, G/D, and hence the scour depth:

$$\frac{S}{D} = f(KC, \frac{G}{D}) \qquad (4.2)$$

Fig. 4.8 illustrates the scour hole in the equilibrium stage for the two-pile, side-by-side arrangement with $G/D = 0.4$, and $KC = 6$. (Note that, in Sumer and Fredsøe's (1998) experiments, the bed was live, $\theta > \theta_{cr}$).

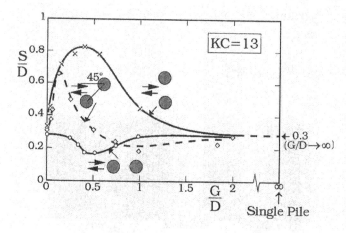

Figure 4.9: Equilibrium scour depth plotted against pile spacing. Sumer and Fredsøe (1998).

Fig. 4.9 presents the maximum equilibrium scour depth normalized by the pile diameter, S/D, versus the normalized pile spacing, G/D, while Fig. 4.10 presents the plan-view extent of the scour hole, L_x and L_y, normalized by D, versus G/D, both for $KC = 13$. (L_x and L_y are both determined in the study by visual observations, and are therefore subject to some uncertainty).

4.2. PILE GROUP OF IN WAVES

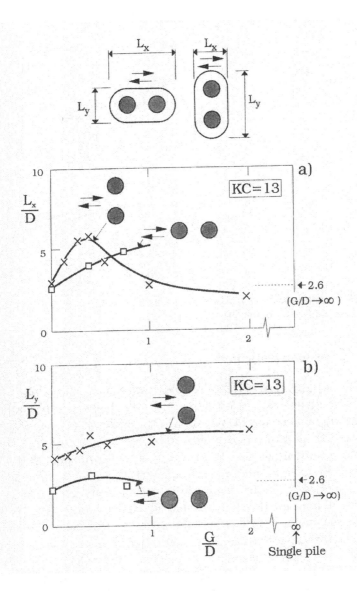

Figure 4.10: Plan view extent of scour hole plotted against pile spacing. Sumer and Fredsøe (1998).

252 CHAPTER 4. SCOUR AROUND A GROUP OF SLENDER PILES

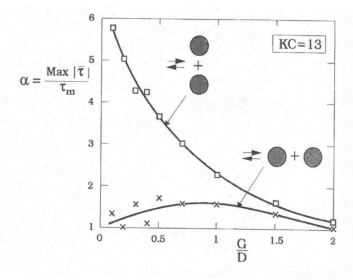

Figure 4.11: Amplification in the bed shear stress. $KC = 13$. Plus sign: measurement location. Sumer and Fredsøe (1998).

Side-by-side arrangement

First of all, Sumer and Fredsøe's (1998) experiments show that the location of the maximum scour depth in the scour hole always is at the midway between the two piles, except for very small gap ratios, namely for $G/D < 0.1$, and for very large gap ratios, namely for $G/D > 2$. For $G/D < 0.1$, the maximum scour depth occurs at the outside edges of the piles; this is linked to the fact that, for such small gap ratios, the piles act as a "single" pile. For $G/D > 2$, on the other hand, the interference effect between the piles practically disappears, therefore each individual pile acts as a single pile, and hence the maximum scour depth occurs at the edge of the individual pile.

Fig. 4.9 shows that the maximum scour depth first increases with increasing G/D and reaches a maximum value, about 0.85, when $G/D \simeq 0.3$, and then it begins to decrease and finally reaches its single-pile value for large values of G/D. The increase in the scour depth with decreasing pile spacing is due partly to the high sediment transport induced by the gap flow (the "jet" effect), and partly to the increased presence of the lee-wake vortices emanating from the outer edges of the pile.

4.2. PILE GROUP OF IN WAVES

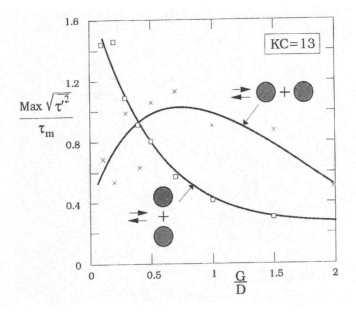

Figure 4.12: The normalized r.m.s. value of the bed shear stress. $KC = 13$. Plus sign: measurement location. Sumer and Fredsøe (1998).

Figs. 4.11 and 4.12 present the results of the bed shear-stress experiments carried out by the same authors; Fig. 4.11 presents the variation of the amplification in the bed shear stress, α, while Fig. 4.12 presents the r.m.s. value of the bed shear stress fluctuations (normalized by the undisturbed bed shear stress) as a function of the normalized pile spacing, G/D. The measurements were taken at the midway between the piles, as indicated in the figure. As seen from Fig. 4.11, for small G/D values, the mean bed shear stress in the gap between the piles can be a factor of $\alpha = 4 - 6$ larger than the undisturbed bed shear stress. Its instantaneous values can be even larger, as implied by Fig. 4.12. This obviously results in a very high sediment transport induced by the gap flow, and hence contributes substantially to the scour depth.

Williamson (1985) studied the influence of cylinder proximity on the vortex wakes in oscillatory flows. Williamson's results for the KC range $7 < KC < 15$ are reproduced in Fig. 4.13.

Figs. 4.13 a and b depict the results regarding the side-by-side arrange-

254   CHAPTER 4. SCOUR AROUND A GROUP OF SLENDER PILES

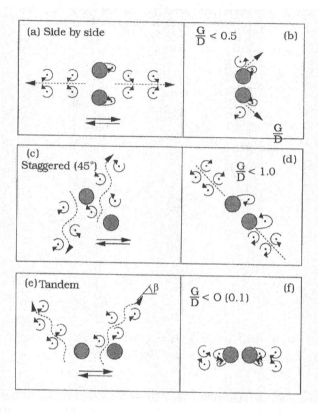

Figure 4.13: Influence of cylinder proximity on the vortex wakes for $7 < KC < 15$. Williamson (1985).

ment. As seen, while the synchronous shedding occurs from the gap sides of the cylinders in every half period of the motion (Fig. 4.13 a), the cylinder pair behaves as a larger solid body for small gaps ($G/D < 0.5$), with vortex pairs shed at flow reversals from the outside edges of the cylinders (Fig. 4.13 b).

The latter effect (where there exists an increased presence of the lee-wake vortices at outside edges of the piles for small gaps ($G/D < 0.5$)) also contributes to the scour depth. The end result will therefore be a marked increase in the scour depth when the pile spacing is small, as revealed by Fig. 4.9.

4.2. PILE GROUP OF IN WAVES

Sumer and Fredsøe's (1998) experiments indicate that, for $G/D > 2-3$, two very distinct scour holes are identified, whereas, for the gap ratios smaller than 2-3, they appear to merge into one single scour hole. As seen from Fig. 4.9, the interference between the piles becomes practically nil, when $G/D > 2$. This latter figure agrees reasonably well with the critical value of the gap ratio (Fig. 4.3) beyond which no interference occurs for the *flow* in the case of two free, side-by-side cylinders in a steady current (Zdravkovich, 1987).

When G/D is zero, the two piles act as a single pile. Fig. 4.9 shows that the normalized scour depth in this case is $S/D = 0.34$. The value for a single circular pile for the corresponding KC number is $S/D \cong 0.3$ (Fig. 3.35).

Fig. 4.10 presents the corresponding data related to the plan view extent of the scour hole. L_x and L_y are the plan-view dimensions of the scour hole in the offshore-onshore direction, and in the cross-shore direction, respectively. The way in which L_x varies with G/D (the side-by-side arrangement in Fig. 4.10 a) is rather similar to that in Fig. 4.9. The result can be interpreted in a similar way to that in Fig. 4.9. Regarding the cross-shore dimension of the scour hole, Fig. 4.10 b shows that it increases with increasing G/D, as expected.

Tandem arrangement

In this case, the maximum scour always occurs at the side edges of the piles.

Fig. 4.9 shows that the influence of pile proximity on the scour depth is quite opposite to that experienced in the case of the side-by-side arrangement; the scour depth first decreases with increasing G/D and attains a minimum (at $G/D \simeq 0.5$) which is a factor of 2 smaller than the single-pile value, and then it begins to increase to reach its single-pile value for large values of G/D.

Williamson's (1985) flow visualization result with regard to the tandem cylinders shows that, except for very small gaps $(G/D < O(0.1))$, the vortex shedding occurs in the manner as sketched in Fig. 4.13 e, with two vortex streets (for each cylinder), diverging from each other, at an angle with the flow direction, β. Williamson's study shows that β is 0^0 for $O(0.1) < G/D < 1$, and it increases to 90^0, the single-cylinder value (corresponding to the transverse vortex street) when $G/D \geq 4$. Furthermore, Williamson's study shows that, for very small gaps, the vortex shedding in each half period takes place only in the rear of the downstream cylinder (i.e., the cylinder pair behaves as a single body) (Fig. 4.13 f).

Now, returning to Fig. 4.9, the decrease in the scour depth here is due to the partial suppression of vortex shedding; for the pile spacing in the range $O(0.1) < G/D < 1$; the shedding behind the upstream pile is partially, or totally suppressed ($\beta = 0^0$), and therefore the end result is a decrease in the scour. On the other hand, Fig. 4.11 suggests a moderate increase (of a factor 1-1.5) in the bed shear stress. These two mechanisms counteract each other; however, Fig. 4.9 shows that the end result is a decrease in the scour.

For $G/D < O(0.1)$, the vortex shedding is restored; the two piles act as a single body, and the shedding occurs in much the same way as in the single-pile case. Hence the scour depth recovers its single-pile value, as revealed by Fig. 4.9.

Likewise, for $G/D > 1$, the piles act like two "individual" piles with practically no interference. Thus, the scour depth recovers its single-pile value again (Fig. 4.9).

Fig. 4.10 illustrates how the plan-view extent of the scour hole varies with G/D. The dimension of the scour hole in the offshore-onshore L_x/D direction increases with increasing G/D, as anticipated (Fig. 4.10 a). The cross-shore dimension of the scour hole, L_y/D, on the other hand, appears to be not very much influenced by the pile spacing. This may be attributed to the fact that the lee-wake vortices are convected away in-line with the pile ($\beta = 0^0$).

Effect of angle of attack

To see the influence of the angle of attack, scour experiments in Sumer and Fredsøe's (1998) study were carried out with the pile group with the 45^0 angle of attack (Fig. 4.7 c) for the same KC number as in the previous experiments, namely for $KC = 13$. Of particular interest is the small-gap ($G/D < 0.5$) behaviour. Williamson's (1985) flow-visualization result (Fig. 4.13 d) shows that the vortices shed into the flow are convected away from the cylinder. Therefore, as in the case of the other arrangements (Figs. 4.13 a-b, and 4.13 e-f), a convection mechanism does exist, to carry the eroded sediment away from the pile. This effect obviously contributes to the scour. The second contribution comes from the gap-flow effect. However, this effect is apparently not as strong as in the case of the side-by-side arrangement; the scour depth in the present case is increased only by a factor of about 2 (Fig. 4.9).

4.2. PILE GROUP OF IN WAVES

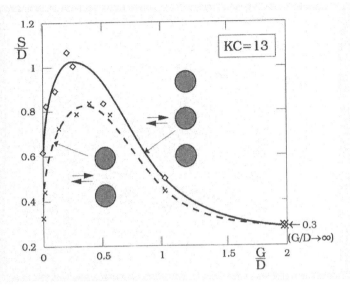

Figure 4.14: Equilibrium scour depth plotted against pile spacing. The single-pile ($G/D \to \infty$) value: from Eq. 3.20. Sumer and Fredsøe (1998).

4.2.2 Three-pile group

Figs. 4.14 and 4.15 compare the scour depth obtained for the two- and three-pile, side-by-side, and the tandem arrangements, respectively (for $KC = 13$). (The two-pile results are reproduced from Fig. 4.9). In the case of the side-by-side arrangement, the scour depth somewhat increases with respect to that of the two-pile arrangement for gaps $G/D < 0.5$, by about 20-30 %. This can be attributed to the increase in the extent of the bed area exposed to the lee-wake vortices. In the case of the tandem arrangement, on the other hand, the scour depth does not change significantly in magnitude. However, (1) the dip experienced at $G/D = 0.5$ in the case of the two-pile case is now experienced at about $G/D = 0.2$; and (2) the scour for $G/D = 0$ in the three-pile case is smaller than in the two-pile case. The decrease in the scour depth in the three-pile case for $G/D = 0$ is due to the decrease in the effective KC number for the single body (the longitudinal dimension of the body is now $3D$).

Fig. 4.16 depicts the scour-depth data plotted against the pile spacing G/D for the triangular group (Fig. 4.7 f) for the same KC number, $KC = 13$.

258   CHAPTER 4. SCOUR AROUND A GROUP OF SLENDER PILES

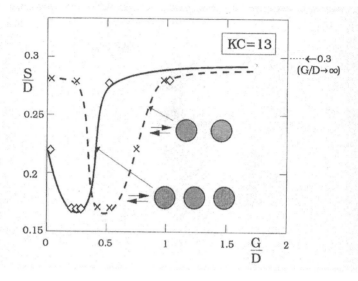

Figure 4.15: Equilibrium scour depth plotted against pile spacing. The single-pile $(G/D \to \infty)$ value: from Eq. 3.20. Sumer and Fredsøe (1998).

As seen, the triangular group gives a picture somewhat similar to the side-by-side arrangement (cf. Fig. 4.9), i.e., the depth of scour at the two side-by-side piles (Piles 2 and 3, Fig. 4.16) in the group increases with increasing G/D, and attains a maximum of $S/D \cong 0.5 - 0.6$ at $G/D \cong 0.3$, and then it decreases to reach a value of about 0.2 at $G/D = 3$. The scour depth for Pile 1 (Fig. 4.16) behaves similarly, but with moderately smaller scour depths. The decrease in the scour depth for the side-by-side piles (Piles 2 and 3) with respect to that depicted for the side-by-side arrangement in Fig. 4.9 is mainly due to the "shielding" effect caused by the presence of the third pile (Pile 1) in the present case.

4.2.3   Four-pile group

Fig. 4.17 displays the scour-depth information obtained in the case of the 4×4 square-pile group (Fig. 4.7 g) for two KC numbers, $KC = 13$ and 37. (The figure also indicates the single-pile $(G/D \to \infty)$ values for the two KC numbers from Fig. 3.35). The maximum scour depth occurred always at the corner piles in the first row of the pile group.

4.2. PILE GROUP OF IN WAVES

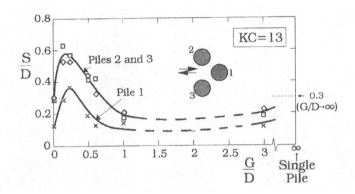

Figure 4.16: Equilibrium scour depth plotted against pile spacing. The single-pile ($G/D \to \infty$) value: from Fig. ?????. Sumer and Fredsøe (1998).

Of particular interest is the case of $KC = 37$. For this KC, the way in which the scour depth changes with G/D is rather similar to the side-by-side arrangements in conjunction with the two-pile and three-pile cases (Figs. 4.14).

For the zero gap situation, the single body has a KC number of $KC = 37/4 = 9.3$. From Fig. 3.37 for the square-section pile, it is found that the normalized scour depth for this KC is $S/(4D) = O(0.1)$, or $S/D = O(0.4)$. This is not radically different from the result shown in Fig. 4.17, $S/D \cong 0.6$, for $G/D = 0$. The difference may be attributed to: (1) The corners of the present single-body square section are rounded; (2) The surface of the present single body is not a plane surface, but rather a "wavy" surface; and (3) The scour in the present case may be influenced by the presence of a slight constant streaming, as will be described in Chapter 6.

The increase in the scour depth with increasing values of G/D from zero is due partly to the previously mentioned gap-flow effect, and partly to the increase in the extent of the shed vortices. The scour depth attains its maximum value, $S/D \cong 2$, at about $G/D \cong 0.4$, and then it begins do decrease, and eventually approaches its single-pile value, $S/D = 0.67$, for G/D values larger than about unity.

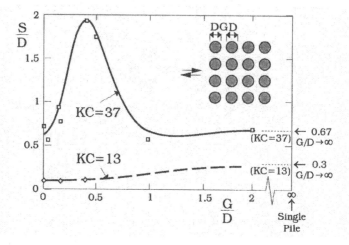

Figure 4.17: Equilibrium scour depth plotted against pile spacing. The single-pile ($G/D \to \infty$) values: from Eq. 3.20. Sumer and Fredsøe (1998).

4.2.4 Effect of the KC number

Fig. 4.18 illustrates the influence of the Keulegan-Carpenter number, KC, on the scour depth for various pile groups for a given pile spacing, namely for $G/D = 0.4$. The figure also includes the single pile result (from Fig. 3.35, the solid curve in the figure) as a reference line.

First of all, it is apparent that the way in which the scour varies with the KC number is much the same as in the single pile case. This implies that, as in the single-pile case, while the scour is mainly governed by the lee-wake vortex flow processes for small KC numbers, it is basically dominated by the horseshoe-vortex flow for large KC numbers (Section 3.2.2). From Fig. 4.18, the KC number beyond which the scour is dominated by the horseshoe-vortex flow is $KC = O(100)$ for the single pile (Section 3.2.2), $O(50)$ for the two-pile, side-by-side arrangement, $O(70)$ for the 45^0 staggered arrangement, and $O(300)$ for the 4×4 square group.

Second, the scour depth in the case of pile groups undergoes substantial changes with respect to the single-pile value; this change can be by an order of magnitude (or even larger) in some cases, particularly for small KC numbers. This can be explained in the same manner as in Section 4.2.1, basically in terms of the gap-flow effect, and the increase in the extent of the lee-wake

4.2. PILE GROUP OF IN WAVES

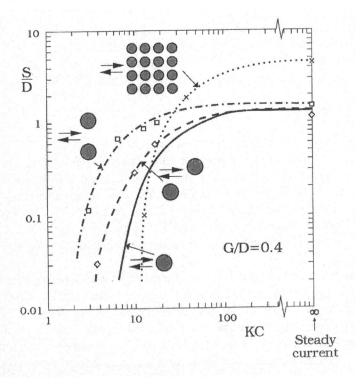

Figure 4.18: Equilibrium scour depth plotted against KC. Sumer and Fredsøe (1998).

vortex.

Third, the onset of scour in the case of the two-pile groups occurs at KC numbers smaller than that in the single-pile case. While the onset of scour occurs at $KC \cong 6$ in the single-pile case, it occurs at $KC \cong 2$ for the two-pile, side-by-side arrangement, and at $KC \cong 3$ for the two-pile staggered arrangement. This is mainly linked with the presence of the gap flow in the two-pile cases.

Fourth, for the 4×4 square pile arrangement, the onset of scour occurs at $KC \cong 12$, a much larger KC value than that for the single pile case. This is mainly due to the rather weak gap flow (cf. Fig. 4.17). As the KC number is increased, however, the horseshoe-vortex flow begins to take over, and this occurs at $KC = O(300)$, as pointed out in the preceding paragraphs. This causes an "explosive" growth in the scour depth; as seen from Fig. 4.18, S/D can be as much as 5, a factor of 3-4 increase in the scour depth with respect to the single-pile value. (cf. $S/D = 8$ found from $S/(4D) = 2$, the steady current value for a single square pile with the width $4D$, Fig. 3.37).

Carreiras et al. (2000) also have conducted experiments on the scour produced by waves around groups of piles, consisting of a single transverse range (a single row). The initial flow conditions and the location of the piles in the wave flume were kept constant - the mean water depth and the wave period were 0.15 m and 2.17 s, respectively. The gap ratio and the KC number (calculated with Eq. 3.19, Chapter 3, using the near-bed velocity measured between the piles) varied in the range 0.2-2.0 and 7.5-17.4, respectively.

In general, the horseshoe vortex generated by the presence of each pile induced a thin local scour hole around each pile, while the jet effect was responsible for the deeper erosion between the piles. Mory, Larroudé, Carreiras and Seabra Santos (2000) argued that the most significant local scour conditions are obtained when KC is between 9 and 25. In this case, the scour was initially dominated by the gap flow that induced a strong and rapid erosion on the bed between the piles and a high sediment transport to both sides of the pile range.

In most of the cases, in particular with multiple transverse ranges, the bed topography in the pile zone was influenced by the ripple dynamics that results from the sediment transport. On the other hand, the presence of the piles induced a global scour around the group of piles in a number of cases.

For groups of piles in a single transverse range, Eq. 3.20 was found to provide a good scaling of local scour when the KC number was determined using the measured flow conditions in the gap zone, i.e., the near-bed velocity measured between the piles, and when the coefficient $m = 0.06$ is used.

The scour produced by regular waves inside a pile group has been measured in a laboratory wave basin (Larroude and Mory , 2000, Mory et al. , 2000); the 9 m by 30 m wave basin.

The pile group consisted of two rows of ten piles each. It was located in a sand beach sloping initially at 1 in 40. Different cases were investigated in which the spacing between piles and the orientation of the pile group with regard to the direction of wave propagation were varied. Two different pile

4.2. PILE GROUP OF IN WAVES

diameters were considered, for which the Keulegan-Carpenter number were $KC = 7.9$ and $KC = 16.5$, respectively. Global scour was observed to be more significant than local scour for the tested range of the KC number.

Among the results obtained in Mory et al.'s study are

1. Large-scale morphological bed changes are very significant, and the pattern and magnitude of change depend both on the orientation of the pile group and on the KC number (the wave condition was the same for all cases). For the pile orientation $\varphi = 45^0$, for example, a deep global erosion is observed for $KC = 16.5$, whereas deposition is observed for $KC = 7.9$. For the pile orientation $\varphi = 90^0$, there was also a strong difference in the morphological changes between the two KC numbers.

2. Regular ripples were observed for all experiments. The amplitude of small scale bed changes inside the pile groups was significantly smaller than the ripple heights. Ripples were damped inside the pile group.

3. Local scour was limited in these experiments. This is not surprising, since the largest KC number investigated was 16.5. The tendency that increasing the KC number increases the scour depth is clearly observed.

For more details see Mory et al. (2000).

Bayram and Larson (2000 a) report some interesting field results. Their analysis of regularly surveyed bed profiles along a 200 m long pier at Ajigaura Beach in Ibaraki Prefecture on the Pacific coast of Japan have shown scour-hole formations at the pile groups (spaced about 30 m apart) which support the pier. Each pile group comprises four piles arranged in an approximately square configuration. The gap-to-diameter ratio is about 4-5. Bayram and Larson report that the scour hole is apparently one single hole rather than four local scour holes developing at the individual piles. The scour depth, plotted in the normalized form S/D, was found to increase with KC in which D is the diameter of the individual pile and KC is defined on the basis of D. The values of S/D range from 1 to 4 with KC in the interval of 7-22. Although no information is given in Bayram and Larson (2000 a), effects such as littoral currents, breaking waves and the turbulence generated by the structure itself may have contributed to the end result. (Posey and Sybert (1961) report that the presence of a littoral current combined with waves in a physical model test made it possible for a single scour hole to

264 CHAPTER 4. SCOUR AROUND A GROUP OF SLENDER PILES

develop beneath an offshore platform, while the waves alone caused small holes to develop around each pile. The mechanism responsible with regard to the combined wave and current climate is that the waves stir up the sediment and put it into suspension, and the current carries the sediment away from the structure). In a follow-up paper, Bayram and Larson (2000 b) have demonstrated that the scour depth correlates with the longshore current velocity.

4.3 Global and local scour at pile groups

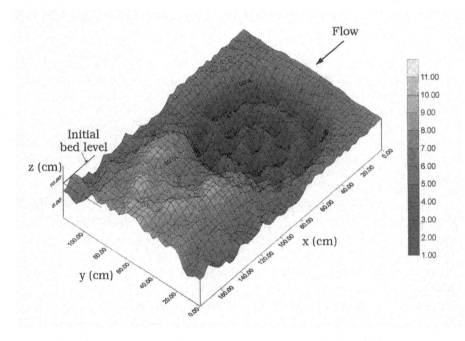

Figure 4.19: Local and global scour. Sumer et al. (2002 a).

As mentioned in Chapter 1, scour at pile groups may occur as local scour and global scour (Fig. 4.19). The studies outlined in the preceding sections viewed the scour process without regard to local and global scour processes. To the authors' knowledge, Sumer, Bundgaard and Fredsøe (2002 a) were the first to study in detail these processes separately. To this end, experiments

4.3. GLOBAL AND LOCAL SCOUR AT PILE GROUPS

have been carried out, employing six different pile groups in *steady currents*. The pile groups used in the tests are: side-by-side pile group (with two piles), tandem pile group (with two piles), 2×2 square pile group, 3×3 square pile group, 5×5 square pile group and an equivalent circular-configuration pile group, along with a single pile as the reference case. In the tests, the global scour could be differentiated from the local scour. The latter has enabled Sumer et al. (2002 a) to correlate the results with the mean and turbulence characteristics of the flow past the structure. The following paragraphs summarize the results of this study.

When a group of piles is exposed to flow action, two kinds of scour patterns emerge: (1) the local scour around the individual piles, and (2) the global scour in the form of a saucer-shaped depression (the general lowering of the bed level over the entire area of the pile group). (Fig. 4.19).

The local scour is caused by the horseshoe vortex, the vortex shedding, and the contraction of streamlines (Fig. 3.1, Chapter 3), associated with the individual pile (as described in Chapter 3). (Note that the resulting local scour characteristics may not be the same as those in the case of a single pile due to the possible influence of the global scour).

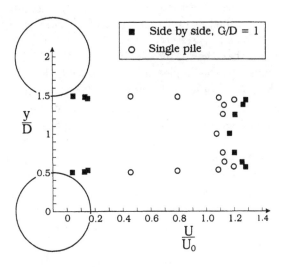

Figure 4.20: Mean velocity at 14.5 cm from the bed in Sumer et al.'s (2002 a) rigid-bed tests. U_0 is the undisturbed velocity at the same level.

266 CHAPTER 4. SCOUR AROUND A GROUP OF SLENDER PILES

The global scour, on the other hand, is caused by the following two effects: (1) the change in the flow velocity in the gap between the piles, and (2) the turbulence generated by the individual piles.

The former effect can be described by reference to Sumer et al.'s (2002 a) velocity measurements depicted in Fig. 4.20 in the case of the side-by-side configuration with two piles. As seen from Fig. 4.20, the velocity in the gap between the piles increases with respect to the case of the single pile. (Similar increases have been observed for the other values of the gap-to-diameter ratio, G/D, tested. Even for $G/D = 4$, an increase in the order of magnitude of 5% has been measured). This implies that the bed will be scoured, and the scouring process will continue until the bed shear stress between the piles is gradually reduced to the value of the undisturbed bed shear stress (also taking into account the local bed slope). This scour is the global scour generated by the increase of the velocity between the piles.

The second effect (namely, the turbulence generated by the individual piles), will be described by reference to velocity measurements of the same authors in the case of the 5 × 5 square group configuration, with $G/D = 4$. Fig. 4.21 displays the mean velocity profiles for four sections indicated in the figure, Sections A, B, C and D, while Fig. 4.22 displays the measured turbulence profiles at the same sections. Here, u is the streamwise component of the velocity, u' the fluctuating component of u, and the overbar denotes the time averaging.

Fig. 4.21 shows that the mean velocities do not exhibit any significant increase with respect to the single-pile velocities (cf. Fig. 4.20). (The velocities inside the structure even decrease at places, apparently due to the "shielding" effect). By contrast, Fig. 4.22 reveals that the turbulence inside the structure is increased quite substantially. (Compare the turbulence profiles at Sections B, C and D with that at Section A. The turbulence inside the structure increases by as much as a factor of 3). The measured turbulence is generated mostly by the wakes of the piles at the upstream sections. (Note that the flow in this test is measured at 14.5 cm from the bed. Although no measurements were undertaken, it may be expected that the turbulence generated by the upstream piles will also be contributed by the horseshoe vortices in areas near the bed).

The implication of the presence of extensive turbulence inside the structure is that the sediment transport underneath/around the structure will increase substantially (Sumer et al., 2002 b), leading to "global" lowering of the general bed level, the global scour generated by turbulence.

4.3. GLOBAL AND LOCAL SCOUR AT PILE GROUPS

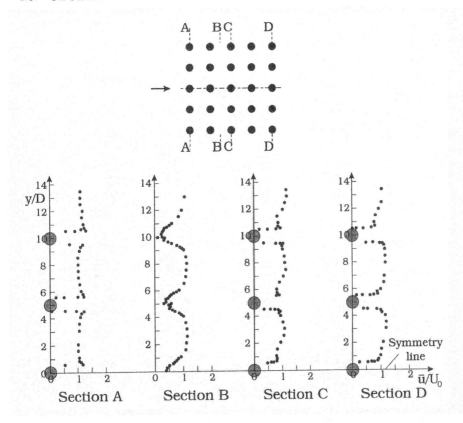

Figure 4.21: Mean velocity at 14.5 cm from the bed in Sumer et al.'s (2002 a) rigid-bed tests. $G/D = 4$. U_0 is the undisturbed velocity at the same level.

Fig. 4.23 shows an example of the measured scour profiles (in the case of the 5 × 5 square group with $G/D = 4$). Fig. 4.23 reveals that the global scour can be differentiated very clearly from the local scour.

The maximum equilibrium global scour depth, S_G, is plotted in Fig. 4.24 for the case of the square group arrangements with $N \times N$ piles with $G/D = 4$. Here, the scour depth is normalized by the pile diameter D, considering its role in the global-scour process through the governing mechanisms, namely the change in the mean and turbulence properties of the flow inside the structure.

As seen from Fig. 4.24, the global-scour depth increases with increasing

268 CHAPTER 4. SCOUR AROUND A GROUP OF SLENDER PILES

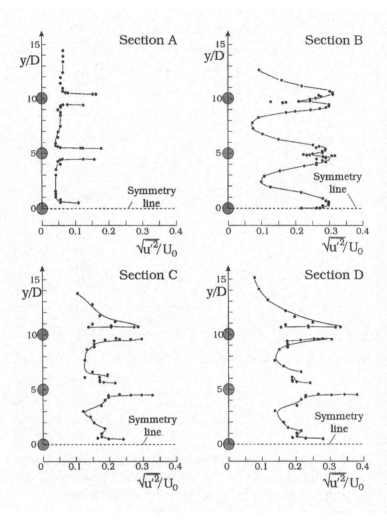

Figure 4.22: Turbulence profiles at 14.5 cm from the bed in Sumer et al.'s (2002 a) rigid-bed tests in the previous figure. $G/D = 4$. U_0 is the undisturbed velocity at the same level.

4.3. GLOBAL AND LOCAL SCOUR AT PILE GROUPS

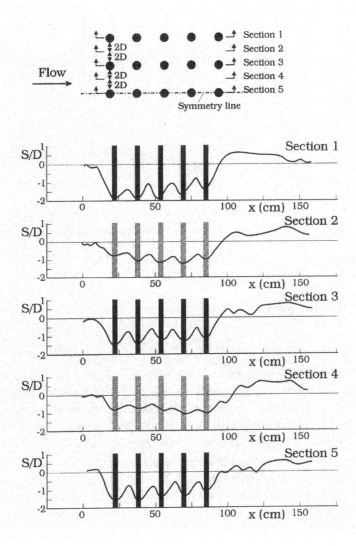

Figure 4.23: Equilibrium scour profiles for the 5×5 square group. Live bed. $G/D = 4$. Sumer et al. (2002 a).

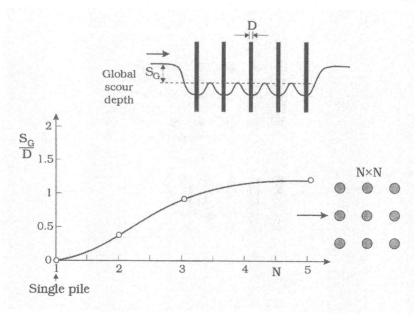

Figure 4.24: Maximum equilibrium global-scour depth for $N \times N$ square group. Live bed. $G/D = 4$. Sumer et al. (2002 a).

N for $N \lesssim 4 - 5$. For further increase in the number of piles, however, the scour depth seems to remain uninfluenced by N. This may be attributed to the effect of turbulence. As seen from Fig. 4.22, no significant change occurs in the turbulence from Section C (the third row of piles) to Section D (the fifth row of piles); indeed, the maximum value of turbulence at these two sections is more or less the same and equal to $\sqrt{\overline{u'^2}}/U_0 = 0.3 - 0.4$. This suggests that the turbulence generated by the upstream piles in a square group will attain its maximum value when $N \to 3 - 5$, and will remain practically constant when $N > 3 - 5$. This evidently supports the finding displayed in Fig. 4.24.

Fig. 4.25 shows the maximum total equilibrium scour depth, S_T, (i.e., the maximum equilibrium global scour depth plus the maximum equilibrium local scour depth), plotted against the number of piles in the $N \times N$ pile square group with $G/D = 4$. The quantity S_0 is the maximum equilibrium scour depth in the case of the single pile measured under exactly the same

4.3. GLOBAL AND LOCAL SCOUR AT PILE GROUPS

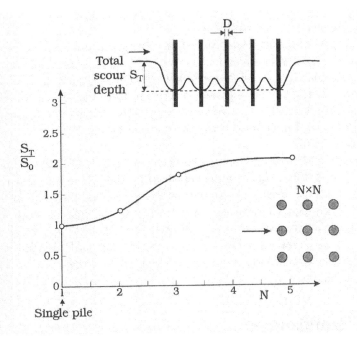

Figure 4.25: Maximum equilibrium total scour depth for $N \times N$ square group. Live bed. $G/D = 4$. Sumer et al. (2002 a).

flow conditions. As seen, the scour depth with the presence of the global scour increases quite substantially. This increase can be as much as a factor of 2 or more for 4×4 and 5×5 pile groups, as revealed by Fig. 4.25.

Sumer et al. (2002 a) note that the maximum local scour depth occurs at the first upstream row of the piles (Fig. 4.23). This behaviour has been observed in the tests for the other square configurations as well. Fig. 4.23 further shows that the corner pile in the first upstream row experiences the maximum scour depth. This may be attributed to the fact that the front row and possibly the corner pile in this row are the most exposed piles to the flow. For the circular group (Fig. 4.19), Sumer et al.'s (2002 a) tests showed that the very first pile at the upstream side of the group experiences the maximum local scour, again due to the largest exposure to the flow.

The data plotted in Figs. 4.24 and 4.25 are for the square group arrangements. The results for the other configurations, the side-by-side and circular group arrangements, are as follows.

The global scour depth for the side-by-side arrangement with two piles is found somewhat larger than that for the 2 × 2 square arrangement; $S/D = 0.78$ for the side-by-side pile group versus $S/D = 0.37$ for the 2 × 2 square pile group. This can be explained by the marked increase in the velocity between the piles in the case of the side-by-side arrangement (Sumer et al., 2002 a). It may be mentioned that these results agree well with the results of Hannah (1978).

Regarding the circular group arrangement (Fig. 4.19), the global scour depth is found to be slightly larger than that for the 5×5 square arrangement; $S/D = 1.5$ for the circular group versus $S/D = 1.2$ for the 5 × 5 square pile group. (Note that the circular group tested in the present study has the same pipe "density" as in the case of the 5 × 5 square group, Sumer et al., 2002 a). This increase may be attributed to the increased turbulence in the circular group, apparently due to the "staggered" arrangement of the piles in the circular group.

4.4 References

1. Bayram, A. and Larson, M. (2000 a): Analysis of scour around a group of vertical piles in the field. J. Waterway, Port, Coastal and Ocean Engineering, ASCE, vol. 126, No. 4, 215-220.

2. Bayram, A. and Larson, M. (2000 a): Analysis of scour due to breaking and non-breaking waves around a group of vertical piles in the field. Proceedings of Coastal Structures 99, Santander, Spain, 7-9 June, 1999, vol. 2, 763-771, Balkema.

3. Breusers, H.N.C. and Raudkivi, A.J. (1991): Scouring. A.A. Balkema, Rotterdam, viii + 143 p.

4. Carreiras, J., Larroudé, Ph.,Santos, F.J. Seabra and Mory, M. (2000): Wave scour around piles. In: Proc. 27th Coastal Engineering Conference, Sydney, Australia, vol. 2, 1860-1870.

5. Chow, W.Y. and Herbich, J.B. (1978): Scour around a group of piles. Proc. Offshore technology Conference, Houston, Texas, Paper No. 3308.

4.4. REFERENCES

6. Eadie, R.W. IV, and Herbich, J.B. (1986): Scour about a single, cylindrical pile due to combined random waves and current. Proc. 20th Coastal Engineering Conference, Taipei, Taiwan, ASCE, New York, N.Y., vol. 3, 1858-1870.

7. Fredsøe, J. (1984): Turbulent boundary layers in wave-current motion. J. Hydraulic Engineering, ASCE, vol. 110, No. HY8, 1103-1120.

8. Gormsen, C. and Larsen, T. (1984). Time development of scour around offshore structures. Master's Study, ISVA, Technical University of Denmark, 139 p. (in Danish).

9. Hannah, C.R. (1978). Scour at pile groups. University of Canterbury, N.Z., Civil Engineering, Research Report, No. 78-3, 92 p.

10. Herbich, J.B., Schiller, R.E., Jr., Watanabe, R.K. and Dunlap, W.A. (1984): Seafloor Scour. Design Guidelines for Ocean-Founded Structures, Marcell Dekker, Inc., New York, NY, xiv + 320 p.

11. Kobayashi, T. and Oda, K. (1994): Experimental study on developing process of local scour around a vertical cylinder. Proc. 24th International Conference on Coastal Engineering, Kobe, Japan, vol. 2, Chapter 93, 1284-1297.

12. Larroudé, P. and Mory, M. (2000): "Erosion autour de structures côtières. VIème journées nationales Génie Civil - Génie Côtier, Caen.

13. Mory, M., Larroudé, P., Carreiras, J. and Seabra Santos, F. J. (2000): Scour around pile groups. Proceedings of Coastal Structures 99, Santander, Spain, 7-9 June, 1999, vol. 2, 773-781, Balkema.

14. Posey, C.J. (1961): Erosion protection of production structures. Proc. 9th Congress of International Association for Hydraulic Research, IAHR, 4-7. September, 1961, Dubrovnik, Yugoslavia, 1157-1162.

15. Sumer, B.M. and Fredsøe, J. (1998): "Wave scour around group of vertical piles". J. Waterway, Port, Coastal and Ocean Engineering, ASCE, vol. 124, No. 5, 248-256.

16. Sumer, B.M., Bundgaard, K. and Fredsøe, J. (2002 a): Global and local scour at pile groups. Manuscript in preparation.

17. Sumer, B.M., Christiansen, N. and Fredsøe, J. (1993): Influence of cross section on wave scour around piles. J. Waterway, Port, Coastal and Ocean Engineering, ASCE, vol. 119, No. 5, 477-495.

18. Sumer, B.M., Christiansen, N. and Fredsøe, J. (1997): The horseshoe vortex and vortex shedding around a vertical wall-mounted cylinder exposed to waves. J. Fluid Mech., vol. 332, 41-70.

19. Sumer, B.M., Chua, L.H.C., Cheng N.-S. and. Fredsøe, J. (2002 b): The influence of turbulence on bedload sediment transport. Manuscript in preparation.

20. Sumer, B.M., Fredsøe, J. and Christiansen, N. (1992): Scour around vertical pile in waves. J. Waterway, Port, Coastal and Ocean Engineering, ASCE, vol. 117, No. 1, 15-31.

21. Wang, R.-K. and Herbich, J.B. (1983): Combined current and wave-produced scour around a single pile. COE Rep. No. 269, Texas A&M University.

22. Williamson, C.H.K. (1985): Sinusoidal flow relative to circular cylinders. J. Fluid Mechanics, vol. 155, 141-174.

23. Zdravkovich, M. (1987): The effects of interference between circular cylinders in cross flow. J. Fluids and Structures, vol. 1, 239-261.

Chapter 5

Examples of more complex configurations

When a structure with a complex configuration (such as a piled steel platform comprising horizontal, vertical and inclined members) is exposed to flow action, two kinds of scour take place, as pointed out in Chapter 1 (Section 1.5); one is the local scour around the individual structural elements such as that around the supporting piles, and the other is the global scour which takes place beneath and around the structure in the form of a saucer-shaped depression, (see Fig. 1.4, for a *conceptual* picture, Angus and Moore, 1982). The global scour here is due to the combined action of all the flow effects generated by the individual elements. These include essentially the effect of the contraction of flow, and that of the "turbulence" generated by the structural elements. (The latter is basically due to the lee-wake vortices formed behind the individual elements, horizontal, vertical, or inclined, and may be regarded as a field of externally generated turbulence). The effects inducing the global scour may vary, depending on the particular configuration of the structure, and the particular flow environment. See also the discussion and analysis in Section 4.3.

The present chapter will review the knowledge of scour at structures with complex geometries. This review will be mostly in the form of examples.

276 CHAPTER 5. EXAMPLES OF MORE COMPLEX CONFIGURATIONS

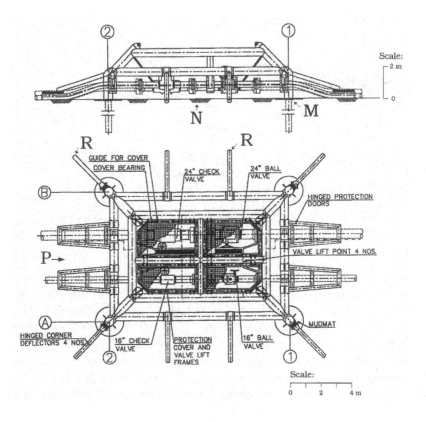

Figure 5.1: Harald-A Valve Station. Plan view of general arrangement (by courtesy of LICENGINEERING A/S, and Mærsk Olie & Gas).

5.1 Scour at pile-supported offshore structures

Fig. 5.1 displays an offshore sub-sea structure, namely a valve station, a structure with complex configuration. The purpose of this structure is to protect a valve station (the Harald valve station in the Danish sector of the North Sea). A 3-D view of the installation is shown in Fig. 5.2. Basically, the structure is supported by four vertical piles (Fig. 5.1). (The mud mat seen in the bottom panel of Fig. 5.1 is to provide the structure initial support before the piles are driven and grouted. The hinged deflectors, both at the corners and at the sides (Fig. 5.1), on the other hand, are to deflect the fishing

5.1. SCOUR AT PILE-SUPPORTED OFFSHORE STRUCTURES

gears, thus preventing the latter from entering the structure and damaging the valves, Fog and Jønsson 1997).

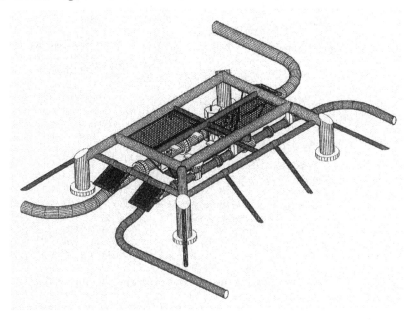

Figure 5.2: 3D view of the valve station and valve spools in the previous figure. (by courtesy of LICENGINEERING A/S, and Mærsk Olie & Gas).

The potential scour areas for this sub-sea structure can be identified as follows

1. The areas around the vertical piles (marked M in Fig. 5.1);

2. The areas below the pipelines (marked N in Fig. 5.1);

3. The areas between the elevated pipelines at the entrance and exit sections (marked P, Fig. 5.1);

4. The areas around the corner and side deflectors (marked R, Fig. 5.1);

While Items 1 and 4 are typical local-scour, Items 2 and 3 may be regarded as global scour. The joints where the vertical, horizontal, and inclined

278 CHAPTER 5. EXAMPLES OF MORE COMPLEX CONFIGURATIONS

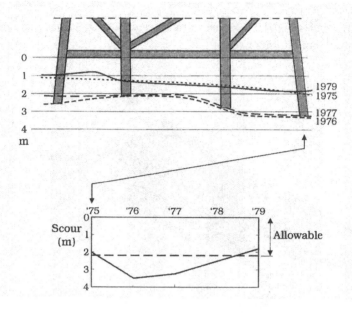

Figure 5.3: Observed scour at an offshore platform. Dahlberg (1983).

members of the structure meet (Figs. 5.1 and 5.2), and also the hinged protection doors (Figs. 5.1 and 5.2), and other secondary structural elements (Figs. 5.1 and 5.2) act as a source of external-turbulence generation. This externally generated turbulence will obviously influence the local scour, and also it will contribute to the global scour as well.

Fog and Jønsson (1997) report that scour considerations have been an important parameter in the design of the aforementioned valve station. They also emphasize that an open design is used, to minimize the scour development.

Dahlberg (1983) reviews observations made in the North Sea of scour around pile-supported offshore oil platforms. Dahlberg (1983) gives several examples of the observed scour profiles. These observations show that the scour depth (the global-plus-local-scour depth) can be as much as 3 m. (Incidentally, scour depths as much as 12 m have been observed in the North Sea, as reported by van Dijk, 1981). Fig. 5.3 displays an example from Dahlberg (1983) where the seabed soil is fine sand. The allowable scour was 2 m. Apparently, the design criteria were exceeded by up to 1.5 m at two occasions before the gravel dumping in 1978 improved the situation.

Our knowledge of scour around structures with complex geometries is rather limited. Only a few cases have been investigated in a systematic manner (DHI/Snamprogetti, 1992). In the DHI/Snamprogetti (1992) study, the steady-current scour has been the focus of the investigation in the majority of the cases. The scour characteristics such as the scour depth and the plan-view extent of the scour have been obtained from the experiments. The following section will present as examples three of the configurations investigated in the DHI/Snamprogetti work (1992).

5.2 Scour characteristics

Example 14 *Scour around a T-shaped structure*

The T-shaped structure is sketched in Fig. 5.4; it comprises two circular pipe members, one is horizontal and the other is vertical. It is exposed to a steady current. The contour plots of the normalized scour depth S/D, obtained for three different values of the gap between the horizontal member and the seabed, are displayed in Fig. 5.4 a, b and c. These contour plots correspond to the equilibrium state of the scour process. Fig. 5.5, on the other hand, presents the contour plot of the scour hole in the case of a single pile with a height equal to D, and exposed to the same steady current. The scour displayed in Fig. 5.5 corresponds to the equilibrium state of the scour process, and the scour picture in Fig. 5.5 is included here as a reference case.

From Figs. 5.4 and 5.5, it can be seen that the scour hole in the case of the T-shaped structure is quite large with respect to that experienced in the case of the single pile (Fig. 5.5) (the scour width is a factor of 2-3 larger in the case of the T-shaped structure). Clearly, this is due to the presence of the horizontal member. Also the scour depth at the pile (the "local scour") is generally larger than that in the case of the single pile; for example, the scour depth in the case of the T-shaped structure for $e/D = 1$ (Fig. 5.4 c) is 30 % larger than in the case of the single pile (Fig. 5.5). The scour picture is quite similar to that described at the beginning of this section, namely the local scour around the pile, and the global scour for the entire structure.

Example 15 *Scour around a structure representing a typical joint*

280 CHAPTER 5. EXAMPLES OF MORE COMPLEX CONFIGURATIONS

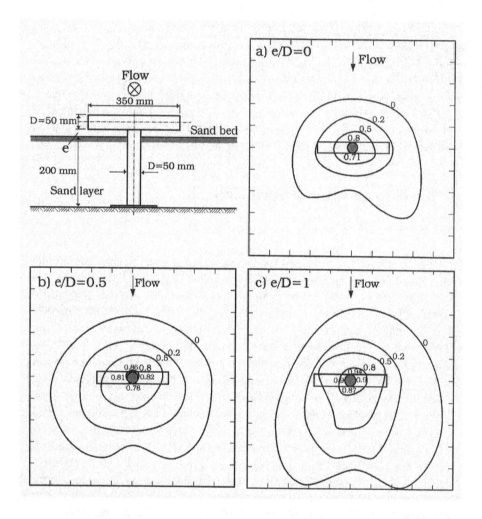

Figure 5.4: Contour plots of S/D, the normalized scour depth. Shields parameter at the critical value. Circular cross-section pipes. By courtesy of DHI and Snamprogetti.

5.2. SCOUR CHARACTERISTICS

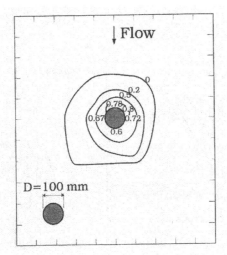

Figure 5.5: Contour plots of S/D, the normalized scour depth. Shields parameter at the critical value. Circular cross-section pipes. By courtesy of DHI and Snamprogetti.

This structure is sketched in Fig. 5.6. It represents a typical joint where the vertical, horizontal, and inclined structural elements join. The structure is exposed to a steady current. Fig. 5.6 illustrates the measured scour around the structure (the contour plots of the normalized scour depth S/D), corresponding to the equilibrium state.

The local scour around the vertical element is quite clear from Fig. 5.6. The global scour is also clear from the figure (the saucer-shaped scour hole extending over a very large area of the bed, caused by the structure-induced turbulence) (cf. Fig. 5.4).

Example 16 *Scour around a porous structure*

Fig. 5.7 displays a porous structure. The porous structure is achieved by a "square" group of vertical rods (the base is not porous, however, Fig. 5.7). The porosity of the structure is characterized by the transparency of the structure, which is defined by

$$\text{Transparency} = \frac{\text{Open Cross-Sectional Area}}{\text{Total Cross-Sectional Area}} \quad (5.1)$$

282CHAPTER 5. EXAMPLES OF MORE COMPLEX CONFIGURATIONS

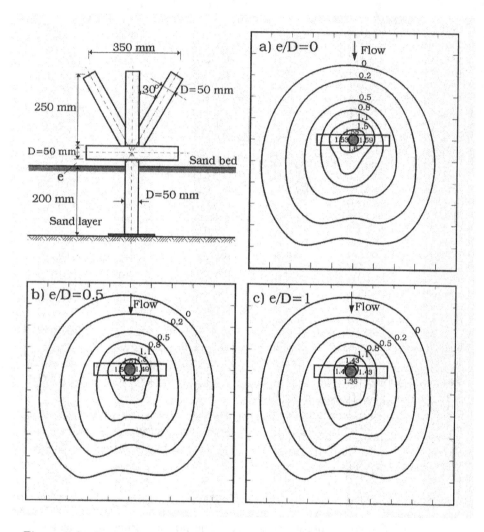

Figure 5.6: Contour plots of S/D, the normalized scour depth. Shields parameter at the critical value. Circular cross-section pipes. By courtesy of DHI and Snamprogetti.

5.2. SCOUR CHARACTERISTICS

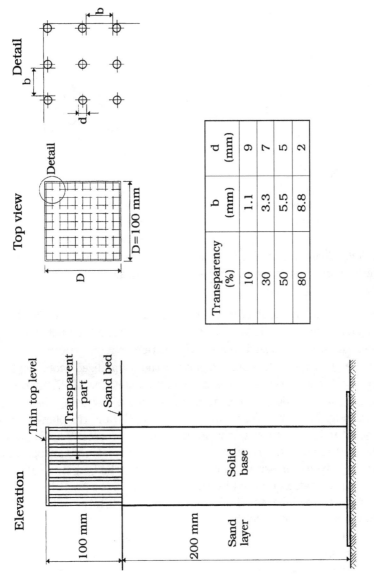

Figure 5.7: Porous structure. By courtesy of DHI and Snamprogetti.

CHAPTER 5. EXAMPLES OF MORE COMPLEX CONFIGURATIONS

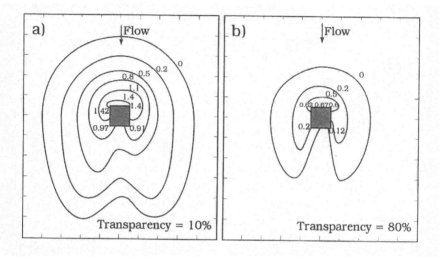

Figure 5.8: Contour plots S/D, the normalized scour depth. Porous structure. The Shields parameter at the critical value. By courtesy of DHI and Snamprogetti.

The small box in the figure depicts the tested transparency values in terms of the spacing between the vertical rods, b, and the rod diameter, d.

The structure is exposed to a steady current.

Fig. 5.8 illustrates the contour plots of the normalized scour depth S/D obtained in the two extreme cases tested; namely for 10 % transparency (Fig. 5.8 a), and for 80 % transparency (Fig. 5.8 b), D being the size of the structure (Fig. 5.7). It is clear that the scour depth (and the plan-view extent of the scour hole) decreases as the transparency increases. This is due to the decrease in the adverse pressure gradient with increasing transparency; the smaller the adverse pressure gradient, the smaller the bed area subjected to the flow concentration (caused by the horseshoe vortex, and the contraction of streamlines, see Chapter 3, Sections 3.1.1 and 3.1.4), and therefore the smaller the scour depth/width.

Fig. 5.9 a depicts the scour depth as a function of the transparency. As seen, the scour depth decreases with increasing transparency. For example, the scour depth decreases by more than 30 % when the transparency is changed from zero to 50 %.

Fig. 5.9 b depicts the variation of the average plan-view extent of the scour hole with respect to the transparency, a picture similar to that in the

5.3. REFERENCES

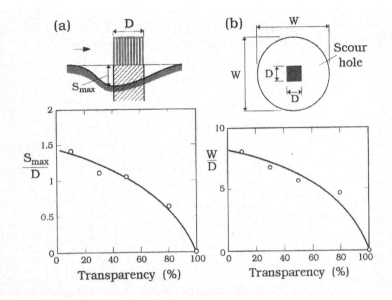

Figure 5.9: Porous structure. By courtesy of DHI and Snamprogetti.

previous figure. Note that although the streamwise dimension of the scour hole is slightly larger than the transverse dimension, an average value, W, is used in Fig. 5.9 b for simplicity.

5.3 References

1. Angus, N.M. and Moore, R.L. (1982): Scour repair methods in the Southern North Sea. Proc. 14th Annual Offshore Technology Conference, OTC, Houston, Texas, May 3-6, 1982, Paper No. 4410, 385-399.

2. Dahlberg, R. (1983): Observations of scour around offshore structures. Can. Geotech. J., vol. 20, 617-628.

3. Fog, N.G. and Jønsson, P.H. (1997): Harald valve station - a Danish subsea installation in the North Sea. Proc. Seventh International Offshore and Polar Engineering Conference, ISOPE-97, Honolulu, Hawaii, USA, vol. I, 387-393.

4. van Dijk, R.N. (1981): Experience of scour in the Southern North Sea. J. Society for Underwater Technology, March Issue, 18-22.

Chapter 6

Scour around large piles

Scour around large bodies *under waves* may be of interest with regard to its application to large marine structures such as offshore gravity platforms, platform legs, bridge piers, breakwater heads, etc. The body size would be so large that the flow is in the so-called diffraction (and hence unseparated flow) regime. Therefore, the vortex shedding and the horseshoe vortex (the mechanisms responsible for scour around slender bodies, Chapter 3) are not present, as will be detailed in the next section. However, observations do show that scour occurs also in this regime. Clearly, the scour in this case must be related to mechanisms other than the previously mentioned vortex-flow processes.

Although much research has been carried out on wave scour around small vertical piles over the past decade or so (see Chapter 3), comparatively little is known about the flow and scour processes around large structures subjected to waves. Rance (1980), Katsui and his co-workers (Katsui and Toue (1988), Toue et al. (1992), Katsui (1992), and Katsui and Toue (1993)) and recently Sumer and Fredsøe (2001) studied the scour in the laboratory around large vertical piles subjected to waves. Their work indicated that the wave-induced steady streaming near the pile may be a possible candidate responsible for the transport of the sediment away from the pile, eventually leading to the formation of scour holes.

The present chapter will study the flow and scour processes around a large vertical pile exposed to waves. We shall first address various aspects of the flow processes; namely, the flow regime in the case of the large pile; the phase-resolved flow; and the steady streaming around the pile, and then we shall turn our attention to the scour processes.

6.1 Large pile. Diffraction regime

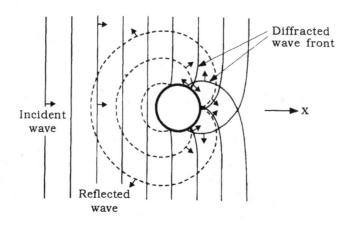

Figure 6.1: Sketch of the incident, diffracted and reflected wave fronts around a pile.

In Chapter 3, attention was concentrated on scour around small piles where the pile diameter, D, is assumed to be much smaller than the wave length. In this case, the presence of the pile does not influence the wave. In the case when D becomes relatively large, however, the body will disturb the incident waves. Consider, for example, a large vertical circular pile placed on the bottom (Fig. 6.1). As the incident wave impinges on the pile, a reflected wave moves outward. On the sheltered side, there will be a "shadow" zone where the wave fronts are bent around the pile (called the diffracted waves). The presence of the pile will therefore disturb the incident waves by the generation of reflected and diffracted waves. This process is generally termed **diffraction**, and the reflected and diffracted waves, combined, are usually called the scattered waves.

It is generally accepted that the diffraction effect becomes important when the ratio D/L becomes larger than 0.2 (Isaacson, 1979).

Now, the amplitude of the horizontal component of water-particle motion at the sea surface, according to the sinusoidal-wave theory, is (Appendix A)

$$a_S = \frac{H}{2} \frac{1}{\tanh(kh)} \qquad (6.1)$$

6.1. LARGE PILE. DIFFRACTION REGIME

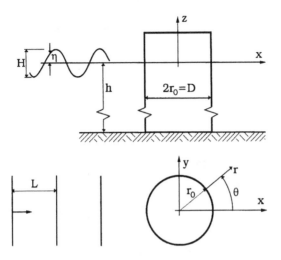

Figure 6.2: Definition sketch.

in which H is the wave height, h the water depth, and k the wave number

$$k = \frac{2\pi}{L} \qquad (6.2)$$

(Fig. 6.2). The Keulegan-Carpenter number (Eqs. 3.4 and 3.6, Chapter 3) for a vertical circular pile (calculated at the sea surface) will then be

$$KC_S = \frac{2\pi a_S}{D} = \frac{\pi(H/L)}{(D/L)\tanh(kh)} \qquad (6.3)$$

The largest Keulegan-Carpenter number is obtained when the maximum wave steepness is reached, namely when $H/L = (H/L)_{max}$. The latter may be given by the following approximate expression (Isaacson, 1979)

$$\left(\frac{H}{L}\right)_{max} = 0.14\tanh(kh) \qquad (6.4)$$

(The waves with the wave steepness larger than $(H/L)_{max}$ will break). Therefore, the largest Keulegan-Carpenter number that the pile experiences may, from the preceding two equations, be written as

$$KC_S = \frac{0.44}{D/L} \qquad (6.5)$$

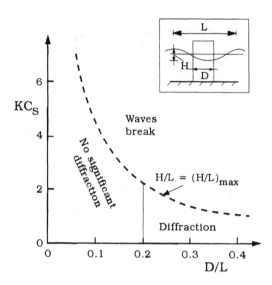

Figure 6.3: Different flow regimes. Adapted from Isaacson (1979).

and for KC_S larger than this limiting value, the waves will break.

Eq. 6.5 is plotted in Fig. 6.3 (the dashed line), representing the boundary between the two regions, namely Waves Break and Waves Do Not Break. Also included in the figure is the previously mentioned value of D/L, 0.2, the value beyond which the diffraction effect becomes significant (the vertical line at $D/L = 0.2$). Now, the following conclusions can be drawn from the preceding analysis:

1. As seen from Fig. 6.3, the Keulegan-Carpenter numbers experienced in the diffraction flow regime (i.e., in the region below the dashed line, and for $D/L > O(0.1)$) are extremely small, namely $KC_S < O(2-3)$. Considering the Reynolds numbers involved (normally extremely large; large compared with $O(10^3)$ in any event), such small Keulegan-Carpenter numbers indicate that the flow around the pile is unseparated. (See, for example, Sumer and Fredsøe, 1997 a, Fig. 3.15). So, no separation vortices and no vortex shedding will exist in the diffraction regime.

2. Likewise, such small Keulegan-Carpenter numbers, again, indicate that

6.2. PHASE-RESOLVED FLOW AROUND THE PILE

the flow in front of the pile will not separate, and therefore no horseshoe vortex will develop. (See Chapter 3, Section 3.1.2).

3. The analysis also implies that an additional parameter, namely D/L, termed the **diffraction parameter**, emerges with regard to the flow (and therefore scour) processes.

6.2 Phase-resolved flow around the pile

When a large pile is subjected to a progressive wave, a reflected wave moves outward from the pile, and a diffracted wave forms on the sheltered area, as described in the preceding paragraphs. These wave fields (the incident wave, the reflected wave and the diffracted wave), combined, create two kinds of flow around the pile. One is the phase-resolved flow, and the other is the steady streaming (the latter is induced by the nonuniform oscillatory motion around the pile, as will be detailed in the following paragraphs). Let us first consider the phase-resolved flow.

Fig. 6.4 illustrates how the flow around the pile evolves as the wave progresses. In the figure, the plan-view component of the velocity around the pile is displayed at phase values $\omega t = 0^0$, 90^0, 180^0, 225^0 and 270^0. The velocity profiles are shown only in the lower half of the plane due to symmetry. The phase $\omega t = 0^0$ corresponds to that where the wave crest is just at the offshore edge of the pile. The sequence of the velocity profiles to the left in Fig. 6.4 is from Sumer and Fredsøe's (1997 b) laboratory measurements, while that to the right is from the theory of MacCamy and Fuchs (1954). The bed in Sumer and Fredsøe's (1997 b, 2001) velocity measurements was a plane rigid bed, and the wave conditions were: the water depth = 40 cm, the wave height = 12 cm, the wave period = 3.5 s, and the velocities depicted in Fig. 6.4 were measured at a level of 5 cm from the bed. The preceding wave conditions correspond to $KC = 1.1$ and $D/L = 0.15$ in which KC is defined by

$$KC = \frac{U_m T_w}{D} \quad (6.6)$$

in which U_m is the maximum value of the undisturbed orbital velocity of water particles at the bed, T_w is the wave period.

The MacCamy and Fuchs theory is basically a linear theory, and the solution is obtained by superposing the incident-wave potential function and

292 CHAPTER 6. SCOUR AROUND LARGE PILES

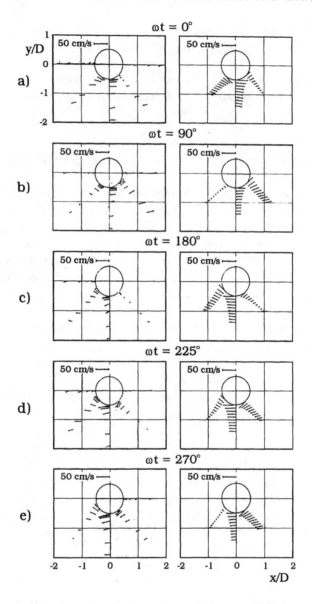

Figure 6.4: Vector diagram of the phase-resolved velocities at 5 cm distance from the bed. Left column: Measurements. Right column: MacCamy and Fuchs (1954) solution. $KC = 1.1$, $D/L = 0.15$. Sumer and Fredsøe (1997b).

6.2. PHASE-RESOLVED FLOW AROUND THE PILE

the scattered-wave potential function with the appropriate boundary conditions, namely the zero vertical velocity at the bed, the constant pressure at the free surface, and the zero normal velocity at the surface of the pile. (A full account of MacCamy and Fuchs' theory can be found in Sarpkaya and Isaacson, 1981, or in Sumer and Fredsøe, 1997 a). The plan view velocity components are

$$u_r = -\frac{\partial \phi}{\partial r} \text{ and } u_\theta = -\frac{1}{r}\frac{\partial \phi}{\partial \theta} \tag{6.7}$$

in which u_r and u_θ are the velocity components in the $r-$ and $\theta-$directions (Fig. 6.2), respectively, and ϕ is the potential function which is given by the MacCamy and Fuchs theory by

$$\phi = -i\frac{gH}{2\omega}\frac{\cosh(k(z+h))}{\cosh(kh)}\sum_{p=0}^{\infty}\varepsilon_p i^p \left[J_p(kr) - \frac{J_p'(kr_0)}{H_p^{(1)'}(kr_0)}H_p^{(1)}(kr)\right]\cos(p\theta)e^{-i\omega t} \tag{6.8}$$

in which i is the imaginary unit $i = \sqrt{-1}$, ω is the angular frequency, $\omega = 2\pi/T_w$, and the derivative terms are

$$J_p'(kr_0) = \left(\frac{dJ_p(\alpha)}{d\alpha}\right)_{\alpha=kr_0} \tag{6.9}$$

and

$$H_p^{(1)'}(kr_0) = \left(\frac{dH_p^{(1)}(\alpha)}{d\alpha}\right)_{\alpha=kr_0} \tag{6.10}$$

where α is a dummy variable. ε_p in Eq. 6.8 is defined by

$$\varepsilon_p = 1 \text{ when } p = 0 \tag{6.11}$$
$$\varepsilon_p = 2 \text{ when } p \geq 1$$

Here $J_p(kr)$ is the Bessel function of the first kind, order p, and $Y_p(kr)$ is the Bessel function of the second kind, order p. The function $H_p^{(1)}$, on the other hand, is the Hankel function of the first kind defined by

$$H_p^{(1)} = J_p(kr) + iY_p(kr) \tag{6.12}$$

The Bessel functions are given in tabulated forms in mathematical handbooks (e.g., Abramowitz and Stegun 1965, Chapter 9), and also in various mathematical softwares as built-in functions (e.g., Mathsoft 1997).

Now, returning to Fig. 6.4, although there exists a general agreement between the measured velocity distributions and the computed ones from the MacCamy and Fuchs solution, there are two major differences: (1) Fig. 6.4 shows that there is a small phase difference between the measured velocities and the computed ones; the latter is leading by $O(30^0)$ over the former; (2) The measured velocity data indicates that there exists a pronounced radial flow, particularly in the neighbourhood of the pile at the side edges. The theoretical solution does not give this. Nevertheless, the observed general agreement implies that the outer flow around the pile is due to the previously mentioned interacting wave fields: namely, the incident wave and the reflected wave in front and at the sides of the pile, and the diffracted wave and the reflected wave in the "shadow" zone behind the pile. The boundary layer over the bed will obviously respond to these interacting wave fields leading to the generation of a steady streaming, as will be described in the next section.

The implication of the phase-resolved flow with regard to scour is that clearly this phase-resolved flow will stir up the sediment on the bed, and put it into suspension. As will be seen later in the chapter, this aspect is an important "ingredient" of the mechanism regarding scour around large piles.

6.3 Steady streaming around the pile

The second kind of flow around the pile is the steady streaming caused by the nonuniform oscillatory motion. The nonuniformity is caused by the presence of the pile itself. We shall describe this process by reference to Sumer and Fredsøe's (2001) velocity measurements (also partly reported in Sumer and Fredsøe, 1997 b).

Figs. 6.5-6.7 show the vector diagram of the *period-averaged*, plan-view component of the velocity measured at three different depths, namely at distances 0.4 cm, 5 cm, and 25 cm from the bed. Fig. 6.8, on the other hand, gives a 3-D illustration of the vertical profiles of the same quantity very close to the pile. (The wave conditions are the same as those given in conjunction with Fig. 6.4, namely the water depth is 40 cm, the wave height 12 cm and the wave period 3.5 s, and the bed is a plane rigid bed).

The period-averaged velocities are defined by

$$U_r = \frac{1}{T_w} \int_0^{T_w} \overline{u_r} \, dt, \text{ and } U_\theta = \frac{1}{T_w} \int_0^{T_w} \overline{u_\theta} \, dt \qquad (6.13)$$

6.3. STEADY STREAMING AROUND THE PILE

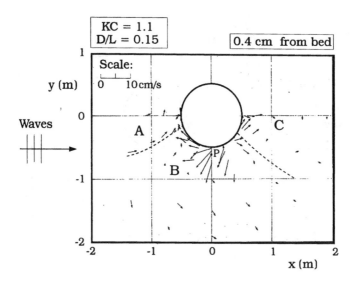

Figure 6.5: Vector diagram of the period-averaged velocities (the plan-view component), the *steady streaming*. Sumer and Fredsøe (2001).

and the velocity vectors plotted in Figs. 6.5-6.7 and 6.8 correspond to the resultant velocity, namely $(U_r^2 + U_\theta^2)^{1/2}$.

The flow field illustrated in Figs. 6.5-6.7 and 6.8 reveals very clearly that a wave-induced steady flow, called the **steady streaming**, takes place around the pile. The magnitude of the velocity of this flow can reach values as high as 20-25 % of the maximum value of the orbital wave velocity, which was $U_m = 30$ cm/s at the bed. This streaming can have a significant effect on scour around the pile when the bed is erodible, as will be demonstrated in the next section.

(It may be noted that Sumer and Fredsøe (1997 b, 2001) also measured the undisturbed (i.e., in the absence of the pile) period-averaged flow velocity at these levels under exactly the same wave conditions. Comparison of these undisturbed period-averaged velocities with those measured in the presence of the pile (Figs. 6.5-6.7) indicates that the measured streaming velocities in the neighbourhood of the pile are essentially due to the presence of the pile itself).

Of particular interest is the steady streaming near the bed (Fig. 6.5). It

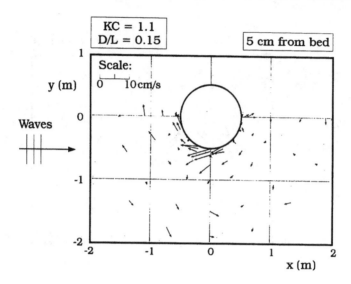

Figure 6.6: Vector diagram of the period-averaged velocities (the plan-view component), the *steady streaming*. Sumer and Fredsøe (2001).

appears that there are three distinct regions in Fig. 6.5, marked A, B and C. In Region A, the streaming is towards the pile, while in Regions B and C, it has a component directed outward. While in Region B the x−component of the streaming is in the direction opposite to the wave propagation, the converse is true for Region C. (The flow pattern in Figs. 6.5-6.7 implies that the streaming must, for continuity reasons, have a 3-D structure with a strong vertical component).

The following features of the streaming in the neighbourhood of the pile in Region B (Fig. 6.5) are noteworthy:

1. The presence of the large non-zero radial velocities in this region suggests that the bed boundary layer is mainly responding to the reflected waves in the radial direction. (Note that no such velocities exist in the absence of the pile).

2. Fig. 6.9 displays the time series of the two velocity components measured at Point P in Region B (Fig. 6.5). The time series for the individual wave cycles are plotted on top of each other for all the wave

6.3. STEADY STREAMING AROUND THE PILE

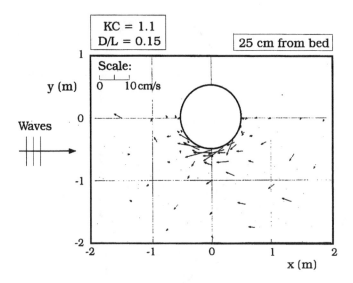

Figure 6.7: Vector diagram of the period-averaged velocities (the plan-view component), the *steady streaming*. Sumer and Fredsøe (2001).

cycles sampled. Fig. 6.9 a shows that the aforementioned radial velocities in the outward direction occur over the phase interval from about $\omega t = 135^0$ to about 360^0. This phase interval coincides largely with the time period during which the water in Region B is "receding" (Fig. 6.10). This motion obviously generates radial velocities near the bed, directed outward.

3. Fig. 6.5 further shows that there is a steady streaming near the pile in the tangential direction opposite to the direction of wave propagation. This is due to the response of the bed boundary layer to the interacting incident and reflected waves. Fig. 6.9 b shows that the negative tangential velocities are experienced over a longer period of time, presumably resulting in the measured non-zero period-averaged tangential velocity in the direction opposite to the wave propagation.

(It is interesting to note that the velocity time series in Fig. 6.9 show a clear cycle-to-cycle variation, as the individual cycles of velocity variation do not collapse on a single curve. This is due to the turbulence generated in

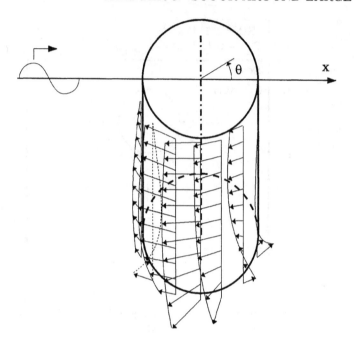

Figure 6.8: 3-D illustration of the period-averaged velocities (the plan-view component), the *steady streaming*.

the boundary layer over the pile surface in Sumer and Fredsøe's experiment. Their measurements show, however, that this behaviour simply disappears completely, and the individual cycles of the velocity variation do collapse on a single curve further away from the pile, for distances larger than $O(50$ cm$)$ from the side edge of the pile).

Sumer and Fredsøe (2001) also report the results of some particle-tracking experiments. The obtained trajectory patterns confirm the streaming picture presented in Fig. 6.5.

It is known that a 2-D steady streaming (Fig. 6.11) emerges when a free circular cylinder is exposed to an oscillatory flow (Schlichting, 1979, p. 428). As seen, the steady streaming in the present case (Fig. 6.5) is completely different from the 2-D case displayed in Fig. 6.11. To see whether or not the inner recirculation cells experienced in the 2-D case (Fig. 6.11) occur in the present case, velocity measurements were made by Sumer and Fredsøe

6.3. STEADY STREAMING AROUND THE PILE

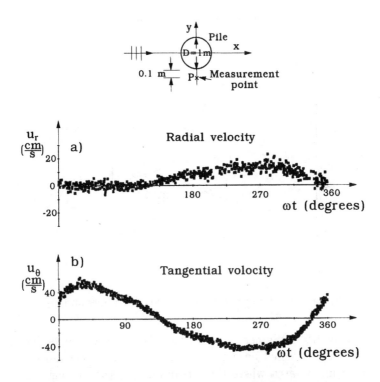

Figure 6.9: Time series of the radial and tangential components of the velocity at Point P. The individual cycles are plotted on top of each other to demonstrate the degree of turbulence. At 0. 4 cm from the bed. Sumer and Fredsøe (2001).

(2001) extremely close to the surface of the pile (at radial distances $O(0.1$ mm)). No such cells were detected. It appears that the entire process of the response of the bed boundary layer in the present pile case overshadows the 2-D streaming which would otherwise be present.

Influence of KC

Fig. 6.12 illustrates the influence of the KC number on the steady streaming. KC is calculated at the bed. The data is for $D/L = 0.15$. In the figure, U_r is the radial component of the steady-streaming velocity at the measurement point with the coordinates $(x;y) = (0;-60$ cm$)$ at the level

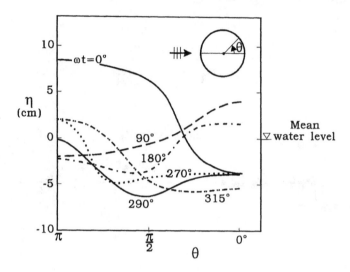

Figure 6.10: Surface elevation profiles around the pile. Sumer and Fredsøe (2001).

0.4 cm from the bed in Sumer and Fredsøe's (2001) experiments. The latter location has been selected on grounds that (1) it has a certain symmetry, and (2) it is in the area where the streaming is rather strong. Hence, it may constitute a good location to study the effect of KC on the streaming process. In Fig. 6.12, the KC number beyond which the waves break, namely $KC = 0.44/(D/L) = 2.9$ (Isaacson, 1979) (Eq. 6.5), is also marked.

Fig. 6.12 shows that the steady streaming, characterized by U_r, increases with increasing KC. Given the pile diameter, the water depth, and the wave period, the larger the wave height, the larger the KC number. On the other hand, the larger the wave height, the larger the steady streaming. Therefore, the streaming should increase with increasing KC.

Influence of D/L

Fig. 6.13 illustrates the influence of D/L, the diffraction parameter, on the steady streaming. The radial velocity is measured at the same point as in the previous figure, and the data is for $KC = 0.4$ (KC being calculated at the bed). The figure shows that the steady streaming increases with increasing D/L. Obviously, the larger the value of D/L, the more pronounced the reflected and diffracted waves, and the more pronounced the response of the

6.4. SCOUR AROUND THE PILE

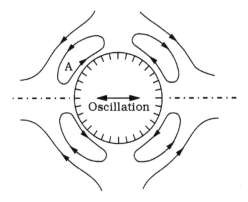

Figure 6.11: Steady streaming around a cylinder subject to a plane oscillatory flow. 2-D case.

bed boundary layer to the reflected and diffracted waves. Since the steady streaming is a consequence of the response of the bed boundary layer to the reflected and diffracted waves, it should increase with increasing D/L.

6.4 Scour around the pile

Although much research has been carried out on wave scour around slender piles over the past decade or so (Wang and Herbich, 1983, Herbich, Schiller, Dunlap and Watanabe, 1984, Eadie and Herbich, 1986, Sumer et al., 1992, 1993, Kobayashi, 1992, and Kobayashi and Oda, 1994; See Chapter 3 for a full account of scour around slender piles), there have been comparatively few investigations of scour around large piles subjected to waves, as already mentioned at the beginning of this chapter. Rance (1980), Katsui and Toue (1988), Toue et al. (1992), Katsui (1992) and Katsui (1993) studied the scour in the laboratory around large vertical piles subjected to waves. On the theoretical side, Saito, Sato and Shibayama (1990), Saito and Shibayama (1992) along with Katsui (1992) and Katsui and Toue (1993) made attempts to simulate the scour process numerically for the case of a circular pile, and Kim, Iwata, Miyaike and Yu (1994) for the case of two circular piles.

Recently, Sumer and Fredsøe (2001; also partly reported in 1997 b) made a study of scour around large vertical circular piles exposed to progressive

302   CHAPTER 6. SCOUR AROUND LARGE PILES

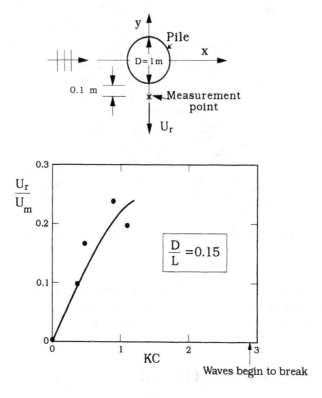

Figure 6.12: Radial component of the steady streaming velocity. At 0.4 cm distance from the bed. Sumer and Fredsøe (2001).

waves by making detailed velocity measurements very close to the bed (as well as away from the bed) in the vicinity of the pile, and subsequently related them to the scour process. The results of Sumer and Fredsøe's velocity measurements are outlined in the previous two sections. The following analysis is essentially based on the scour measurements and analysis presented in Sumer and Fredsøe (2001).

6.4.1 Mechanism of streaming-induced scour

Sumer and Fredsøe (2001) made their scour experiments in the same wave basin as that used in their velocity measurements depicted in Figs. 6.4-6.9

6.4. SCOUR AROUND THE PILE

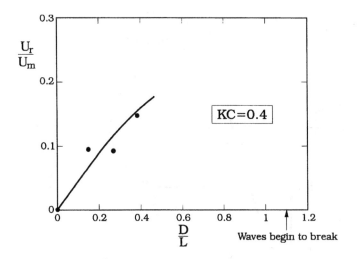

Figure 6.13: Radial component of the steady streaming velocity. At 0.4 cm distance from the bed. Sumer and Fredsøe (2001).

and 6.12-6.13.

Fig. 6.14 displays the bed topography around the pile at the equilibrium stage in one test (after 8 hours' test time). In this particular test, the wave conditions were maintained exactly the same as those in the rigid-bed velocity measurements where the steady-streaming velocity was measured (depicted in Figs. 6.5-6.7, 6.12 and 6.13). Fig. 6.15 gives a 3-D illustration of the bed topography around the pile at the equilibrium stage.

Comparison of Figs. 6.5 and 6.14 suggests that, in the case of the erodible bed (Fig. 6.14), the sediment stirred up by the waves and brought up into suspension is transported by the wave-induced steady streaming near the bed from Region B (Fig. 6.5) outward. This will eventually result in the formation of a scour hole in this area near the pile. The scour picture in Fig. 6.14 reveals this.

(Note that the phase-resolved velocities induced by the waves near the pile can be quite large, as much as a factor of 2 larger than the maximum undisturbed wave velocity, Fig. 6.4, meaning that the temporal value of the Shields parameter will be very large, like $\theta = O(0.4)$, much larger than the undisturbed value of the Shields parameter, which was $\theta = 0.125$ in the test.

304   CHAPTER 6. SCOUR AROUND LARGE PILES

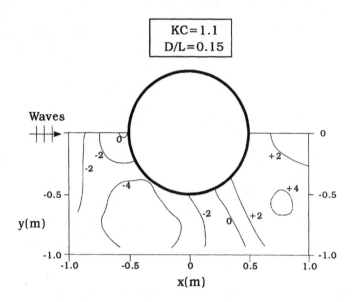

Figure 6.14: Scour/deposition picture after 8 hours of test time (the equilibrium state). Live bed. The wave conditions are the same as those in the rigid-bed tests (Fig. 6.5). The numerical figures indicate scour (-) and deposition (+) in cm. Sumer and Fredsøe (2001).

This implies that the sediment will be stirred up by the waves and brought up into suspension fairly easily. See Eq. 6.15 below for the definition of θ).

It may be noted that the flow picture illustrated in Fig. 6.5 corresponds to the initial, plane-bed situation. Obviously, as the scour/deposition process continues, the interaction between the wave-boundary layer and the outer flow due to the interacting waves will change, therefore the streaming will undergo a constant adjustment. As a matter of fact, the streaming may cease to exist when the scour/deposition process attains its equilibrium state. However, no measurements are yet available, revealing this aspect of the problem.

Fig. 6.14 further shows that, in the "shadow" zone (in Region C, Fig. 6.5), deposition occurs. This can be explained by the steady-streaming picture in Fig. 6.5. The sediment which is transported from Region B towards Region C will end up in this latter region, causing the observed deposition

6.4. SCOUR AROUND THE PILE

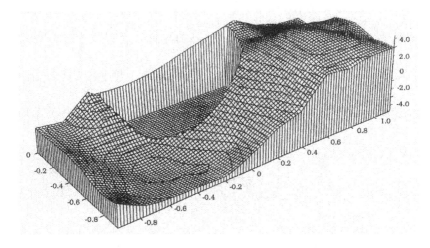

Figure 6.15: 3-D view of Scour / deposition picture in the previous figure.

pattern.

As seen from Fig. 6.14, the scour depth can reach values as high as 4% of the pile diameter. This value is in accord with the scour-depth values reported in Katsui and Toue (Fig. 4, 1993) and Rance (1980) for the streaming-induced scour around a circular pile.

The present result is also in accord with the scour-depth measurements reported in Sumer et al. (1994) and Fredsøe and Sumer (1997) for the streaming-induced scour around the head of a breakwater.

6.4.2 Influence of KC and D/L

From dimensional considerations, the scour characteristics (for example, the maximum scour depth) in the case of the streaming-induced scour around a pile depend mainly on the following parameters

$$\frac{S}{D} = f(KC, \frac{D}{L}, \theta) \tag{6.14}$$

The parameters KC and D/L are related to the process of wave-induced steady streaming, while θ is related to the sediment and its motion caused

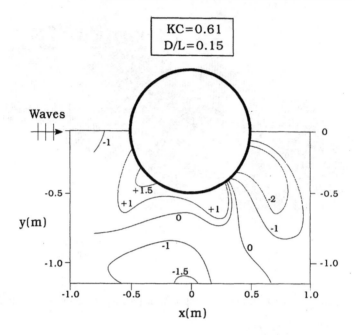

Figure 6.16: Contour plot of bed topography in the equilibrium stage. The numerical figures indicate scour (-) and deposition (+) in cm. Sumer and Fredsøe (2001).

by the phase-resolved component of the wave-induced flow in which θ is the Shields parameter defined by

$$\theta = \frac{U_{fm}^2}{g(s-1)d_{50}} \tag{6.15}$$

Here g is the acceleration due to gravity, s is the specific gravity of the sand grains, d_{50} is the grain size and U_{fm} is the maximum value of the undisturbed bed shear stress velocity, defined by

$$U_{fm} = \sqrt{\frac{f_w}{2}} U_m \tag{6.16}$$

in which U_m is the maximum value of the undisturbed orbital velocity of water particles at the bed, and f_w is the wave friction coefficient. In the

6.4. SCOUR AROUND THE PILE

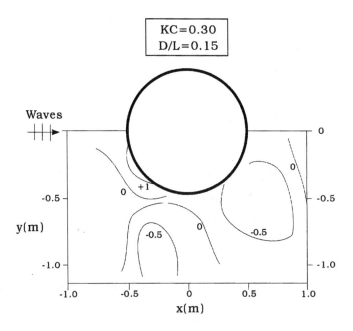

Figure 6.17: Contour plot of bed topography in the equilibrium stage. The numerical figures indicate scour (-) and deposition (+) in cm. Sumer and Fredsøe (2001).

case of a live bed scour ($\theta > \theta_{cr}$), the dependence on θ is expected to be weak. However, if θ is very large, the sediment (stirred up by the waves) will be brought up into suspension at much higher elevations (see Sumer and Fredsøe, 2000, for a discussion on mode of sand transport). In this case, some influence of the parameter θ may be expected. This subsection focuses on the variations with KC and D/L.

Figs. 6.16 and 6.17 present the bed topography around the pile for two other KC numbers, $KC = 0.61$, and 0.30 (KC calculated at the bed). The parameter D/L is maintained constant, at $D/L = 0.15$.

As seen from Figs. 6.14, 6.16 and 6.17, two changes occur, as the KC number decreases from 1.1 to 0.3: (1) the scour area extends towards the shoreward side of the pile, replacing the deposition area on the onshore side; and (2) deposition begins to occur in front and at the side of the pile. Furthermore, the scour depth decreases substantially with decreasing KC.

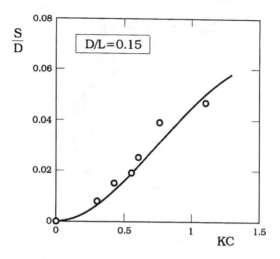

Figure 6.18: Maximum scour depth versus D/L. Live bed. Sumer and Fredsøe (2001).

Fig. 6.18 depicts the maximum scour depth obtained in Sumer and Fredsøe's (2001) experiment plotted against KC (for $D/L = 0.15$). The figure indicates that there is a good correlation between the scour depth and the KC number. The larger the KC number, the larger the scour depth. This is related to the steady streaming; the larger the KC number, the larger the steady streaming (Section 6.3), and therefore the larger the scour depth.

Fig. 6.19 illustrates the influence of the diffraction parameter D/L on the maximum scour depth. The data is for $KC = 0.4$. The figure shows that the scour depth increases with increasing D/L. This is, again, related to the steady streaming; the larger the value of D/L, the larger the steady streaming (Section 6.3); hence the scour should increase with increasing D/L. However, it can be expected that there may be a limit to the increase in S/D for large values of D/L (see the discussion in the next subsection).

Scour at the periphery of the pile base. Fig. 6.20 displays the variation of the scour/deposition depth at the periphery of the pile base for the previously mentioned three KC numbers (Figs. 6.14, 6.16 and 6.17). Fig. 6.20 shows that the maximum scour depth moves towards the shoreward side of the pile. Fig. 6.20 further shows that the scour depth decreases, as the KC number decreases, in agreement with Fig. 6.18. Note that the maximum

6.4. SCOUR AROUND THE PILE

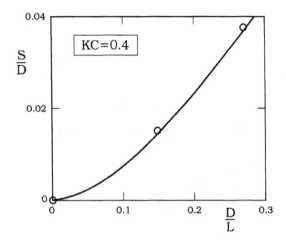

Figure 6.19: Maximum scour depth versus D/L. Live bed. Sumer and Fredsøe (2001).

scour depth at the periphery is somewhat smaller than that measured over the scour-hole area (cf. Fig. 6.18 and Fig. 6.20).

The maximum scour depth at the periphery of the pile base is plotted in Fig. 6.21. Also plotted in Fig. 6.21 is the maximum scour-depth data obtained in Sumer et al.'s (1992) small-pile ($D/L \to 0$) experiments (Fig. 3.35 in Chapter 3).

Regarding this latter case, (1) the maximum scour depth is always at the periphery of the pile; (2) The scour depth is mainly governed by the Keulegan-Carpenter number, KC; and (3) For the variation of the scour depth with KC, Sumer et al. (1992) give the following empirical expression: $S/D = 1.3(1 - \exp(-0.03(KC - 6)))$; $KC > 6$, $D/L \to 0$ (Eq. 3.20). This expression is also plotted in Fig. 6.21.

There are three different regions in the diagram in Fig. 6.21:

1. For $KC < O(1)$, the scour is caused by the steady streaming, as detailed in the preceding paragraphs.

2. For $KC > 6$, the scour is caused by the vortex structures around the pile, as described in Chapter 3 (mainly by the vortex-shedding

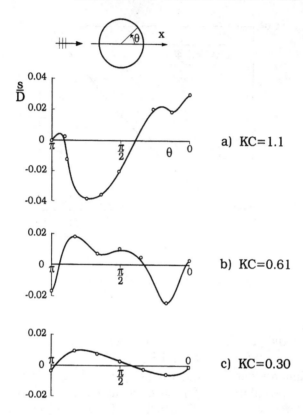

Figure 6.20: Scour at the perimeter of the pile. $D/L = 0.15$. Sumer and Fredsøe (2001).

mechanism when $KC \lesssim 30$, and by the horseshoe-vortex mechanism when $KC \gtrsim 30$).

3. For $O(1) < KC < 6$, it is expected that the separation vortices may have an influence on the steady streaming, and therefore on the scour.

As far as the streaming-induced scour is concerned, S/D increases with KC and D/L. However, although the increase in the scour depth is very clear when D/L is increased from 0.08 to 0.15, it seems that the data collapses on a single curve for further increase in D/L. Although no clear explanation has been found for this behaviour, Sumer and Fredsøe (2001) link this to

6.4. SCOUR AROUND THE PILE

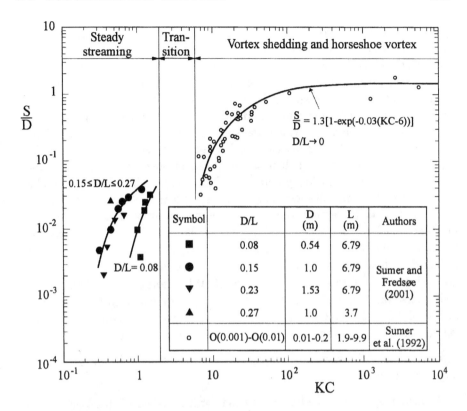

Figure 6.21: Maximum scour depth at the *periphery* of the pile base. Live bed. Sumer and Fredsøe (2001).

the decrease in the phase-resolved velocity for very large values of D/L (The latter can readily be seen from MacCamy and Fuchs', 1954, solution, see Section 6.2 above). When the phase-resolved velocity decreases (for very large values of D/L), the sand will not be brought as far away from the bed as it is for relatively smaller D/L, and therefore it will not be exposed to the same streaming as in the case of smaller D/L; this may, in turn, influence the end scour.

Remarks on scale effects. There may be several scale effects: The wave boundary layer on the bed may be different in the field (the Re number effect (Re here being the Reynolds number for the wave boundary layer,

$U_m a/\nu$) and/or the effect of bed roughness); the Shields parameter θ may be so small that the scour will not be in the live-bed regime (the effect of θ, the clear-water scour); or conversely, θ may be so large that the mode of the phase-resolved sediment transport may be in the suspension regime (the effect of mode of sand transport). The precise implications of these effects are unknown today. Therefore Sumer and Fredsøe (2001) urge that caution must be exercised when implementing the results of their study for field conditions. Sumer and Fredsøe (2001) discuss another issue; namely, only one sand size ($d_{50} = 0.2$ mm) was used in their study. As mentioned previously, the sand size appears in the Shields parameter (Eq. 6.15). Therefore the Shields parameter rather than the actual sand size is important when the sand size is considered. It is interesting to note that the maximum scour depth obtained by Rance (1980) (i.e., $S/D = 0.032$) for a similar set of wave conditions and under live-bed conditions, but with an entirely different sediment, namely Bakelite with a specific gravity of the sediment grains 1.4 and a median grain size of 0.39 mm $< d_{50} < 0.83$ mm (R. Whitehouse of HR Wallingford, 2000, personal communication) is not radically different from that obtained in Fig. 6.21, $S/D = 0.038$. Also, the bed topographies obtained in the two experiments are quite similar (cf. Fig. 6.17 and Fig. 6.25 a).

Example 17 *A numerical example for the prediction of scour depth at a large structure*

Consider a 70 m diameter gravity platform installed offshore on a sand bed with the sand size $d_{50} = 0.2$ mm. Predict the scour depth when this platform is exposed to waves with a period of $T_w = 15$ s and a wave height of $H = 20$ m. The water depth is $h = 104$ m.

1. Calculate the quantity L_0 (the deep-water wave length)

$$L_0 = \frac{gT_w^2}{2\pi} = \frac{9.81 \times 15^2}{2\pi} = 351 \text{ m}$$

2. Calculate the parameter h/L_0

$$\frac{h}{L_0} = \frac{104}{351} = 0.3$$

3. From the wave tables for sinusoidal waves

$$\sinh(kh) = 3.48 \text{ for } \frac{h}{L_0} = 0.3$$

6.4. SCOUR AROUND THE PILE

4. Calculate the amplitude of the orbital motion of water particles at the seabed, assuming that the small-amplitude sinusoidal wave theory is applicable (Appendix A)

$$a = \frac{H}{2}\frac{cosh(k(z+h))}{sinh(kh)} = \frac{H}{2}\frac{cosh(k(-h+h))}{sinh(kh)} = \frac{H}{2}\frac{1}{sin(kh)} = \frac{20}{2}\frac{1}{3.48}$$
$$= 2.9 \text{ m}$$

in which z = the vertical distance measured from the mean water level which is put equal to $z = -h$ (the seabed). The maximum value of the velocity at the bed is (Appendix A)

$$U_m = \frac{\pi H}{T_w}\frac{cosh(k(z+h))}{sinh(kh)} = \frac{\pi \times 20}{15}\frac{1}{3.48} = 1.2 \text{ m/s}$$

5. Check if the sinusoidal theory is applicable

$$U \text{ (the Ursell parameter)} = \frac{HL^2}{h^3} < 15$$

in which L, the wave length, is $L = L_0 \; tanh(kh)$. (Otherwise, use the cnoidal theory). From the wave tables

$$tanh(kh) = 0.961 \text{ for } \frac{h}{L_0} = 0.3$$

Therefore

$$L = 351 \times 0.961 = 337 \text{ m}$$

and then

$$U = \frac{HL^2}{h^3} = \frac{20 \times 337^2}{104^3} = 2$$

which is smaller than 15. Therefore we may assume that the sinusoidal theory is applicable.

6. Calculate the Keulegan-Carpenter number at the seabed

$$KC = \frac{2\pi a}{D} = \frac{2 \times \pi \times 2.9}{70} = 0.3$$

7. Calculate the diffraction parameter

$$\frac{D}{L} = \frac{70}{337} = 0.21$$

8. Predict the scour depth from Fig. 6.21, for $KC = 0.3$ and $D/L = 0.21$,

$$\frac{S}{D} = 0.005, \text{ or } S = 0.005 \times 70 = 0.35 \text{ m} \qquad (6.17)$$

9. Calculate the Shields parameter

$$\theta = \frac{U_{fm}^2}{g(s-1)d} = \frac{\frac{f_w}{2}U_m^2}{g(s-1)d} = \frac{\frac{0.003}{2} \times 1.2^2}{9.81 \times (2.65-1) \times 0.0002} = 0.67$$

where f_w is calculated from

$$f_w = 0.035 RE^{-0.16}$$

as 0.003 (Fredsøe and Deigaard, 1992, p. 29), assuming that the bed is acting as a smooth wall ($dU_{fm}/\nu \lesssim 10$). Here, $RE = aU_m/\nu$, the wave-boundary-layer Reynolds number. As seen, the Shields parameter, $\theta = 0.67$, is larger than the critical value, $\theta_{cr} \simeq O(0.05)$, i.e., the bed is live, therefore the graph used to calculate the scour depth (Fig. 6.21) is valid.

Example 18 *Time scale of scour around large piles*

Sumer and Fredsøe (2002), in a separate study, investigated the time scale of scour at large piles. Fig. 6.22 illustrates how the scour depth develops with time.

As seen from Fig. 6.22, the time development of the scour depth can be represented by the same kind of exponential relation as in Eq. 1.2 where T is the time scale during which substantial scour develops (Fig. 1.1, Chapter 1). This time scale can be predicted from the scour-depth-versus-time information (e.g., Fig. 6.22), for example, by calculating the slope of the line tangent to the $S_t(t)$ curve at $t = 0$ (as described in Chapter 2, Section 2.3.4).

Five tests were carried out, and the time scales from the $S_t(t)$ curves were obtained using the previously described method. The results are summarized in the following table:

$D(m)$	KC	θ	D/L	T(min)	T^*
1.00	0.30	0.035	0.15	26.2	0.018
1.00	0.61	0.072	0.15	32.0	0.022
0.54	0.55	0.035	0.08	14.5	0.034
0.54	1.13	0.072	0.08	12.7	0.029
0.40	1.75	0.082	0.06	5.6	0.024
1.00	1.1	0.126	0.15	31.9	0.022

6.4. SCOUR AROUND THE PILE

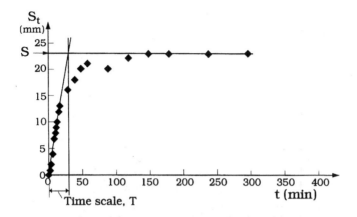

Figure 6.22: Time development of the scour depth. $KC = 0.61$, $D = 1$ m. At the point where the scour is maximum. (see Fig. 6.16).

In the table, KC is calculated at the bed (Eq. 6.6), and the Shields parameter is calculated in the same way as in Eq. 6.15. T^* is the familiar normalized time scale, Eq. 2.44. In the experiments, the sand size was $d_{50} = 0.2$ mm, and the wave period was $T = 3.5$ s with the water depth 40 cm.

This data is plotted in Fig. 6.23. The figure shows that the time scale decreases with increasing Shields parameter θ, consistent with the result obtained in the case of slender piles (Chapter 3, Section 3.3). T^* decreases with θ, because the larger the Shields parameter, the larger the sediment transport, and therefore the smaller the time period during which substantial scour develops. Fig. 6.23 further shows that, given $D/L = 0.06 - 0.15$, the time scale increases with increasing KC for a given value of θ. This is related to the fact that the volume of sediment to be scoured obviously increases with increasing KC (Fig. 6.18). Therefore, the time scale should increase with KC.

The range of D/L for the data is too narrow to reveal any pattern as regards the variation of time scale with this parameter. However, it may be expected that, given the θ value and the KC number, the time scale may increase with increasing D/L. This is because, again, the volume of sediment to be scoured will increase with increasing D/L (Fig. 6.19), and therefore the time scale should increase with D/L.

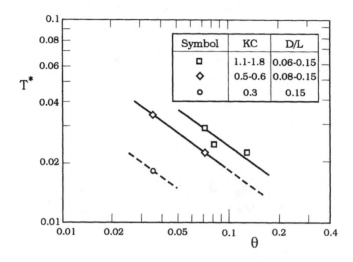

Figure 6.23: Normalized time scale for scour around a large pile.

6.4.3 Influence of combined waves and current, and cross-sectional shape

The influence of combined waves and current and cross-sectional shape can be described by reference to Figs. 6.25-6.29, which are reproduced from Rance (1980). Rance reports that the pile was placed on a bed of Bakelite (with a specific gravity 1.4 and a median grain size of 0.39 mm $< d_{50} <$ 0.83 mm, R. Whitehouse of HR Wallingford, 2000, personal communication) and subjected to waves of sufficiently large amplitude to make sure that the whole bed was in motion. The pile-diameter-to-wave-length ratio was $D/L = 0.2$. The bed topographies depicted in the figures correspond to the equilibrium stage. The top diagrams in Figs. 6.25-6.29 are obtained for waves alone, while the bottom diagrams are obtained for combined waves and current, as indicated in the figures. The current in the latter case was in the same direction as the wave propagation, and had a magnitude of 40% of the maximum oscillatory velocity. The symbol D in Figs. 6.25-6.29 is the equivalent diameter, namely the diameter of a circular pile having the same cross-sectional area as the angular pile. The equivalent diameters of the piles used in Rance (1980) experiments are depicted in Fig. 6.24.

From Fig. 6.25, the overall pattern of scour and deposition remains much

6.4. SCOUR AROUND THE PILE

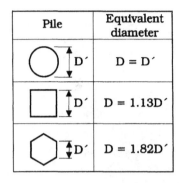

Figure 6.24: Equivalent diameter of pile in Rance's (1980) study.

the same, but, in the case of the combined waves and current, the maximum scour depth increases by a factor of 2. The lateral extent of scour increases too.

From Figs. 6.26-6.29, it is seen that the maximum scour depth always increases by a factor of 2 when the current is superimposed, irrespective of the cross-sectional shape of the pile except in the case of the hexagonal pile with leading corner. In the latter case (Fig. 6.26), the increase is only by about 20%. However, note that the end result must be a function of the velocity ratio, U_c/U_m where U_c is the current velocity.

Comparison of Figs. 6.25-6.27 shows that, by and large, the picture of scour and deposition around a hexagonal pile is not so very different from that around a circular pile.

Finally, scour around a square pile is rather more severe (Figs. 6.28-and 6.29).

Rance (1980) notes that the variability of current and wave direction will mitigate the scouring effect, adding that it is to be expected that there would be no net deposition or scour in the long term; from storm to storm the mean direction of the waves will change and with it the location of the areas of deposition and scour, meaning that previously eroded areas will be filled in.

However, it must be stated that the problems regarding large piles are those due to transient scour, the maximum amount of scour that will take place in any given storm.

Example 19 *Observations of scour around offshore gravity structures and*

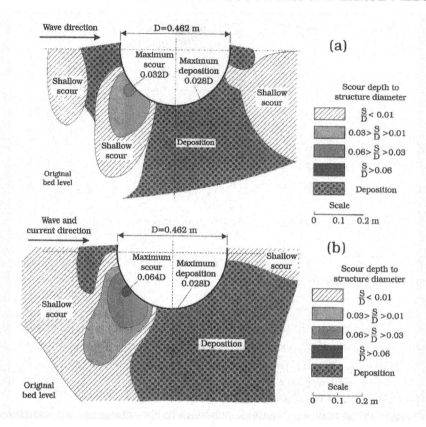

Figure 6.25: Large circular pile. (a) Wave; (b) Combined wave and current with $U_c/U_m = 0.4$. Rance (1980).

protection (Dahlberg, 1983)

Offshore gravity structures are installed as production structures (for the production of oil from proven reserves). In contrast to pile-supported offshore platforms, the gravity platforms remain on the site by virtue of their massive weight, bearing against the sea bottom. Because of their large volumes, these types of platforms are used as oil storage structures as well. They are built in waters up to 150 m depth. Fig. 6.30 gives some examples of the shape and dimensions of foundation base of such structures installed in the North Sea.

6.4. SCOUR AROUND THE PILE

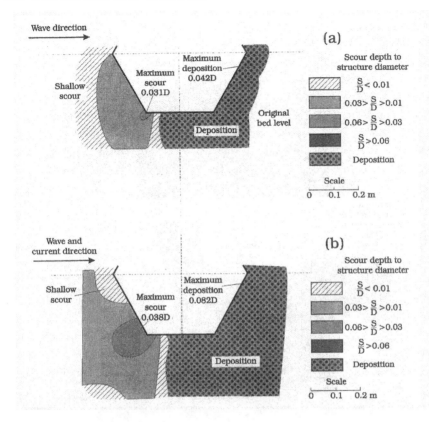

Figure 6.26: Large hexagonal pile with leading corner. (a) Wave; (b) Combined wave and current with $U_c/U_m = 0.4$. Rance (1980).

As seen from Fig. 6.30, the size of the structure is typically very large, and therefore scour occurs in the diffraction regime. Dahlberg (1983) compiled observational data on scour around 17 such structures. The following paragraphs will give the highlights of the latter work.

Frigg TP1 platform (Fig. 6.30 a). The water depth is 104 m, and the soil conditions close to the sea bottom are: dense fine sand over clay. Scour took place at two corners of the platform with a scour depth of about 2 m, mainly during the summer months of 1976 (cf. Figs. 6.28 and 6.29). Scour protection by gravel bags and gravel fill (placed by divers in the spring of 1977) effectively stopped further development of scour.

320   CHAPTER 6. SCOUR AROUND LARGE PILES

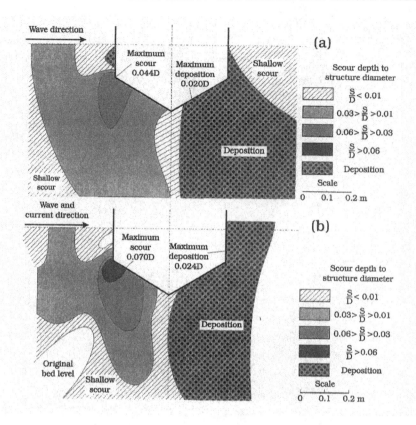

Figure 6.27: Large hexagonal pile with leading face. (a) Wave; (b) Combined wave and current with $U_c/U_m = 0.4$. Rance (1980).

Dahlberg (1983) notes that because nearby Frigg TCP2 (only 40 m away) has experienced no scour, it has been concluded that the square base of TP1 is more "sensitive" to scour than the approximately circular base of TCP2 (cf. Fig. 6.25 and Figs. 6.28-6.29).

It may be noted that, for a similar set of conditions, the scour depth for a circular shaped platform, the scour depth is estimated to be $S = 0.35$ m in Example 17 in Section 6.4.2 (Eq. 6.17). Comparison of Fig. 6.24, Fig. 6.25 a and Fig 6.29 a indicates that the latter estimate should be increased by a factor of $(0.082/0.032) \times 1.13 \approx 3$ to account for the cross-sectional shape of the Frigg TP1 platform (i.e., the square shape). Hence, an estimate of

6.4. SCOUR AROUND THE PILE

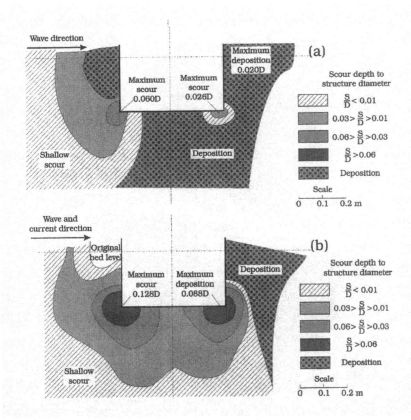

Figure 6.28: Large square pile with leading face. (a) Wave; (b) Combined wave and current with $U_c/U_m = 0.4$. Rance (1980).

the scour depth for the Frigg TP1 platform may be given as $S = 3 \times 0.35$ m $\approx O(1 \text{ m})$. This latter figure is not radically different from the observed value, 2 m.

Brent D platform (Fig. 6.30 g). The water depth is 140 m, and the soil conditions close to the sea bottom are: silty fine sand over stiff clay with sand layers. Along the periphery of the base of the platform, a few local scour holes were observed (the deepest hole having a depth of about 0.5 m), but otherwise no scour was present. The deepest hole was filled with sand bags.

Brent B platform (Fig. 6.30 g). The water depth is 140 m, and the soil

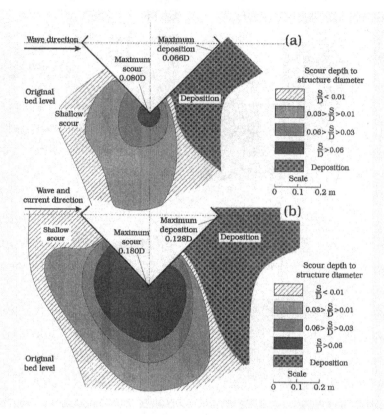

Figure 6.29: Large square pile with leading corner. (a) Wave; (b) Combined wave and current with $U_c/U_m = 0.4$. Rance (1980).

conditions close to the sea bottom are: thin layer of fine sand over stiff clay with sand layers. No protection against scour. No scour was observed. It is interesting to note that a calculation exercise similar to Example Example 17 in Section 6.4.2 for this platform gives practically zero scour depth, consistent with the previously mentioned observation. This is simply because of a significant decrease in the Keulegan-Carpenter number (Fig. 6.21), mainly due to the increase in the water depth (the water depth is in the present case 40 m larger than that in the case of the Brent D platform); even with a platform size of 70 m, the scour will be practically nil.

Dahlberg (1983) reports that, for gravity platforms in the North Sea, the

6.4. SCOUR AROUND THE PILE

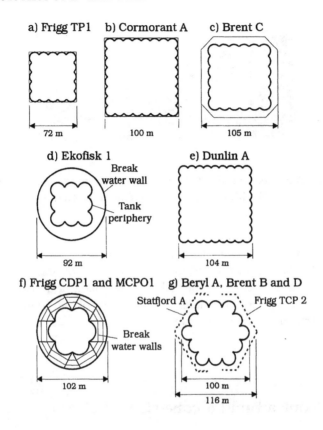

Figure 6.30: Shape of foundation base for gravity structures. Dahlberg (1983).

following types of scour protection have been used:

1. Perforated breakwater wall (used for Ekofisk 1, Frigg CDP1, Frigg MCPO1 and Ninian platforms), (Zaleski-Zamenhof, 1976).
2. Extended base slab.
3. Well-graded gravel over a width of a minimum of 10% of the base diameter along the periphery.
4. Sand, gravel, or concrete bags.
5. Skirts. These may be vertical steel "sheets" installed to the foundation of the structure, as sketched in Fig. 6.31. The idea is

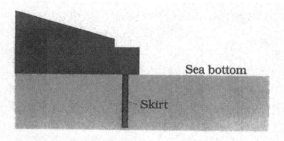

Figure 6.31: Skirt used to prevent the rocking motion of the platform which may lead to liquefaction. The latter may result in removal of sand, leading to scour.

to prevent the structure from rocking, and therefore to avoid the liquefaction of the soil due to this cyclic rocking motion. Without these elements, the soil may be liquefied, and therefore may be easily removed along the periphery of the foundation of the structure, leading to scour.

6.4.4 Scour around a cone-shaped object

Fredsøe and Sumer (1997) measured scour around a cone-shaped structure. Fredsøe and Sumer's (1997) results are presented in Fig. 6.32. The objective of these scour tests was to study the influence of side slopes in conjunction with scour processes at the head of rubble-mound breakwaters. As will be seen in Chapter 7, one of the mechanisms related to the scour in this latter case is the steady streaming. To observe the similar effect (i.e., the steady streaming) in the cone tests, Fredsøe and Sumer (1997) used a relatively large cone (86 cm in diameter at the base). The results revealed the presence of the streaming. Fig. 6.32 shows that scour decreases with decreasing side slopes. The details of the measurements and the results of a parallel study of the bottom shear stress are given in the original publication, Fredsøe and Sumer (1997). See also Section 3.2.3 in Chapter 3 for further discussion on the bottom shear stress around a cone-shaped object.

6.5. REFERENCES

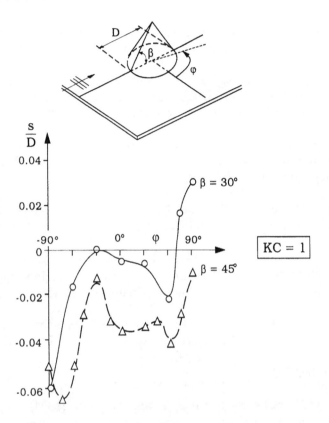

Figure 6.32: Depth of scour around the periphery of a cone-shaped structure. $KC = 1$. Fredsøe and Sumer (1997).

6.5 References

1. Abramowitz, M. and Stegun, I.A. (1965): Handbook of Mathematical Functions. Dover Publications, Inc., New York.

2. Dahlberg, R. (1983): Observations of scour around offshore structures. Can. Geotech. J., vol. 20, 617-628.

3. Eadie, R.W. IV, and Herbich, J.B. (1986): Scour about a single, cylindrical pile due to combined random waves and current. Proc. 20th Coastal Engineering Conference, Taipei, Taiwan, ASCE, New York, N.Y., vol.. 3, 1858-1870.

4. Fredsøe, J. and Sumer, B.M. (1997): Scour at the round head of a rubble-mound breakwater. Coastal Engineering, vol. 29, 231-262.

5. Herbich, J.B., Schiller, R.E., Jr., Watanabe, R.K. and Dunlap, W.A. (1984): Seafloor Scour. Design Guidelines for Ocean-Founded Structures, Marcell Dekker, Inc., New York, NY, xiv + 320 p.

6. Isaacson, M. (1979): Wave-induced forces in the diffraction regime. In: Mechanics of Wave-Induced Forces on Cylinders, (Ed. T.L. Shaw). Pitman Advanced Publishing Program, 68-89.

7. Katsui, H. (1992): Study on scouring and scour protection around offshore structures. Doctoral Dissertation, University of Tokyo, 382 p.

8. Katsui, H. and Toue, T. (1988): Inception of sand motion around a large obstacle. Proc. 21st International Coastal Engineering Conference, ASCE, Costa del Sol- Malaga, Spain, vol.. 2, 1280-1294.

9. Katsui, H. and Toue, T. (1993): Methodology of estimation of scouring around large-scale offshore structures. Proc. Third International Offshore and Polar Engineering Conference, Singapore, 6-11 June 1993, vol. I, 599-602.

10. Kim, C.J., Iwata, K., Miyaike, Y. and Yu, H.-S. (1994): Topographical change around multiple large cylindrical structures under wave actions. Proc. 24th International Coastal Engineering Conference, ASCE, Kobe, Japan, vol.. 2, 1212-1226.

11. Kobayashi, T. (1992): Three-Dimensional analysis of the flow around a vertical cylinder on a scoured seabed. Proc. 23rd International Conference on Coastal Engineering, Venice, Italy, vol. 3, Chapter 99, 3482-3495.

12. Kobayashi, T. and Oda, K. (1994): Experimental study on developing process of local scour around a vertical cylinder. Proc. 24th International Conference on Coastal Engineering, Kobe, Japan, vol. 2, Chapter 93, 1284-1297.

13. MacCamy, R.C. and Fuchs, R.A. (1954): Wave forces on piles: A diffraction theory. U.S. Army Corps of Engineers, Beach Erosion Board, Tech. Memo No. 69, 17 p.

14. Mathsoft (1997): Mathcad 7. User's Guide. Mathsoft Inc., Cambridge, MA.

15. Rance, P.C. (1980): The potential for scour around large objects. In: Scour Prevention Techniques Around Offshore Structures. Society for Underwater Technology, London, 41-53.

16. Saito, E., Sato, S. and Shibayama, T. (1990): Local scour around a large circular cylinder due to wave action. Proc. 22nd International Coastal Engineering Conference, ASCE, Delft, the Netherlands, vol. 2, 1795-1804.

17. Saito, E. and Shibayama, T. (1992): Local scour around a large circular cylinder on the uniform bottom slope due to waves and current. Proc. 23rd International Coastal Engineering Conference, Venice, Italy, ASCE, vol. 3, 2799-2810.

18. Sarpkaya, T. and Isaacson, M. (1981): Mechanics of Wave Forces on Offshore Structures. Van Nostrand, Reinhold, xv + 651 p.

19. Schlichting, H. (1979): Boundary-Layer Theory. 7th Edition, McGraw-Hill Co.

20. Sumer, B.M. and Fredsøe, J. (1997 a): Hydrodynamics Around Cylindrical Structures. World Scientific, xviii + 530 p.

21. Sumer, B.M. and Fredsøe, J. (1997 b): Scour around a large vertical circular cylinder in waves. Proc. 16th International Conference on Offshore Mechanics and Arctic Engineering, 13-18. April, 1997, Yokohama, Japan OMAE 1997, ASME, volume I-A, 57-64.

22. Sumer, B.M. and Fredsøe, J. (2000): Experimental study of 2D scour and its protection at a rubble-mound breakwater, Coastal Engineering, vol.. 40, Issue 1, 59-87.

23. Sumer, B.M. and Fredsøe, J. (2001): "Wave scour around a large vertical circular cylinder". J. Waterway, Port, Coastal and Ocean Engineering, ASCE, vol. 127, No. 3, 125-134.

24. Sumer, B.M. and Fredsøe, J. (2002): Time scale of scour around a large vertical cylinder in waves. Manuscript in preparation.

25. Sumer, B.M., Fredsøe, J. and Christiansen, N. (1992): Scour around a vertical pile in waves. J. Waterway, Port, Coastal and Ocean Engineering, ASCE, vol.. 118, No. 1, 15-31.

26. Sumer, B.M., Christiansen, N. and Fredsøe, J. (1993): Influence of cross section on wave scour around piles. J. Waterway, Port, Coastal and Ocean Engineering, ASCE, vol. 119, No. 5, 477-495.

27. Sumer, B.M., Fredsøe, J., Christiansen, N. and Hansen, S.B. (1994): Bed shear stress and scour around coastal structures. Proc. 24th International Coastal Engineering Conference, ASCE, Kobe, Japan, vol.. 2, 1595-1609.

28. Toue, T., Katsui, H. and Nadaoka, K. (1992): Mechanism of sediment transport around a large circular cylinder. Proc. 23rd International Coastal Engineering Conference, Venice, Italy, ASCE, vol.. 3, 2867-2878.

29. Wang, R.-K. and Herbich, J.B. (1983): Combined current and wave-produced scour around a single pile. COE Rep. No. 269, Texas A&M University.

30. Zaleski-Zamenhof, L.C. (1976): Antiscour protection by means of perforated wall. Proc. Behaviour of Offshore Structures, BOSS '76, Trondheim, vol. 2, 553-555.

Chapter 7
Scour around breakwaters

A breakwater is a structure protecting an area from wave action. In the case of shore-connected breakwaters, the protected area may be a shore area, a harbour, or a basin, while in the case of offshore breakwaters, it will be an area located offshore.

Scour is one of the failure modes of breakwaters.

Soundings reveal large potential risks. For example, it is not unusual to observe scour holes with depths larger than 10 m at the head of a breakwater at an exposed location (Fredsøe and Sumer, 1997).

Breakwater failures (for vertical-wall breakwaters) due to scour have been reported in a review paper by Oumeraci (1994 a). A report from the US Army Corps of Engineers (Lillycrop and Hughes, 1993) gives numerous examples of scour hole formations at the tip and along the sides of rubble-mound breakwaters and jetties, and the resulting failures monitored in twenty-one case studies in the U.S. The latter report also gives the costs involved for repair of the failed parts of the structures where the failure was due to scouring. This cost typically runs at the level of $ 2-10 million.

Fig. 7.1 displays two field measurements from the Danish West Coast (Fredsøe and Sumer, 1997); Fig. 7.1 a presents a three-dimensional picture of a scour formation monitored near the head of a groin, Groin Nr. 57, located in Thyborøn, while Fig 7.1 b presents a contour plot of bed topography monitored near the head of the Hirtshals harbour breakwater. Although the precise environmental conditions which generated these scour formations are not known, the general pattern of scour displayed in Fig. 7.1 resembles rather well that observed in the laboratory investigations, as will be seen later in the chapter.

330   CHAPTER 7. SCOUR AROUND BREAKWATERS

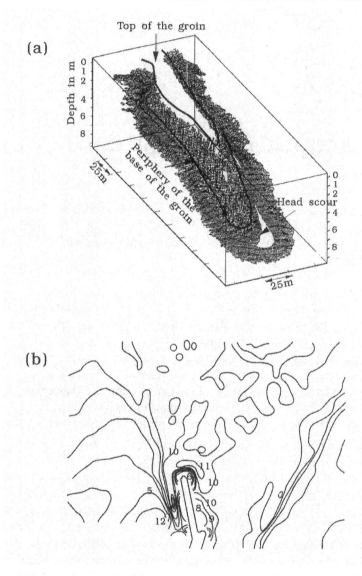

Figure 7.1: (a) Scour formation near the head of Groin No. 57, located in Thyborøn on the Danish West Coast (by courtesy of John Jensen, Danish Coastal Authority). (b) Bed topography near the head of Hirtshals harbour breakwater, on the Danish West Coast; numerical figures indicate the depth in metres (by courtesy of Hans Kjær, State Authority, Denmark). Fredsøe and Sumer (1997).

Figure 7.2: Damage to breakwater. Barels et al. (2000).

Gunbak, Gokce and Guler (1990) report damage to the "bend" section of the main rubble-mound breakwater of Samandag Fishing Port near Iskenderun, Turkey, apparently due to extensive scour at the toe; the scour was measured to be 4 to 5 m.

Fig. 7.2 displays the damage to a section of the main breakwater of a small-craft harbour near Cape Town, South Africa. This also is due to extensive scour at the toe of the breakwater (Bartels, Kloos and Phelp, 2000).

Fig. 7.3 from the same study gives the results of two surveys, one in June 1989 just after the completion of the breakwater construction, and the other in April 1994. An example of scour at the toe (of about 4 m) is seen in this figure.

Fig. 7.4, on the other hand, displays the results of photographic analysis of the top row of Accropode units that are used for the side wall of the breakwater in the study of Bartels et al. (2000). The subsidence of the top row of Accropodes (apparently due to scour) is clear from the figure.

Silvester and Hsu (1997) cite field evidence of scour from Europe, Japan, United States and Africa.

The present chapter will study the scour at breakwaters.

There are two kinds of scour processes: one is the scour process in front of the structure along the length of the trunk section (A, in Fig. 7.5) and the other is that around the head of the breakwater (B, in Fig. 7.5). When the waves attack at right angles to the breakwater, the scour in front of the

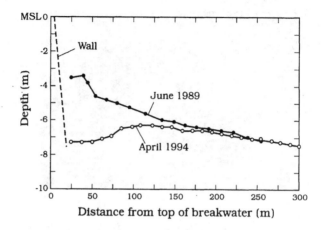

Figure 7.3: Section seaward of breakwater. Bartels et al. (2000).

breakwater will be a two-dimensional process. The head scour, on the other hand, is always a three-dimensional process.

We shall start off with the scour at the trunk section of a breakwater. Both the vertical-wall breakwater and the rubble-mound breakwater will be covered. This will be followed by the scour at the head section. Similar to the trunk section scour, both cases (namely, the vertical-wall and rubble-mound breakwaters) will be covered in the latter analysis.

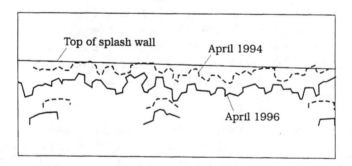

Figure 7.4: Photographic analysis of top row of accropode. Bartels et al. (2000).

7.1. SCOUR AT THE TRUNK SECTION OF A BREAKWATER

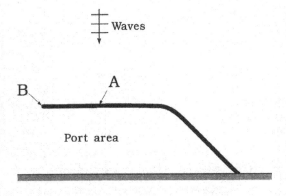

Figure 7.5: Definition sketch. A: Trunk section. B: Head.

7.1 Scour at the trunk section of a breakwater

7.1.1 Scour at the trunk section of a vertical-wall breakwater

Description of the scour process

Consider a vertical, rigid wall (Fig. 7.6), subjected to a progressive wave (the incident wave). As the incident wave impinges on the wall, a reflected wave moves in the offshore direction. The superposition of these two waves results in a standing wave. The surface elevation of this standing wave (Fig. 7.6) is given by

$$\eta(x,t) = \frac{(2H)}{2} \cos(kx) \cos(\omega t) \tag{7.1}$$

$$\omega^2 = gk \tanh(kh) \tag{7.2}$$

in which ω is the angular frequency ($\omega = 2\pi/T_w$), k is the wave number ($k = 2\pi/L$), T_w is the wave period, L is the wave length, and h is the water depth. The quantity $2H$ is the height of the standing wave, and is twice the wave height (H) of each of the two progressive waves forming the standing wave (Dean and Dalrymple, 1984, Section 4).

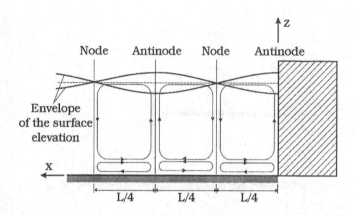

Figure 7.6: Steady streaming in the vertical plane in front of a breakwater.

Now, this standing wave apparently generates a field of steady streaming, a system of recirculating cells (consisting of bottom and top cells), as illustrated in Fig. 7.6 (Carter, Liu and Mei, 1973; see also Mei, 1989). The formation of the bottom cells is related to the boundary layer over the bed.

The sediment on the bed essentially responds to these recirculating cells. If the sand size is relatively small, the sand stirred up by the waves (by the phase-resolved component of the wave-induced flow) and brought up into suspension will be carried to higher elevations, and therefore it will respond mostly to the top cells. The end result will then be the scour-and-deposition pattern sketched in Fig. 7.7 a (as demonstrated experimentally by Xie, 1981). If the sand size is relatively coarse, on the other hand, the sand will be transported in the no-suspension regime, and therefore will respond mostly to the bottom cells, and the end result will be the scour-and-deposition pattern shown in Fig. 7.7 b (Xie, 1981).

As seen from Fig. 7.7, the scour/deposition pattern in front of a vertical-wall breakwater emerges in the form of alternating scour and deposition areas lying parallel to the breakwater. Carter, Liu and Mei (1973) were the first to recognize the scour and deposition process in a field of standing waves in conjunction with the development of sand bars parallel to the shore line. The 2-D scour in front of a vertical-wall breakwater has been investigated by de Best, Bijker, Wichers (1971), Xie (1981, 1985), Irie and Nadaoka (1984) and Hughes and Fowler (1991).

7.1. SCOUR AT THE TRUNK SECTION OF A BREAKWATER 335

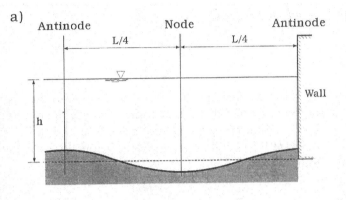

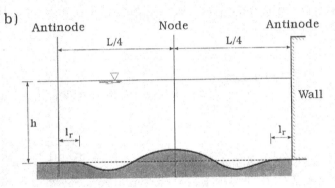

Figure 7.7: Scour/deposition pattern. (a) Suspension mode of sand transport. (b) No-suspension mode of sand transport. Xie (1981).

The preceding description implies that the scour-and-deposition pattern in front of a breakwater is heavily dependent on the mode of sand transport. The following subsection will address this issue.

Mode of sand transport

There are two modes of sand transport: the suspension-mode sand transport, and the no-suspension-mode sand transport.

The mode of sand transport is governed basically by two parameters, namely the Shields parameter, θ, and the fall-velocity-to-friction-velocity ratio, w/U_{fm} in which θ is defined by

$$\theta = \frac{U_{fm}^2}{g(s-1)d} \tag{7.3}$$

in which U_{fm} is the maximum value of the friction velocity, d is the sand size, s is the specific gravity of the sand grains, and g is the acceleration due to gravity. For the sand grains to be suspended from the bed, θ should be larger than $\theta_s(dU_{fm}/\nu)$, the Shields parameter required for the initiation of *suspension* from the bed:

$$\theta > \theta_s\left(\frac{dU_{fm}}{\nu}\right) \tag{7.4}$$

(see Example 1 in the following paragraphs for an empirical expression for the function $\theta_s(dU_{fm}/\nu)$).

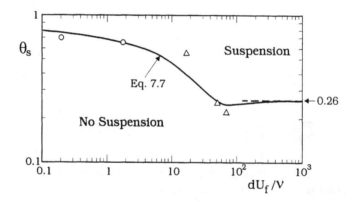

Figure 7.8: Criterion for initiation of suspension from bed. ○, Mantz (1978, Table 1 Run 2, Table 2, Run 2). △, Sumer (1986, Fig. 8). Dashed line, Shields (1936).

This is a necessary condition for the initiation of suspension of sand grains from the bed. However, it is not sufficient for the suspension mode of sand transport. For the sand grains to be maintained in suspension, there is an additional condition, namely:

$$\frac{w}{U_{fm}} < 1 \tag{7.5}$$

7.1. SCOUR AT THE TRUNK SECTION OF A BREAKWATER

(Batchelor, 1965, and Bagnold, 1966).

Hence, if these two conditions (Eqs. 7.4 and 7.5) are satisfied, then the sand transport will be in the **suspension mode**. Otherwise, the transport will be in the **no-suspension mode**.

Example 20 *Initiation of suspension from the bed in steady current*

The criterion for the initiation of suspension of sediment grains from a flat bed exposed to a *steady current* may be given in terms of the Shields parameter

$$\text{Suspension occurs from the bed when } \theta > \theta_s(dU_f/\nu) \tag{7.6}$$

(Sumer, 1986, see also Nezu and Nakagawa, 1993, p. 258). Here θ is the Shields parameter defined as in Eq. 7.3 with U_{fm} replaced with U_f, the friction velocity. Fig. 7.8 displays the available experimental data regarding the initiation of suspension in the case of steady currents. The data in the figure can be represented by the following empirical expression

$$\theta_s = (\frac{dU_f}{\nu})^{-0.05} \left[0.7\exp(-0.04\frac{dU_f}{\nu}) \right] + 0.26 \left[1 - \exp(-0.025\frac{dU_f}{\nu}) \right] \tag{7.7}$$

In the case of waves, no data are available. However, the preceding expression can, to a first approximation, be used, provided that U_f is replaced with U_{fm}.

For the mode of sand transport, Xie (1981) introduced the following criterion. The suspension mode of sand transport occurs when

$$\frac{U_m - U_{cr}}{w} \geq 16.5 \tag{7.8}$$

in which U_m is the maximum value of the orbital velocity at the bed, and U_{cr} is the critical velocity corresponding to the incipient sediment transport. Xie (1981) termed this kind of sand as fine sand. By contrast, when

$$\frac{U_m - U_{cr}}{w} \leq 16.5 \tag{7.9}$$

the mode of sand transport will be in the no-suspension regime; Xie (1981) termed this latter kind of sand as coarse sand. Xie's criterion essentially corresponds to that mentioned in the preceding paragraphs.

An approach similar to Xie (1981) has also been adopted in the study of Irie and Nadaoka (1984).

Scour depth

As seen from the above analysis, scour is caused by the steady streaming induced by standing waves (Fig. 7.6). The steady streaming depends on the following quantities: the incident-wave height, H, the incident-wave period, T_w (or, alternatively, the incident-wave length, L), the water depth, h, the bed roughness characterized by the sand-grain size, d, the water density, ρ, and the water viscosity, μ. Scour may also depend on the sediment properties, namely the grain size, d, and the submerged specific weight of the sand grains, $\gamma_s - \gamma$.

From dimensional analysis, the scour characteristics for the case of the *no-suspension-mode transport* (coarse sand) can be found to depend on the following parameters

$$\frac{S}{H} = f(\frac{h}{L}, \theta, \frac{L}{d}, RE) \tag{7.10}$$

in which $RE = aU_m/\nu$ where a is the amplitude of the undisturbed orbital motion of water particles, and U_m is the maximum value of the undisturbed orbital velocity at the bed. In most engineering problems, the bed under storm conditions acts as a rough wall, and therefore RE in the preceding equation drops out. Furthermore, when the live-bed scour ($\theta > \theta_{cr}$) is considered, the influence of θ is not very important and may be ignored to a first approximation. Therefore, the scour characteristics in this case will vary only with h/L, and L/d. It may be noted that the wave period may also be involved in the form of a Keulegan-Carpenter number, $T_w U_m/L_r$, as another parameter responsible for the process of the steady streaming (here, U_m is the maximum value of the orbital motion of water particles at the bed, and L_r is the sediment-ripple length). However, the present analysis focuses on the variation of the scour depth with the parameters h/L and L/d.

Xie (1981) investigated experimentally the scour (for both coarse sand and fine sand) in front of a vertical concrete wall. Fig. 7.9 displays his coarse sand data. S in the figure is the maximum scour depth which occurs at about $L/8$ from the breakwater (the no-suspension-mode transport, Fig. 7.7 b). The data shows the following:

1. There is a remarkable correlation between the scour depth S/H and the parameter h/L. The larger the value of h/L, the smaller the scour

7.1. SCOUR AT THE TRUNK SECTION OF A BREAKWATER

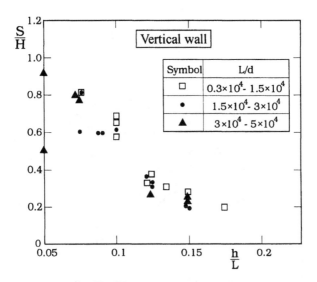

Figure 7.9: Data for maximum scour depth at vertical-wall breakwater for no-suspension mode of sand transport. Recast from Xie (1981).

depth. This is in fact expected. Clearly, for very large water depths, waves may not even induce any motion at the bed. Therefore, $S/H \to 0$, as $h/L \to \infty$.

2. Although the size of the data is not very large, the plot does show that the scour depth slightly increases as L/d decreases. This can be attributed to the bed roughness characterized by d. The streaming becomes more pronounced with increasing bed roughness (or alternatively with decreasing L/d). The larger the streaming, the larger the scour. Therefore the scour should increase with increasing bed roughness, or with decreasing L/d. Nevertheless, this variation in the scour depth with L/d is not very extensive, and for all practical purposes may be ignored.

Similar to the preceding analysis, the scour characteristics in the case of the *suspension-mode transport* (fine sand) can be found to depend on the previous nondimensional parameters, namely h/L, and L/d. Xie's (1981) fine sand data reveal that, similar to the previous case, the dependence on

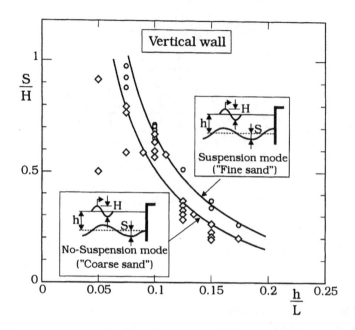

Figure 7.10: Maximum scour depth at vertical-wall breakwater. Live bed. Xie (1981).

L/d can, to a first approximation, be ignored, and therefore the maximum scour depth S/H is a function of only h/L. (Recall that, in the case of the suspension-mode transport, the maximum scour depth occurs at $L/4$ from the breakwater, Fig. 7.7 a).

Fig. 7.10 displays Xie's (1981) results. The data corresponding to the no-suspension mode transport (from Fig. 7.9) is also included for comparison.

The figure indicates that the fine-sand scour process produces larger scour depths than those experienced in the coarse-sand case.

Xie (1981) gives the following empirical expression for the maximum scour depth corresponding to the fine-sand scour (the top curve in Fig 7.10):

$$\frac{S}{H} = \frac{0.4}{\left[\sinh(\frac{2\pi h}{L})\right]^{1.35}} \qquad (7.11)$$

For the maximum scour depth in the case of the "coarse sand" (no-suspension-mode sand transport), a similar empirical expression can be obtained (the

7.1. SCOUR AT THE TRUNK SECTION OF A BREAKWATER   341

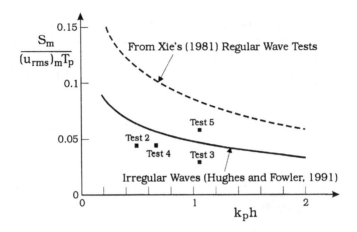

Figure 7.11: Maximum scour depth at vertical-wall breakwater. Comparison.

bottom curve in Fig. 7.10):

$$\frac{S}{H} = \frac{0.3}{\left[\sinh\left(\frac{2\pi h}{L}\right)\right]^{1.35}} \tag{7.12}$$

An empirical expression similar to Eq. 7.11 for the case of normally incident, **irregular waves** was given by Hughes and Fowler (1991):

$$\frac{S_m}{(U_{rms})_m \, T_p} = \frac{0.05}{[\sinh(k_p h)]^{0.35}} \tag{7.13}$$

in which T_p is the wave period of the spectral peak, k_p the wave number associated with the spectral peak by linear wave theory, $(U_{rms})_m$ the root-mean-square of horizontal velocity, which is given by (Hughes, 1992)

$$\frac{(U_{rms})_m}{g \, k_p \, T_p \, H_{m0}} = \frac{\sqrt{2}}{4\pi \, \cosh(k_p h)} \left[0.54 \, \cosh\left(\frac{1.5 - k_p h}{2.8}\right)\right] \tag{7.14}$$

in which H_{m0} is the zeroth-moment wave height defined as 4 times the standard deviation of sea surface elevation, g the acceleration due to gravity. The preceding equation is empirically based, and the range for which the equation can be applied is given as $0.05 < k_p h < 3.0$.

Fig. 7.11 compares the preceding relation with that of Xie's regular-wave result (Eq. 7.11). As seen, the scour depth in the case of irregular waves is reduced considerably. Therefore the scour generated by irregular waves, particularly in the case of the fine sand, may be a less threat to the structure in many cases.

Influence of angle of attack. Obliquely incident waves

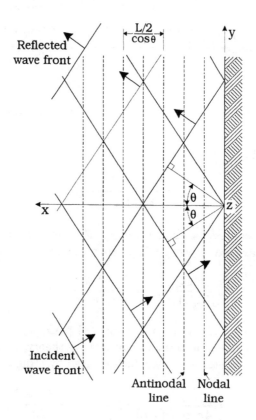

Figure 7.12: Waves incident obliquely at a vertical-wall breakwater.

In the case of obliquely incident waves, the flow and scour processes undergo substantial changes.

7.1. SCOUR AT THE TRUNK SECTION OF A BREAKWATER

Flow process. As the incident wave impinges on the wall, a reflected wave moves in the offshore direction (Fig. 7.12). Unless the wall is porous, almost all the energy reaching the breakwater will be reflected, meaning that the reflected wave will have the same properties as the incident wave. The superposition of these two waves results in a new wave. The surface elevation of this "resultant" wave is given by (see, for example, Hsu, Tsuchia and Silvester, 1979)

$$\eta(x, y, t) = \frac{(2H)}{2} \cos[k(\cos\theta)x] \cos[\omega t - k(\sin\theta)y] \qquad (7.15)$$

$$\omega^2 = gk \tanh(kh) \qquad (7.16)$$

in which θ is the incident angle (Fig. 7.12), ω is the angular frequency ($\omega = 2\pi/T_w$), k is the wave number ($k = 2\pi/L$), T_w is the wave period, L is the wave length (all corresponding to the incident wave), and h is the water depth. Note that Eq. 7.15 reduces to Eq. 7.1 when $\theta = 0$.

Eq. 7.15 indicates that, along $y = constant$ lines, the resultant wave evidently resembles a standing wave with the wave number equal to $k\cos\theta$ (or, alternatively, with the wave length equal to $L/\cos\theta$) (cf. Eq. 7.1). As in the case of the normally incident waves, $2H$ is the height of this standing wave, and, as seen, it is twice the wave height (H) of each of the two progressive waves (i.e., the incident wave and the reflected wave) forming the resultant wave.

Likewise, Eq. 7.15 indicates that, along $x = constant$ lines, the resultant wave resembles a progressive wave (propagating in the y-direction) with the wave number equal to $k\sin\theta$ (or, alternatively, with the wave length equal to $L/\sin\theta$), and with the celerity $c = \omega/(k\sin\theta)$) (cf. Eq. A.7, Appendix A). Here, too, $2H$ is the height of this progressive wave.

A snapshot of the surface elevation in plan view will show the picture sketched in Fig. 7.13 (as can readily be seen from Eq. 7.15); the surface elevation topography comprises discrete "islands" of crests. For this reason, these waves are called short-crested waves.

The presence of a standing wave in the x-direction implies that a system of steady, recirculating cells, similar to that in Fig. 7.6, will develop in the (x, z) vertical plane. Likewise, the presence of a progressive wave in the y-direction implies that there exists a steady streaming in the y-direction near the bottom (due to the non-uniform wave boundary layer in which the period-averaged velocity becomes different from zero, presumably resulting

344   CHAPTER 7. SCOUR AROUND BREAKWATERS

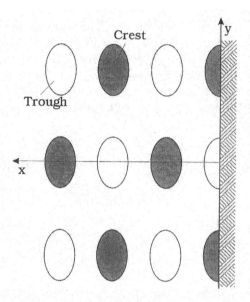

Figure 7.13: Snapshot of the surface elevation in plan view in the case of oblique incidence waves.

in a steady streaming, the so-called mass transport, Longuet-Higgins, 1957; see e.g., Fredsøe and Deigaard, 1992, p. 44 for a full account of this process). The sketch in Fig. 7.14 illustrates how the period-averaged flow (the steady streaming) would look near the seabed.

As for the y−component of the steady streaming, clearly the strength of the streaming should be largest along the antinodal lines, because these lines represent the lines with maximum wave height. Obviously the larger the wave height, the larger the streaming velocity.

An important point regarding the short-crested waves (which has implications with reference to the scour process) is that both the orbital velocities and the steady streaming velocities may be larger than those corresponding to the progressive waves which form the short-crested waves. This is because, the waves now have a wave height twice that of the corresponding progressive waves. Hsu and Silvester (1989) (see also Silvester, 1990) report that

1. the maximum orbital velocities can be as much as a factor of 2 larger, and

7.1. SCOUR AT THE TRUNK SECTION OF A BREAKWATER

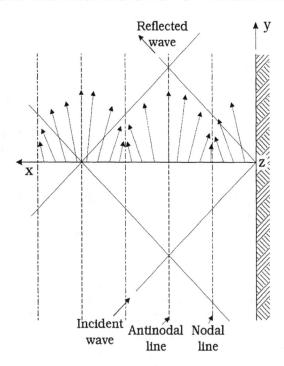

Figure 7.14: Steady streaming under standing waves (with oblique incidence waves).

2. likewise, the steady-streaming velocities can be as much as a factor of 3-4 larger than those corresponding to the progressive incident wave, depending on the incident angle and the x−location.

Obviously, these have significant implications with regard to the scour process, as will be discussed in the following paragraphs.

Papers by Hsu, Tsuchia and Silvester (1979), and Hsu, Silvester and Tsuchia (1980), and the review articles by Hsu (1990) and Silvester (1990) can be consulted for further information on the flow processes in short-crested waves.

Scour process. The scour process occurs in the following way.

First of all, the sediment is stirred up by the orbital motion (phase-resolved component) of the waves, and consequently put into the main body of the water.

Secondly, the sediment put into the main body of the water will respond to the recirculating cells forming in the vertical (x, z) plane, in much the same way as in the case of the normally incident waves (Section 3.1.1). This will presumably lead to the scour/deposition pattern in the form of alternating scour and deposition areas lying parallel to the breakwater. If the sand transport is in the suspension mode, the scour/deposition pattern will be as illustrated in Fig. 7.7 a, if it is in the no-suspension mode, it will be as in Fig. 7.7 b.

Clearly, the presence of the mass-transport generated along the length of the breakwater (y−direction), will induce additional scouring. Therefore, the scour depth will be increased by this latter effect.

In a real-life situation, obviously the wave properties change. Therefore, the mode of transport may also change. This may lead to the replacement of the deposition areas with the scour areas, and vice versa. The latter effect will cause scouring in the previous deposition areas. This process will eventually result in an overall scouring in front of the breakwater (Hsu and Silvester, 1989, Silvester and Hsu, 1989, and Silvester, 1990).

The latter authors report scour measured in the laboratory and in the field, demonstrating various aspects of the scour process in front of breakwaters (including the rubble-mound breakwaters) exposed to obliquely incident waves.

No design relationships yet exist to predict the scour in the case of the obliquely incident waves. The empirical expressions given for the normally incident waves can be implemented to make assessments. However, the previously mentioned features of the short-crested waves (such as the increase in the wave height as a function of the incident angle, and the x−position, and also the presence of the mass-transport velocity along the length of the breakwater) need to be taken into account in making such assessments.

Finally, longitudinal current induced by oblique waves along a vertical-wall porous breakwater and the resulting sediment transport leading to the formation of bars have been studied by Baquerizo and Losada (1998 a, and 1998 b). The time- and depth-averaged equations in a porous medium are used to analyze how dissipation inside the breakwater generates a current along the structure (inside, and, by virtue of diffusion, outside the breakwater at the seaward and leeward sides) (Baquerizo and Losada, 1998 a). With two selected cases, it was demonstrated that, depending on the angle of incidence, scour or deposition may occur at the toe of the structure (Baquerizo and

7.1. SCOUR AT THE TRUNK SECTION OF A BREAKWATER

Losada, 1998 b).

Breaking waves

The waves may break before they reach the breakwater, or they may break on the breakwater. Clearly, the wave breaking will affect the scour. These and other scenarios related to wave breaking will be studied in Chapter 8.

Longshore current

Longshore currents may co-exist with waves. These may be tidal currents, or they may be generated by the obliquely incident breaking waves. See the discussion in Chapter 8, Section 8.1 under 3-D Effects.

7.1.2 Scour at the trunk section of a rubble-mound breakwater

Scour process

In the case of a rubble-mound breakwater, there will be two changes:

1. first of all, the breakwater is made of rubble; and
2. it has a sloping wall, normally in the range 1:1.3 to 1:2, depending on the material used (1:2 for the rock-cube material, while 1:(4/3) for the Accropode material).

These changes will cause the reflection coefficient to be reduced. This will affect the steady streaming (i.e., the recirculating cells, described in the previous section, Fig. 7.6), and consequently the scour.

Sumer and Fredsøe (2000) studied the scour at a rubble-mound breakwater. In this study, a 2-D breakwater model, constructed of crushed stones the size of $D = 4.7$ cm, was placed in a wave flume, and subjected to regular and irregular waves. The breakwater model was a permeable structure (no impermeable core was used). Three kinds of breakwater slopes (1:1.75, 1:1.2, and a vertical-wall breakwater, the latter constructed from a plywood plate) were tested. The sand transport in the tests was in the *no-suspension mode*.

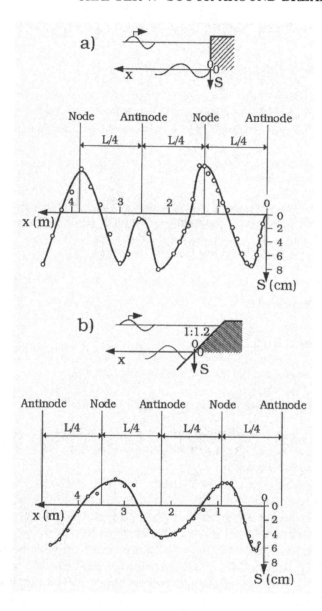

Figure 7.15: Scour/deposition in front of a vertical-wall breakwater (a) and a rubble-mound breakwater (b). Sumer and Fredsøe (2000).

7.1. SCOUR AT THE TRUNK SECTION OF A BREAKWATER 349

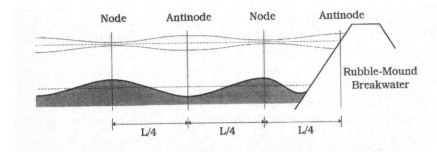

Figure 7.16: Schematic illustration of scour/deposition in front of rubble-mound breakwater. No-suspension mode of sand transport. Sumer and Fredsøe (2000).

Fig. 7.15 displays the scour/deposition profiles from Sumer and Fredsøe's (2000) study for two breakwater slopes, namely the vertical-wall (Fig. 7.15 a), and the slope 1:1.2 (Fig. 7.15 b). The tests were conducted under exactly the same incident wave conditions. Therefore, the results illustrate the influence of the breakwater slope.

First of all, note that the scour/deposition pattern in the case of the vertical-wall breakwater agrees with the pattern depicted in Fig. 7.7 b (recall that the sand transport in the tests occurred in the no-suspension regime).

Secondly, the scour/deposition pattern that emerges from Fig. 7.15 b for the rubble-mound breakwater case is schematically illustrated in Fig. 7.16. Comparison of Fig. 7.16 and Fig. 7.7 b indicates that the scour/deposition process occurs essentially in the same way as in the case of the vertical-wall breakwater. Namely, the recirculating cell system illustrated in Fig. 7.6 is basically responsible for the scour/deposition process; the sand (transported in the no-suspension mode in the tests) responds to the bottom cells, and the end result is the scour/deposition pattern depicted in Fig. 7.15 b, or in Fig. 7.16.

Although the scour/deposition occurs essentially in the same way as in the case of the vertical-wall breakwater, there are differences between the two cases:

1. While the scour just at the breakwater is zero in the case of the vertical-wall breakwater (Fig. 7.15 a), it has a non-zero value in the case of the rubble-mound breakwater (Fig. 7.15 b). This has obviously important practical consequences. The scour at the vertical-wall breakwater is nil

because the steady streaming at this location is practically nonexistent. The fact that there is a substantial amount of scour at the rubble-mound breakwater suggests that the steady streaming at this location must be significant. This is evidently related to the sloping surface of the breakwater (with 1:1.2 slope in the case presented in Fig. 7.15 b). (To the authors' knowledge, no study is yet available, revealing the influence of the breakwater slope on the steady streaming illustrated in Fig. 7.6). Also, it should be noted that there may be a flow from inside the breakwater towards the sand layer, and this may help enhance the scour at the toe region.

2. While no significant scour occurs below the antinode point in the case of the vertical-wall breakwater (Fig. 7.15 a), quite a substantial amount of scour occurs at this location in the case of the rubble-mound breakwater (Fig. 7.15 b). Sumer and Fredsøe (2000) note that no clear explanation has been found for this behaviour.

3. The maximum scour depth in the case of the rubble-mound breakwater is smaller (about 25 %) than that in the case of the vertical-wall breakwater. This is expected because the streaming in the case of the rubble-mound breakwater is weaker (due to the smaller reflection); therefore the resulting scour will be smaller than that of the vertical-wall breakwater case.

4. Finally, it is interesting to note that deposition occurs under the node points (Figs. 7.15 b and 7.16), in much the same way as in the case of the vertical-wall breakwater (Fig. 7.15 a). However, the precise locations of the maximum deposition in the rubble-mound case seem to be shifted slightly in the onshore direction (cf. Fig. 7.15 a). This effect is observed in other tests of Sumer and Fredsøe (2000), too. This shift may be attributed to the phase shift on the reflected waves in the case of a sloping structure/beach (Hughes and Fowler, 1995, Losada, Silva and Losada, 1997, and Sutherland and O'Donoghue, 1998). The latter effect leads to the change in the locations of nodes and antinodes, as described in the preceding publications. In the case of the vertical-wall breakwater, this phase shift is zero, therefore the locations of the maximum deposition coincide with the nodal points at precisely $L/4$ from the breakwater (Fig. 7.15 a).

7.1. SCOUR AT THE TRUNK SECTION OF A BREAKWATER

Maximum scour depth at the breakwater

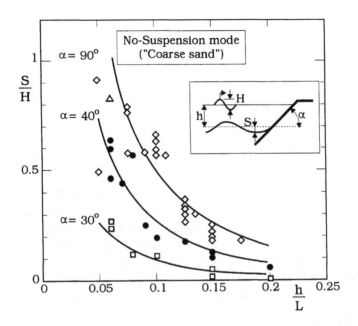

Figure 7.17: Maximum scour depth at rubble-mound breakwater. Sumer and Fredsøe (2000).

Scour is caused by the steady streaming induced by the standing waves. The steady streaming depends on the following quantities: the angle of the breakwater slope, α, the incident-wave height, H, the incident-wave period, T_w (or, alternatively, the incident-wave length, L), the water depth, h, the bed roughness characterized by the sand-grain size, d, the water density, ρ, and the water viscosity, μ. Scour may also depend on the sediment properties, namely the grain size, d, and the submerged specific weight of the sand grains, $\gamma_s - \gamma$.

From dimensional analysis, the scour characteristics for the case of the no-suspension-mode transport (coarse sand) can be found to depend on the following parameters (cf. Eq. 7.10)

$$\frac{S}{H} = f(\frac{h}{L}, \alpha, \theta, \frac{L}{d}, RE) \tag{7.17}$$

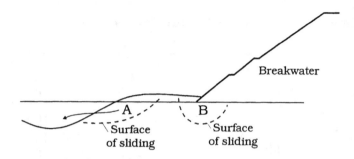

Figure 7.18: Possibility of sand slide in front of a rubble-mound breakwater.

For most engineering problems, the bed under storm conditions acts as a rough wall, therefore RE in the preceding equation drops out. Furthermore, when the live-bed scour ($\theta > \theta_{cr}$) is considered, the influence of θ is not very important and may be ignored to a first approximation. Therefore, the scour characteristics in this case will vary only with h/L, and L/d. Xie's (1981) data for the case of the vertical-wall breakwater, when recast, indicate that the parameter L/d only slightly affects the end results, as mentioned in the preceding subsection in conjunction with Fig. 7.9. Therefore, the scour depth can, to a first approximation, be expected to vary with the parameters h/L and α. It may be noted that the wave period may also be involved in the form of a Keulegan-Carpenter number, $T_w U_m/L_r$, as another parameter responsible for the process of the steady streaming (here, U_m is the maximum value of the orbital motion of water particles at the bed, and L_r is the sediment-ripple length). However, the present analysis focuses on the variation of the scour depth with the parameters h/L and α.

Fig. 7.17 displays the data related to the maximum scour depth near the breakwater as a function of h/L and α. The data obtained by Xie (1981) for the case of the vertical breakwater (from his "coarse sand" experiments, Fig. 7.10) is reproduced in the figure as a reference line.

1. As seen from Fig. 7.17, the scour depth decreases in the case of the rubble-mound breakwater. The milder the slope of the breakwater, the smaller the scour depth. This is simply because the strength of the steady streaming (Fig. 7.6) decreases with decreasing slope.

2. The rubble-mound breakwater data in Fig. 7.17 can be represented by

7.1. SCOUR AT THE TRUNK SECTION OF A BREAKWATER

the following empirical expression with a function f which is dependent on the breakwater slope:

$$\frac{S}{H} = \frac{f(\alpha)}{\left[\sinh(\frac{2\pi h}{L})\right]^{1.35}} \qquad (7.18)$$

in which $f(\alpha)$ is

$$f(\alpha) = 0.3 - 1.77 \exp(-\frac{\alpha}{15}) \qquad (7.19)$$

in which α is the breakwater slope in degrees (Fig. 7.17) in the range $30^0 \leq \alpha \leq 90^0$.

Sumer and Fredsøe (2000) compared their results with the numerical results of Arneborg, Hansen and Juhl (1995 a and b), and found a good agreement. Also, it was found that the results were in qualitative agreement (namely, the scour depth decreases with decreasing slope) with the results of Herbich and Ko (1969; see also Herbich, Schiller, Dunlap and Watanabe, 1984) who conducted laboratory experiments with a sloping wall (the wall in these experiments was made of plexiglas (rather than a rubble-mound), simulating a seawall). For the details of these comparisons, the reader is referred to the original paper.

In Sumer and Fredsøe's (2000) experiments, the breakwater model for the majority of the tests extended into the sand bed down to the base bottom. This kind of set-up enabled them to achieve structural stability under scour. However, they studied scour in three cases where the breakwater was sitting on the sand bottom, to observe the influence of the *bottom-seated breakwater* on scour. The scour at the breakwater in the case of the bottom-seated breakwater was somewhat smaller (about 30% smaller). Sumer and Fredsøe (2000) noted that this may be due partly to the fact that the stones slumped down in the scour hole and formed a protective slope at the toe of the breakwater, resulting in relatively smaller scour depths. It may be noted that, from the preceding discussion, the data presented in Fig. 7.17 for rubble-mound breakwaters give somewhat conservative estimates of the scour depth.

Finally, although no scour is expected at the toe of a rubble-mound breakwater for *suspension-mode sediment transport* (cf. Fig. 7.7 a), and therefore scour is not an immediate threat to the breakwater, soil failure illustrated in Fig. 7.18 may be a risk for the stability of the structure, and hence needs to be considered.

Maximum scour depth for irregular waves

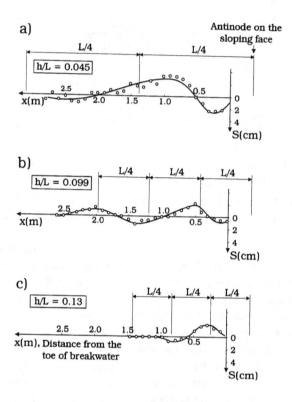

Figure 7.19: Equilibrium bed profiles in the case of irregular waves at rubble-mound breakwaters. Slope = 1:1.2 ($\alpha = 40°$). No-suspension mode of sand transport. Live bed. Sumer and Fredsøe (2000).

In the case of irregular waves, Sumer and Fredsøe (2000) found that the scour/deposition profiles showed distinct features, different from those obtained in the case of regular waves.

Fig. 7.19 presents three equilibrium scour/deposition profiles in the case of the irregular waves for three different values of the parameter h/L in which L is the wave length associated with the peak period.

First of all, it is seen that the scour/deposition "subsides" with increasing distance from the breakwater, in contrast to the results obtained in the case

7.1. SCOUR AT THE TRUNK SECTION OF A BREAKWATER

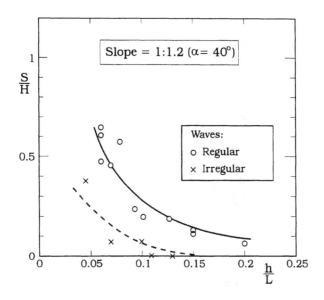

Figure 7.20: Maximum scour depth at rubble-mound breakwaters. Comparison. No-suspension mode of sand transport. Live bed. Sumer and Fredsøe (2000).

of the regular waves (cf. Fig. 7.15 b). This is linked to the fact that, in the case of irregular waves, the wave height only emerges as a standing wave pattern close to the structure, while, in the case of regular waves, a pattern of nodes and antinodes in the wave height extends over the entire area in front of the structure (Klopman and van der Meer, 1999).

Secondly, the location where the deposition is maximum is not always shifted in the onshore direction, in contrast to the regular wave results (cf. Fig. 7.15 b); it may be shifted in the onshore direction (Fig. 7.19 a), or in the offshore direction (Fig. 7.19 b), or it may not experience any significant shift at all (Fig. 7.19 c).

Thirdly, the maximum scour depth in the case of the irregular waves is generally smaller than that of the corresponding regular-wave cases. Fig. 7.20 presents the scour-depth data for the two cases in the case of the steep-slope tests of Sumer and Fredsøe's (2000) study. H in the diagram is taken as the root-mean-square value of the wave heights, H_{rms}, in the case of the irregular waves. This is because this quantity reduces to the conventional wave height, H, for the case of regular waves.

As seen from Fig. 7.20, the scour depth in the case of the irregular waves is reduced by a factor of 2, or even more. The latter result is in good, quantitative agreement with the results of Hughes and Fowler (1991) for the case of the vertical-wall breakwater (Fig. 7.11).

Time scale of scour process

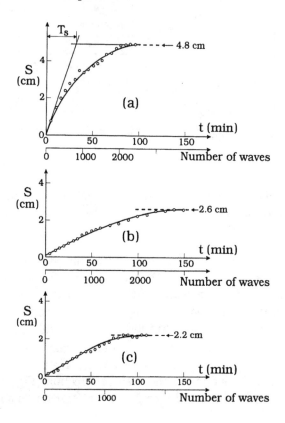

Figure 7.21: Time development of maximum scour depth. Live bed. (a) Regular waves, Slope = 1:1.2. (b) Regular waves, Slope = 1:1.7. (c) Irregular waves, Slope = 1:1.2. Sumer and Fredsøe (2000).

From dimensional analysis, the time scale of the scour process can be expressed in the following nondimensional form (cf. Eq. 7.17):

7.1. SCOUR AT THE TRUNK SECTION OF A BREAKWATER

$$T^* = T^*(\frac{h}{L}, \alpha, \theta, \frac{L}{d}, RE) \qquad (7.20)$$

in which T^* is the nondimensional time scale which may be defined by

$$T^* = \frac{(g(s-1)d^3)^{1/2}}{H^2} T \qquad (7.21)$$

in which T is the time scale, i.e., the time during which a substantial amount of scour occurs. Previous work related to time scale in conjunction with other scour processes such as scour below a pipeline (see Chapter 2, Section 2.3.4, and that around a pile (see Chapter 3, Section 3.3) indicated that the Shields parameter is an important parameter. The subject has been briefly discussed in Sumer and Fredsøe (2000) on the basis of the time development of the maximum scour depth obtained in a limited number of experiments.

Fig. 7.21 depicts the development of the maximum scour depth with respect to time for three different cases; Fig. 7.21 a for the case of the breakwater slope 1:1.2 ($\alpha = 40^\circ$) with regular waves, Fig. 7.21 b for the breakwater slope 1:1.75 ($\alpha = 30^\circ$) with, again, regular waves, and Fig. 7.21 c for the breakwater slope 1:1.2 ($\alpha = 40^\circ$) with irregular waves. A second time axis is added in the figures indicating the number of waves. In the case of the irregular waves (Fig. 7.21 c), the number of waves is obtained by dividing the sampling time by the peak period.

The time scale, T, may be defined by (see the discussion in conjunction with Eq. 1.2)

$$S_t = S\left(1 - \exp(-\frac{t}{T})\right)$$

The T values from the time variations displayed in Fig. 7.21 are depicted in the following table together with the nondimensional time scale values T^*.

h/L	Time scale, T (min)	Number of waves corresponding to T	Normalized time scale, T^*
0.078	33	800	3.3
0.06	85	1700	6.0
0.045	78	1200	15.8

As seen, T^* increases with decreasing values of h/L. This is because, with decreasing values of h/L, the scour depth will increase (Fig. 7.17), and therefore sediment will be removed in a larger and larger quantity, causing the time scale of the scour process to increase. The table also presents the number of waves, n, corresponding to the indicated values of T^*. However, it may be noted that the direct transfer of n values to the prototype may not be suitable since this quantity cannot be substituted for T.

Clearly, the nondimensional T^* values given in the above are at best suggestive. However, they may help assess the time scale for scour processes in the corresponding real-life situations.

Longshore current

Longshore currents may co-exist with waves. It may be tidal currents, or they may be generated by the obliquely incident breaking waves. See the discussion in Chapter 8, Section 8.1 under 3-D Effects.

An interesting physical model study has been undertaken by Hughes and Schwichtenberg (1998) to investigate scour resulting from longshore currents flowing along the lee-side of a detached breakwater. Current-induced scour developed as an unanticipated result of structural modifications that reduced problematic shoaling of the main navigation channel. The model investigation was directed to optimize the design of toe protection intended to prevent lee-side armour layer damage on the detached breakwater.

Finally, scour at the toe of hydraulic and coastal structures such as revetments in rivers and coastal inlet structures (jetties) under currents parallel to the structure has been investigated experimentally by Fredsøe, Sumer and Bundgaard (2001). It was found that the scour primarily is governed by the horseshoe vortices forming in front of the individual stones at the junction between the revetment and the bed. The scour depth normalized by the stone size was found to vary with the Shields parameter and with the stone shape for a given slope of the revetment. All the tests were conducted in the case of live bed. The influence of a falling apron to protect the revetment structure against scour was also investigated.

Remarks on other effects

Sumer and Fredsøe's (2000) study outlined above deals with the case of the no-suspension mode sand transport. Therefore the presented results should only be used for the no-suspension mode transport. In the case of the suspension-mode sand transport, it may be expected that the scour depth may be increased in the same way as in the case of the vertical-wall breakwater (Fig. 7.10). From the vertical-wall breakwater results in Fig. 7.10, this

7.1. SCOUR AT THE TRUNK SECTION OF A BREAKWATER 359

increase in the scour depth may be expected to be in the order of magnitude of 20-40 %, depending on the depth-to-wave-length ratio. However, it may be noted that the scour in the case of the suspension-mode sand transport may not represent as much a threat to the structure as in the case of the no-suspension mode sand transport due to its location, namely at $L/4$ from the breakwater, as already pointed out in the preceding paragraphs.

The analysis given in Sumer and Fredsøe's (2000) study is for the non-breaking waves. Clearly, the scour process in the breaking-wave case is entirely different from that studied in the present investigation. Larger scour depths may be expected when the waves are breaking.

The case of obliquely incident waves, and the case when the waves coexist with a longshore current may increase the scour depth with respect to that experienced in the case of the normally incident waves, since the longshore current in these cases acts as an additional convection mechanism to carry the sand away from the structure, presumably resulting in additional scour.

Finally, the issue of scale effects will be discussed. The nondimensional function which describes the scour depth is given in Eq. 7.17. Sumer and Fredsøe's (2000) study focuses on the variations concerning the two key parameters, namely, h/L and α. For a complete simulation of the scour process, however, the variations regarding the other parameters, namely, θ, L/d and RE, need to be known.

As regards the parameter θ, the influence of this parameter would be to increase the scour depth. However, for the live-bed conditions, the increase in the scour depth with increasing θ would be rather small, as already pointed out. As for the parameter L/d, as mentioned in the preceding paragraphs in relation to Eq. 7.17, Xie's (1981) data for the case of the vertical-wall breakwater, when recast, indicate that the parameter L/d only slightly affects the end results; the scour depth slightly decreases with L/d. The latter can be attributed to the bed roughness characterized by the grain size d; the streaming becomes stronger with increasing bed roughness (or alternatively with decreasing L/d). As regards the $RE-$number effect, this can manifest itself when the sediment size is small, so small that the bed acts as a hydraulically smooth boundary. In this case, the streaming (therefore, the scour depth) may become stronger with decreasing RE. However, no data is available that reveals this behaviour.

Also, no study is yet available that investigates the influence of the porosity of the rubble-mound structure on scour.

Regarding the time scale of scour process, similar considerations apply to this case, too. The nondimensional function given in Eq. 7.20 governs the nondimensional time scale T^*. A complete similarity can be obtained only when this function is determined explicitly. The direction in which each parameter influences the end result can be expected to be similar to that in other scour processes, such as scour below a pipeline (Chapter 2), and that around a pile (Chapter 3).

See Oumeraci (1994 b) for further discussion of the scale effects.

Submerged breakwater

Submerged breakwaters, inspired by the behaviour of natural reefs, also are used in practice. Their purpose is essentially to retard offshore sand movement (caused by storm waves) by introducing a structural barrier. The main difference between a surface-piercing breakwater and a submerged breakwater is that while the former acts to reduce waves, the latter acts as a barrier to shore-normal sediment motion; the effect of submerged breakwaters on waves is relatively small because their crest elevation is at or below the water level (US Army Corps of Engineers, 1994).

In a recent series of work by Sanchez-Arcilla and his co-workers, the wave field over/through/around the submerged structure has been assessed from an observational analysis based on rigid bed (Gironella and Sanchez-Arcilla, 2000) and mobile bed tests (Sanchez-Arcilla, Gironella, Verges, Sierra, Pena and Moreno, 2000) with the same combinations of seabed and breakwater geometry. The sand size was 0.25 mm. The resulting wave pattern showed the co-existence of partial reflection and dissipation due to the rigid/mobile bed where the presence of the breakwater also affected the resulting wave field. The mobile-bed tests (Sanchez-Arcilla et al., 2000) were executed with permeable and near-impermeable breakwaters (i.e. with and without core) in a large-scale wave flume, and paid particular attention to the scouring in front of the structure. The bed evolution around the structure, and particularly the scouring in front of it, has been analyzed from the hydraulic tests. For the 18 tests carried out (see Sumer, Whitehouse and Tørum, 2001, for a brief outline of this study), 3 test series corresponded to submerged permeable breakwaters (the free board being 0.25 m), 5 to permeable structures with crest level at the mean water surface (zero free-board) and 3 to surface-piercing breakwaters (the free-board being -0.75 m). In addition, tests with a nearly impermeable breakwater were conducted; 4 with a submerged structure (the free-board being 0 m) and 3 with a surface-piercing structure (the

7.1. SCOUR AT THE TRUNK SECTION OF A BREAKWATER

free-board being -0.75 m). The scour tests were "designed" to cover a realistic range of possible wave-steepness and were run until an almost equilibrium bed configuration was reached (10,000 to 15,000 waves on average).

Not surprisingly the clearest scouring was found for the more "reflective" structure, corresponding to surface-piercing breakwaters with an impermeable core. Under equivalent wave conditions and structure permeability the maximum scouring moves away from the structure as the freeboard decreases. For a high freeboard and the conditions tested the scour depth is small at the toe whereas for the zero-freeboard case the scour moves in the offshore direction. In the latter case, most of the scouring takes place under the antinode of the standing wave pattern, and there is also a large accretion at the toe of the structure.

Hydrodynamic (numerical) modelling of flow at submerged breakwaters has also been attempted by the same group (Verges and Sanchez-Arcilla, 2000); The (comparatively) simpler time-averaged current field has been determined numerically from the Fluxus model, based on the corresponding 2DV Reynolds equations. The wave and turbulence closure terms are externally imposed by means of algebraic expressions based on the state of art (and fine-tuned with the experimental results). The resulting fluxes show two important features (confirmed by the experimental results): (1) a recirculation cell in front of the structure which is the "degenerated" version of the twin recirculation cells in front of a fully reflective structure (associated with the standing wave pattern, see Fig. 7.6) and (2) a significant (and therefore non-negligible) mass flux through the permeable structure (depending, of course, on porosity). The numerical model solves in a coupled fashion the mass/momentum equations in the fluid/porous media, with corresponding driving/resistance terms. The number of fitting parameters appearing in the equations requires a careful validation in order to develop a robust engineering tool.

7.1.3 Scour protection at the trunk section of a breakwater

It is imperative to construct a protection layer for toe protection. This protection layer may be constructed in the form of a toe berm, or in the form of a berm and an additional apron (Fig. 7.22). The apron must be designed so that it will remain intact under wave and current forces, and it should be "flexible" enough to conform to an initially uneven seabed. With this countermeasure, scour can be minimized, but not entirely avoided. Some scour will occur at the edge of the protection layer, and consequently, armour stones will slump down into the scour hole. This latter process will, however, lead to the formation of a protective slope, a desirable effect for "fixing" the scour. One of the most important design concerns is the determination of the width of the protection layer which should be sufficiently large to ensure that some portion of the berm/apron remain intact, providing adequate protection for the stability of the breakwater. The following paragraphs address this question.

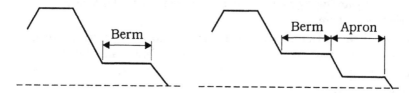

Figure 7.22: Berm and berm with an apron for scour protection.

Sumer and Fredsøe (2000), from their measurements, give the width of the scour hole at the breakwater as follows:

Slope of breakwater	Width of the scour hole at the breakwater, W
Vertical-wall breakwater	$1.0 \times (\frac{L}{4})$
1:1.2	$0.6 \times (\frac{L}{4})$
1:1.75	$0.3 \times (\frac{L}{4})$

In the table, W = the width of the scour hole at the breakwater (measured from the dune crest to the toe of the breakwater) in the absence of scour protection, Fig. 7.16) and L = the wave length.

Sumer and Fredsøe's (2000) experiment for the breakwater slope 1:1.2 indicated that the scour is practically zero when the width of a protection apron (Fig. 7.23), ℓ, is selected to be equal to W depicted in the above table. Xie's (1981) experiments gave the same results for the vertical-wall breakwater.

Sumer and Fredsøe (2000) recommend that the width of the protection apron, ℓ, may be selected as $\ell = W$.

7.1. SCOUR AT THE TRUNK SECTION OF A BREAKWATER

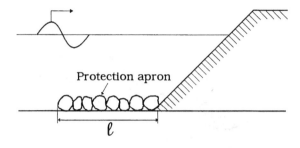

Figure 7.23: Scour protection.

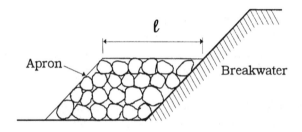

Figure 7.24: Protection apron with several layers of stones.

If ℓ is selected to be smaller than W, clearly some scour will occur. If the protection apron consists of several layers of stones in this case (Fig. 7.24), the stones in the protection apron will slump down into the scour hole (Fig. 7.25), presumably forming a protective slope, a desired effect for scour protection. Sumer and Fredsøe (2000) studied the effect of the number of stone layers on the scour protection.

In their test with a breakwater with a slope of 1:1.2, Sumer and Fredsøe selected ℓ to be $\ell = 0.14 \times (\frac{L}{4})$, a value much smaller than the recommended value for a complete protection for this breakwater slope, namely $0.6 \times (\frac{L}{4})$. Therefore, scour occurred. However, the scour in this case was reduced by employing several layers of stones in the protection apron in which case the stones slumped down into the scour hole. Fig. 7.26 displays the results of Sumer and Fredsøe's tests for this experiment. The figure shows that the scour depth is reduced from $S/H = 0.7$, the scour depth with no protection, to about $S/H = 0.3$. Apparently, the larger the value of the number of stone

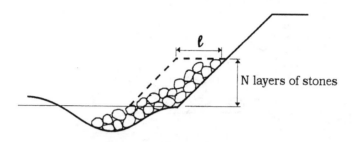

Figure 7.25: Stones slump into the scour hole to form a protective slope.

layers, N, the smaller the scour. This is because the larger number of armour layers leads to a more effective protective slope. However, Fig. 7.26 implies that the scour depth does not change for values of N larger than a certain value, as expected. (The latter value in the present case is $N = 4 - 5$ where the stone size was $D/H = 0.24$).

Finally, Hales (1980) (see also Coastal Engineering Manual, 2001) surveyed scour protection practices in the U.S. and found that the minimum scour protection was typically an extension of the structure bedding layer and filter layers. The following *minimum* rule-of-thumb resulted from this survey:

1. Minimum toe apron thickness: 0.6 m to 1.0 m (1.0 m to 1.5 m in Northwest US);

2. Minimum toe apron width: 1.5 m (3 m to 7.5 m in Northwest US); and

3. Material: Quarry stone to 0.3 m diameter, gabions, mats, etc.

Coastal Engineering Manual (2001) notes that these rules-of-thumb are inadequate when the water depth at the toe is less than two times the maximum nonbreaking wave height at the structure or when the structure reflection coefficient is greater than 0.25 (structures with slopes larger than 1:3). Under these more severe conditions, other scour protection methods are recommended (Coastal Engineering Manual, 2001).

7.1. SCOUR AT THE TRUNK SECTION OF A BREAKWATER

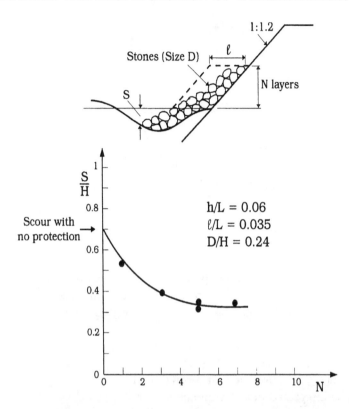

Figure 7.26: Influence of the number of layers on the "performance" of the protection apron. No-suspension mode of sand transport. Sumer and Fredsøe (2000).

7.1.4 Mathematical modelling of scour at the trunk section of a breakwater

Arneborg, Hansen and Juhl's model (1995 a, 1995 b). Arneborg et al. (1995 a) developed a numerical model to simulate the aforementioned 2-D scour process. The model has two modules: the flow module, and the sediment module.

The flow module considers the flow in two regions: the boundary layer over the bed; and the flow outside the boundary layer (Fig. 7.27).

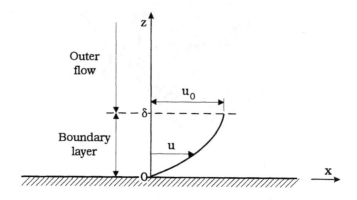

Figure 7.27: Velocity distribution in the wave boundary layer.

For the flow outside the boundary layer, the model uses the linear wave theory. The effect of a varying bed geometry is taken into account by making a conformal mapping of the linear solution for a plane bed into the domain with this varying bed geometry.

For the boundary layer over the bed, the integrated momentum equation is used, similar to Fredsøe (1984) (see also, e.g., Schlichting, 1979, p. 159). The model has been developed for a rough bed.

Now, the integrated momentum equation reads as follows:

$$\int_{k/30}^{\delta+k/30} \frac{\partial u}{\partial t}\,dz + \frac{\partial}{\partial x}\int_{k/30}^{\delta+k/30} u^2 dz - u_0 \frac{\partial}{\partial x}\int_{k/30}^{\delta+k/30} u\,dz = -\int_{k/30}^{\delta+k/30} \frac{1}{\rho}\frac{\partial p}{\partial x} - \frac{\tau_b}{\rho} \quad (7.22)$$

in which k = the Nikuradse equivalent sand roughness of the bed, taken as $2.5 d_{50}$, δ = the boundary layer thickness (Fig. 7.27), u = the velocity in the boundary layer, u_0 = the velocity at the top of the boundary layer (i.e., the orbital velocity at the bed), p = the pressure, z = the vertical distance measured from the bed upwards and τ_b = the bed shear stress. It is important to emphasize that the last term on the left-hand-side of the preceding equation originates from the convection term $w\frac{\partial u}{\partial z}$ in the $x-$component of the equation of motion, and represents the exchange of momentum at the top of the boundary layer between the latter and the outer flow, the key "ingredient" for the steady streaming near the bed.

Next, Arneborg et al. (1995 a) assume that the velocity distribution

7.1. SCOUR AT THE TRUNK SECTION OF A BREAKWATER

across the boundary-layer depth is logarithmic

$$\frac{u}{U_f} = \frac{1}{\kappa} \ln(\frac{30z}{k}) \tag{7.23}$$

in which $U_f = \sqrt{\tau_b/\rho}$, the friction velocity and κ the von Karman constant. Obviously, $u = u_0$ (Fig. 7.27) when $z = \delta + k/30$:

$$\frac{u_0}{U_f} = \frac{1}{\kappa} \ln(\frac{30(\delta + k/30)}{k}) \tag{7.24}$$

from which one obtains

$$\delta = \frac{k}{30}(e^{\kappa u_0/U_f} - 1) \tag{7.25}$$

From Eqs. 7.22, 7.23 and 7.25, the following differential equation is obtained

$$\frac{\partial \xi}{\partial t}[e^\xi(\xi - 1) + 1]\frac{u_0}{\xi^2} = [e^\xi(\xi - 1) + 1]\frac{1}{\xi}\frac{\partial u_0}{\partial t} + [e^\xi(\xi^2 - 3\xi + 4) - \xi - 4] \times$$

$$\times \left[\frac{1}{\xi^2}u_0\frac{\partial u_0}{\partial x} - \frac{\partial \xi}{\partial x}\frac{u_0^2}{\xi^3}\right] \frac{1}{\rho}\frac{\partial p}{\partial x}(e^\xi - 1) + \frac{\kappa^2}{\frac{k}{30}\xi^2}u_0|u_0| \tag{7.26}$$

in which ξ is defined by

$$\xi = \frac{\kappa u_0}{U_f} \tag{7.27}$$

Given the quantities $u_0(x, t)$ and $\frac{\partial p}{\partial x}(x, t)$, Eq. 7.26 is solved for the quantity ξ as a function of x and t, and from Eq. 7.27 the friction velocity U_f is found as a function of x and t. The latter is then used for the calculation of sediment transport in the sediment module.

The sediment module itself consists of a bed load part and a suspended load part. For the former, Engelund and Fredsøe's (1976) bed load formula is employed where the effect of a local bed slope is also taken into consideration. The part related to the suspended load is determined by solving the equation of conservation of mass for suspended sediment (Eq. 2.85). The obtained bed load, q_b, and suspended load, q_s, are then inserted in the equation of continuity for the sediment at the bed, i.e.,

$$\frac{\partial z_b}{\partial t} = -\frac{1}{1-n}\left[\frac{dq_b}{dx} + q_s\right] \tag{7.28}$$

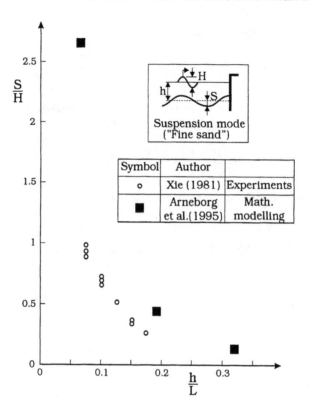

Figure 7.28: Maximum scour depth at vertical-wall breakwater. Suspension mode of sand transport. Comparison.

(cf. Eq. 2.100) in which z_b = the scoured bed level. Eq. 7.28 is then solved for z_b to update the bed morphology.

Fig. 7.28 displays the results of Arneborg et al.'s (1995 a) numerical simulations for vertical-wall breakwaters for fine sand. As seen, the model gives the correct variation of the scour depth with h/L; however, the model results are a factor of 2 larger than the experiments. No explanation for this discrepancy has been offered in Arneborg et al. (1995 a).

Arneborg et al. (1995 b) viewed that the difference between the vertical-wall breakwater case and the rubble-mound breakwater case lies in the reflection coefficient associated with the rubble-mound structure. They considered

7.1. SCOUR AT THE TRUNK SECTION OF A BREAKWATER

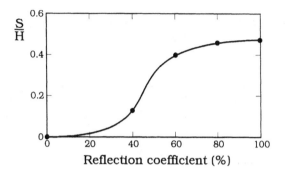

Figure 7.29: Maximum scour depth at breakwaters. Suspension mode of sand transport. $H = 1.5$ m, $kh = 1.2$, $h = 10$ m, $d_{50} = 0.1$ mm, $w = 0.01$ m/s. Arneborg et al. (1995 b).

this latter quantity in order to simulate the scour process at rubble-mound breakwaters; they extended their numerical model (Arneborg et al., 1995 a) so as to cover partially standing waves to account for the variation of scour as function of the reflection coefficient. Fig. 7.29 presents their results for a selected set of input parameters indicated in the figure caption. The scour depth decreases with decreasing values of the reflection coefficient.

Arneborg et al.'s (1995 b) calculations are for $kh = 1.2$, or alternatively, $h/L = 0.2$. The reflection coefficient for the rubble-mound breakwater used in the experiments of Sumer and Fredsøe (2000) for this value of h/L with the slope $1:1.2$ ($\alpha = 40°$), and with $H = 9.0$ cm and $T = 1.08$ s (Sumer and Fredsøe, 2000, Table 1, Test 7) is found to be $c_r = 0.32$, using the empirical expression given by Losada and Gimenez-Curto (1981), which is based on the data of Gunbak (1976, 1979) for reflection on a rip-rap slope,

$$c_r = 1.35(1 - \exp(-0.071\, Ir)) \qquad (7.29)$$

in which Ir is the Irribaren number

$$Ir = \frac{\tan(\alpha)}{\sqrt{2\pi H/(gT^2)}} \qquad (7.30)$$

and $\tan(\alpha)$ is the slope. For this value of the reflection coefficient ($c_r = 0.32$), Arneborg et al.'s (1995 b) numerical simulation of the scour at the breakwater gives $S/H = 0.045$. The latter value is not radically different from the measured value in Sumer and Fredsøe's experiments, namely 0.06 (Sumer and Fredsøe, 2000, Table 1, Test 7).

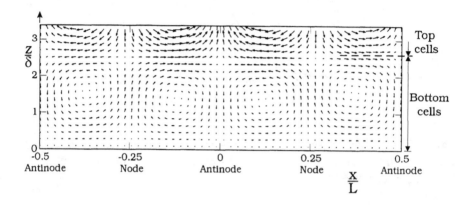

Figure 7.30: Steady streaming in front of a vertical-wall breakwater from mathematical modelling. Gislason et al. (2000).

Gislason, Fredsøe, Mayer and Sumer's (2000) model. Gislason et al. (2000) solve the time-dependent two-dimensional Navier-Stokes (N.-S.) equations with fully non-linear free-surface boundary conditions. The numerical method follows Mayer, Grapon and Sørensen (1998). The geometry of the free surface is described by a height function, and its evolution is tracked by integrating in time the kinematic boundary conditions based on the free-surface volume flux. The fluid domain is discretized by adapting a time-varying curvilinear grid to all boundaries. The N.-S. equations are discretized by a conservative cell-centered finite volume formulation, taking into account the time dependency of the grid.

Gislason et al. (2000) report some early results from their simulation of two-dimensional periodic standing waves for the laminar case in the case of a vertical-wall breakwater. Fig. 7.30 shows an example of the steady-streaming velocities near the bottom. The vertical distance in the figure is normalized by the Stokes length, the length characterizing the thickness of the laminar boundary layer, $\delta = \sqrt{2\nu/\omega}$ in which ω = the angular frequency of the wave and ν = the kinematic viscosity (see, e.g., Fredsøe and Deigaard, 1992, p. 19). The steepness of the wave is $H/L = 0.0963$, the length of the computational domain is equal to the wave length, L, and the depth is equal

7.2. SCOUR AROUND THE HEAD OF A BREAKWATER

to $h/L = 0.375$. The computation is started with a free-surface elevation which is prescribed to the eighth harmonic components, and the fluid velocity is initially set to zero. After the flow field is obtained for some wave periods, the steady-streaming velocities at some fixed points are determined by period averaging. As seen, Fig. 7.30 reveals the presence of the bottom and top cells referred to earlier (Fig. 7.6). It is interesting to note that, from Fig. 7.30, the bottom cell is restricted to a layer with a thickness of $O(2\delta)$. For a wave period of, say, 2 s in the laboratory, this thickness will be $O(1 \text{ mm})$, a very thin layer near the bottom. This makes it difficult to penetrate into this thin layer and measure the streaming-velocity distribution across the depth in the bottom cells. However, the steady streaming in this layer can be made "visible" by sprinkling sand grains (with d small compared with $O(1 \text{ mm})$) onto the bottom and tracking their motion.

Gislason et al. (2000) note that the $k - \omega$ model will be used for modelling the turbulent case. They also note that the calculations will be extended so as to cover the case of a rubble-mound breakwater, including the scour/deposition prediction.

7.2 Scour around the head of a breakwater

7.2.1 Scour around the head of a vertical-wall breakwater

Flow and scour processes

The mechanism regarding the scour around the head of a vertical-wall breakwater is different from that at the trunk section.

In the case of the head scour (a 3-D scour) (where the head is exposed to waves propagating in the direction perpendicular to the breakwater), separation vortices form at the lee side (Fig. 7.31) during each half period of the waves. As will be seen in the following paragraphs, these vortices play the major role in the process of scour (Sumer and Fredsøe, 1997).

Fig. 7.31 depicts the different flow regimes around the head of the vertical-wall breakwater. The Keulegan-Carpenter number, KC, based on the diameter of the breakwater head, B, is the key parameter that governs the flow

CHAPTER 7. SCOUR AROUND BREAKWATERS

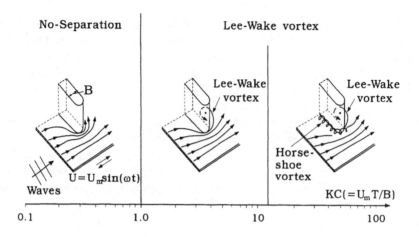

Figure 7.31: Near-bed flow regimes around the head of a vertical-wall breakwater. Sumer and Fredsøe (1997).

processes. Here KC is defined by

$$KC = \frac{U_m T_w}{B} \qquad (7.31)$$

in which U_m is the maximum value of the undisturbed orbital velocity at the bed, and T_w is the wave period.

KC usually is very small in real-life situations. Therefore the flow regime for the KC range $KC > 12$ (where, in addition to the lee-wake vortex, a horseshoe vortex is formed, Fig. 7.31) is of no practical significance with regard to scour problems encountered in practice, unless the wave climate is extremely severe, or when there is a current component perpendicular to the breakwater.

Fig. 7.32 schematically illustrates the way in which the scour develops around the breakwater head (taken from a video record; the video camera views the toe of the head directly). The small sketch in the top left corner of the panels in Fig. 7.32 gives the surface-elevation time variation.

1. The bed at the head is eroded by the high-speed flow (Frame 1).

2. The eroded sand is swept into the lee-wake vortex where the sand grains are lifted up into the upper portion of the lee-wake vortex (A in Frame 2).

7.2. SCOUR AROUND THE HEAD OF A BREAKWATER

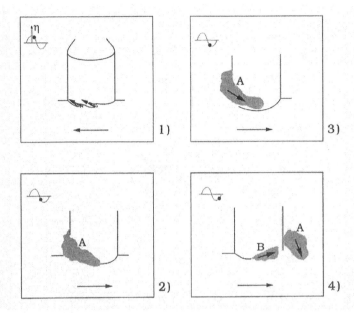

Figure 7.32: Sequence of scour process around the head of a vertical breakwater during a wave cycle. Sumer and Fredsøe (1997).

3. As this vortex is washed around the structure by the flow reversal, the sand trapped in the vortex is carried by this vortex (A in Frame 3), and eventually deposited away from the structure (A in Frame 4).

4. In the next half cycle, the process repeats itself; this time, the flow structure B (Frame 4) plays the same role as A in the previous half cycle.

Fig. 7.33 illustrates the amplification in the bed shear stress, α, around the head (Sumer and Fredsøe, 1997). The quantity α is defined by

$$\alpha = \frac{\max |\vec{\tau}|}{\tau_m} \quad (7.32)$$

in which τ = the bed shear stress, and τ_m = the maximum value of the undisturbed bed shear stress. As seen the amplification factor can be as much as 10-15. Certainly, the large amplification in the bed shear stress

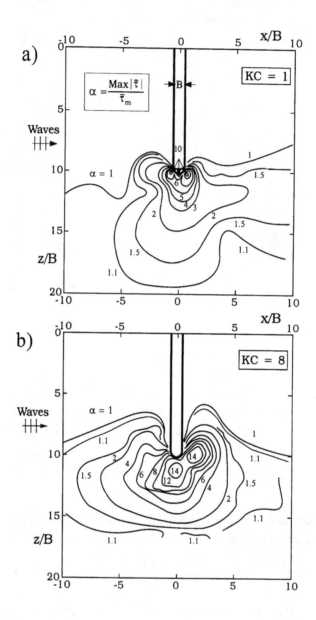

Figure 7.33: Amplification of bed shear stress. Sumer and Fredsøe (1997).

7.2. SCOUR AROUND THE HEAD OF A BREAKWATER

plays a significant role in the erosion process illustrated in Frame 1 in Fig. 7.32.

Scour depth

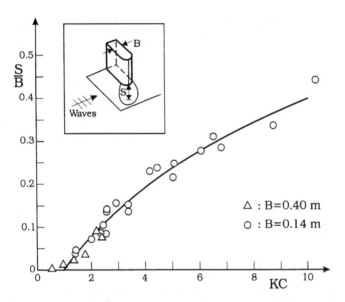

Figure 7.34: Maximum scour depth at the head of a vertical-wall breakwater. Live bed. Sumer and Fredsøe (1997).

Since the formation and the development of the lee-wake vortex are primarily governed by the KC number (as shown by the study of Sumer and Fredsøe, 1997), it might be expected that the resulting scour, too, is mainly governed by this parameter. The scour tests of the latter authors revealed this KC dependence (Fig. 7.34). The scour depth was found to increase with increasing KC number. The scour depth-versus-KC relationship was given by the following empirical expression (for the live-bed situation, $\theta > \theta_{cr}$)

$$\frac{S}{B} = 0.5C[1 - \exp(-0.175(KC - 1))] \qquad (7.33)$$

in which KC = the Keulegan-Carpenter number calculated at the bed (Eq. 7.31), C is an uncertainty factor with a mean value of 1 and a standard deviation of $\sigma_C = 0.6$. The location of the maximum scour was found at the tip of the breakwater head.

Other aspects investigated in Sumer and Fredsøe's (1997) study are among others: the plan view extent of the scour hole, the effects of head shape and the presence of co-directional current.

Regarding the **effect of head shape**, the scour depth was found to increase (by a factor of 2), when the head shape was changed from a round shape to a sharp-edged one. See the original publication for a detailed analysis.

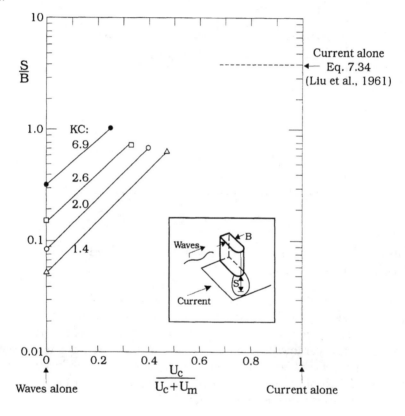

Figure 7.35: Effect of current on scour in combined co-directional wave and current. Live bed. Sumer and Fredsøe (1997). (See the text for details).

Likewise, it was found that the scour depth increased in the presence of

7.2. SCOUR AROUND THE HEAD OF A BREAKWATER

a **co-directional current**. This increase can be very excessive when the current component of the flow is relatively large, Fig. 7.35. Note that Sumer and Fredsøe's (1997) data plotted in Fig. 7.35 do not represent the equilibrium scour, but rather correspond to a scouring time equal to 30 minutes. (This was due to the limited sand layer thickness in the tests). The current scour depth in the same figure (the dashed asymptotic line) represents the empirical relation given by Liu, Chang and Skinner (1961), corresponding to the equilibrium scour depth,

$$\frac{S}{h} = c(\frac{L}{h})^{0.4}(\frac{U_c}{\sqrt{gh}})^{0.33} \qquad (7.34)$$

for the case of a groin exposed to a steady current. Here L = the length of the groin, h = the water depth, U_c = the mean current velocity, g = the acceleration due to gravity, and c is a constant which can be taken as 2.15, as recommended by Liu et al. (1961), when the groin terminates at a vertical wall.

7.2.2 Scour around the head of a rubble-mound breakwater

In the case of a rubble-mound breakwater, the range of the KC number, defined in terms of the *base* diameter of the breakwater head, B, is from practically nil to $O(1)$, meaning that no lee-wake vortices would form in this case (Fig. 7.31), as opposed to the vertical-wall breakwater case. This means that the scour around the head in the present case is caused by mechanisms different from that caused by the lee-wake vortex described in the preceding section.

Two key mechanisms have been identified in the study of Fredsøe and Sumer (1997): (1) The steady streaming which occurs above the bed around the head of the breakwater (Fig. 7.36); and (2) The plunging breaker which occurs locally at the breakwater head (Fig. 7.37).

Streaming-induced scour

The steady streaming illustrated in Fig. 7.36 is caused by the non-uniform wave boundary layer over the bed. This non-uniformity is due to the strong convergent-divergent geometry in plan view of the flow environment. (Note that this phenomenon has been treated experimentally

378   CHAPTER 7. SCOUR AROUND BREAKWATERS

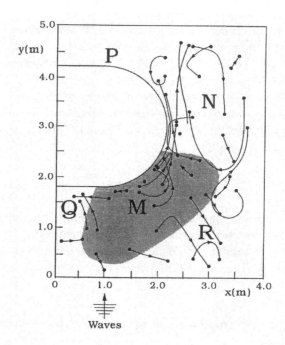

Figure 7.36: Particle trajectories, and scour and deposition at the head of a rubble-mound breakwater. Plan view. M: Scour induced by steady streaming. P: Scour induced by plunging breakwater. N,Q,R: Deposition induced by steady streaming. Fredsøe and Sumer (1997).

in a convergent-divergent channel in an oscillating water tunnel by Sumer, Laursen and Fredsøe, 1993).

This steady streaming generates a scour hole in front of the breakwater head and adjacent to it (Area M in Figs. 7.36 and 7.38). The scour mechanism is rather similar to that in the case of large piles (Chapter 6). The waves (when they are strong enough) will stir up the sediment and bring it into suspension, and the streaming will carry this suspended sediment away from the Area M, and deposit it on the bed over Areas N, Q and R. The end result will be a scour hole emerging in Area M, and the deposition areas over N, Q and R (Figs. 7.36 and 7.38).

Fig. 7.39 displays the amplification factor in the bed shear stress around the head of the breakwater, taken from Fredsøe and Sumer's study (1997).

7.2. SCOUR AROUND THE HEAD OF A BREAKWATER

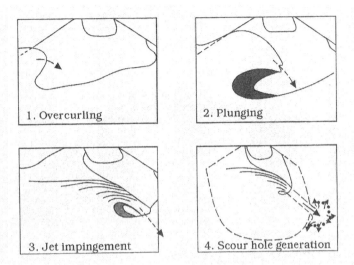

Figure 7.37: Sequence of sketches illustrating the development of 3-D plunging breaker at the head of a rubble-mound breakwater. Fredsøe and Sumer (1997).

In contrast to the vertical-wall breakwater case, the amplification of the bed shear stress in the present case is not very extensive; while, α is $O(10)$ for the vertical-wall breakwater, it is a factor of 4 smaller for $KC = 1$, and a factor of 7 smaller for $KC = 8$ in the case of the rubble-mound breakwater. This large reduction in α is due to the geometry of the rubble-mound structure; quite a substantial amount of water will be diverted over the sloping wall of the breakwater at the head, therefore much less contraction of streamlines will take place at the bed near the tip of the structure, and hence the amplification in the bed shear stress will not be radically large in the rubble-mound case. Yet, the amplification in the order of magnitude of 2-3 around the head will be sufficiently large to stir the sediment and bring it into suspension, and therefore to help develop the streaming-induced scour fairly quickly.

Now, the steady streaming around the head of the breakwater depends on the following quantities: the maximum value of the undisturbed orbital velocity at the bed, U_m, the wave period, T_w, the base diameter of the breakwater head, B, the bed roughness (characterized by the grain diameter, d), the kinematic viscosity, ν, and the side slope of the breakwater, m. From dimensional analysis, the scour and deposition characteristics can be found

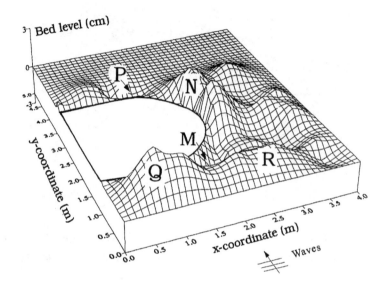

Figure 7.38: 3-D plot of the measured equilibrium bed topography. Fredsøe and Sumer (1997).

to depend on the following nondimensional parameters

$$\frac{S}{B} = f(KC, \theta, m, \frac{a}{d}, Re) \tag{7.35}$$

in which a = the amplitude of the undisturbed orbital motion at the bed and $Re = U_m B/\nu$.

Given the breakwater and the sediment, the scour and deposition will vary only with KC, and θ. When the live-bed conditions are considered ($\theta > \theta_{cr}$), the influence of θ may not be very important, and may be ignored to a first approximation, therefore the scour and deposition characteristics in this case will vary only with KC.

7.2. SCOUR AROUND THE HEAD OF A BREAKWATER

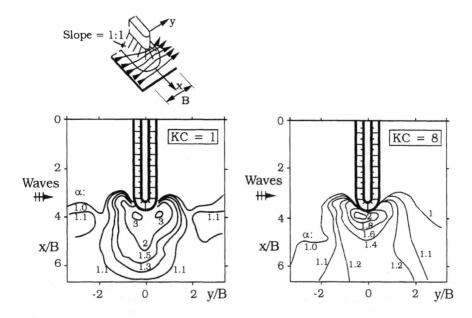

Figure 7.39: Amplification of bed shear stress. Note the side slopes. Fredsøe and Sumer (1997).

Fig. 7.40 displays the scour data presented by Fredsøe and Sumer (1997). As seen, the scour depth correlates rather well with the KC number; the larger the KC number, the larger the scour depth. This is caused by the increasing magnitude of the streaming with increasing KC. Note the agreement between the laboratory results of Fredsøe and Sumer (1997), and the field result of Lillycrop and Hughes (1993) in Fig. 7.40. (Fredsøe and Sumer, 1997, found that the deposition depth also correlates rather well with the KC number; see the original paper for the details).

Finally, the following empirical expression was given in Fredsøe and Sumer's (1997) paper for the scour depth

$$\frac{S}{B} = 0.04 C_1 \left[1 - \exp(-4(KC - 0.05))\right] \qquad (7.36)$$

in which KC = the Keulegan-Carpenter number calculated at the bed and based on the base diameter of the breakwater head, C_1 is an uncertainty factor with a mean value of 1, and the standard deviation $\sigma_{C_1} = 0.2$. The

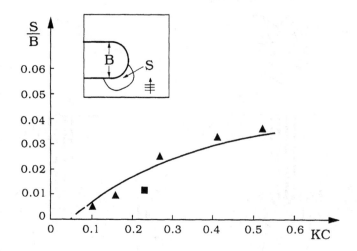

Figure 7.40: Maximum depth of scour hole in front of the rubble-mound breakwater. Scour induced by steady streaming. ▲: Laboratory experiments of Fredsøe and Sumer (1997), Slope = 1:1.5. Live bed. ■: Field data, Lillycrop and Hughes (1993).

waves in Fredsøe and Sumer's experiments (1997) were irregular (this was to avoid large bed undulations which emerge under large-amplitude regular waves in long-duration tests, $O(10) - O(40)$ hrs). The KC number in the preceding equation is defined by

$$KC = \frac{(\sqrt{2}\sigma_U)T_p}{B} \quad (7.37)$$

in which σ_U = the standard deviation of the undisturbed orbital velocity at the bed, and T_p = the wave period corresponding to the peak frequency of the wave spectrum.

Fredsøe and Sumer (1997) give data for the deposition including the plan-view extents of the deposition areas, and also an empirical relation for the deposition depth as a function of the KC number.

Plunging-breaker-induced scour

When the waves are sufficiently large, plunging breakers may occur on the head of the breakwater, as illustrated schematically in Fig. 7.37. Unlike the

7.2. SCOUR AROUND THE HEAD OF A BREAKWATER

familiar two-dimensional breaker, the breaker in the present case is three-dimensional, and it generates a round jet (Frame 3, Fig. 7.37). This jet impinges on the bed and mobilizes the sand grains there, leading to a scour hole (Frame 4, Fig. 7.37). This scour hole is located at the lee-side of the breaker at the junction between the head and the trunk sections (marked P in Fig. 7.38).

By analogy to the scour by a submerged vertical jet (Breusers and Raudkivi, 1991, p. 103), it can be seen that the scour is mainly governed by the following features of the breaker (Fredsøe and Sumer, 1997):

1. the velocity of the jet generated by the breaker (at the point where the jet enters into the main body of the water),

2. the size of the jet (again, at the point of entrance), and

3. the water depth.

The first two features depend on the following parameters: the wave height at breaking, taken as H_s (= the significant wave height) in Fredsøe and Sumer's (1997) analysis; the wave period, T_p; the water depth, h; the side slope of the breakwater, m; and the acceleration due to gravity, g. From dimensional analysis (considering the sediment properties as well), the scour can then be found to depend on the following non-dimensional parameters

$$\frac{S}{H_s} = f(\frac{T_p\sqrt{gH_s}}{h}, \theta, m) \tag{7.38}$$

Given the breakwater and the sediment, and for a live-bed scour ($\theta > \theta_{cr}$), the scour will depend only on the nondimensional parameter $T_p\sqrt{gH_s}/h$.

Fig. 7.41 displays the variation of the scour depth with this parameter. Note the data points from two field studies reported in Lillycrop and Hughes (1993).

Fig. 7.41 shows that the scour depth normalized by H_s correlates with this parameter rather well. The larger the value of this parameter, the larger the scour depth. This can be explained as follows. The numerator, $T_p\sqrt{gH_s}$, actually characterizes the amount of water in the plunging breaker entering into the main body of the water. Clearly, the scour must be directly proportional to this quantity. Regarding the denominator, h, on the other hand, the submerged jet has a penetration distance. If h is large compared with

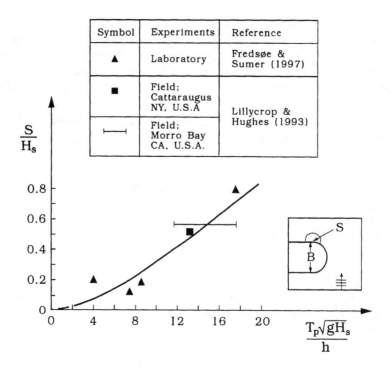

Figure 7.41: Maximum depth of scour hole at the lee side of the rubble-mound breakwater. Scour induced by the plunging breaker. Live bed.

the aforementioned penetration distance, practically no scour would occur. Obviously, the smaller the water depth, the larger the scour. Hence, from the preceding arguments, the scour depth should increase with increasing $T_p\sqrt{gH_s}/h$.

The agreement between the laboratory results of Fredsøe and Sumer (1997) and the field results suggests that the scale effects are not very significant.

The following empirical expression was given for the scour depth, representing the relation depicted in Fig. 7.41

$$\frac{S}{H_s} = 0.01 C_2 \left(\frac{T_p\sqrt{gH_s}}{h} \right)^{1.5} \tag{7.39}$$

in which C_2 is an uncertainty factor with a mean value of 1 and a standard

7.2. SCOUR AROUND THE HEAD OF A BREAKWATER

deviation of $\sigma_{C_2} = 0.34$. Note that the above expression is valid for the live-bed situation ($\theta > \theta_{cr}$).

As mentioned in the original publication, the preceding equations were developed for impermeable, smooth breakwater heads. The porosity and roughness of rubble-mound breakwaters may effectively reduce the scour, and therefore scour estimates based on the preceding equations are somewhat on the conservative side.

7.2.3 Scour protection at the head section

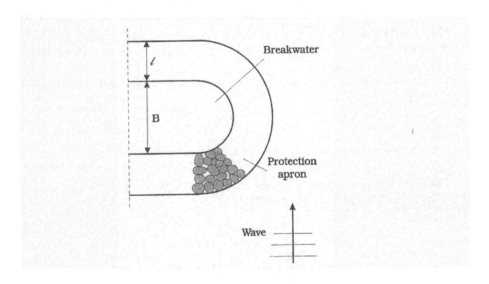

Figure 7.42: Protection apron at the head. B = the width of the vertical-wall breakwater, or the width of the base of the rubble-mound breakwater.

For *vertical-wall breakwaters*, Sumer and Fredsøe (1997) recommend the width of the protection apron (Fig. 7.42) to be selected as

$$\frac{\ell}{B} = 1.75(KC - 1)^{0.5} \tag{7.40}$$

This corresponds to a scour depth at the tip of the protection apron less than 1% of B (i.e., $S/B \leq 0.01$). In Eq. 7.40, KC is defined by Eq. 7.31.

Using the linear wave theory, the preceding equation can be written in the following alternative form

$$\ell = 1.75 B \left[\frac{\pi H}{B \sinh(kh)} - 1 \right]^{0.5} \tag{7.41}$$

in which H = the wave height, k = the wave number and h = the water depth. Sumer and Fredsøe (1997) note that ℓ is virtually the same as the dimension of the lee-wake vortex (Fig. 7.31), the main agent causing scour at the head. This implies that the bed area which is exposed to the action of the lee-wake vortex must be covered for protection. See Sumer and Fredsøe (1997) for further details regarding this issue. Sumer and Fredsøe (1997) also made a sensitivity study with regard to the thickness of the protection apron; this study indicated that the results were insensitive to the height of the protection apron.

For *rubble-mound breakwaters*, the recommended width for the protection apron (Fig. 7.42) is

$$\frac{\ell}{B} = A_1 KC \tag{7.42}$$

in which KC is defined by Eq. 7.37 (Fredsøe and Sumer, 1997). The coefficient $A_1 = 1.5$ for a complete scour protection (i.e., $S/B = 0$) and $A_1 = 1.1$ for $S/B \leq 0.01$. Again, using the linear wave theory, the preceding equation can be written in the following form

$$\ell = \frac{AH_s}{\sinh(kh)} \tag{7.43}$$

in which H_s = the significant wave height and A is

$$A = 3.3 \text{ for a complete scour protection, } S/B = 0, \text{ and} \tag{7.44}$$
$$A = 2.4 \text{ for } S/B \leq 0.01$$

Fredsøe and Sumer (1997) note that the preceding equation is for protection against the streaming-induced scour (Section 7.2.2 above). Their scour-protection tests showed, however, that the implemented widths of the protection layer was able to protect the sand bed completely against the breaker-induced scour. However, they note that, in these tests, scour (damage) occurred in the protection layer itself, despite the extremely small values of the Shields parameter, corresponding to the protection material (θ was a factor of 100 smaller than that of the sand bed).

7.2. SCOUR AROUND THE HEAD OF A BREAKWATER

7.2.4 Influence of finite length of breakwater

Experiments on the scour and deposition around a single detached offshore rubble-mound breakwater have been performed by Sutherland, Chapman and Whitehouse (1999); see also Sutherland, Whitehouse and Chapman (2000). A breakwater with 1:2 front and rear slopes and a crest width of 0.15 m was constructed with 2 to 3 armour layers over a rock core. The model breakwater had a front to back width (B) of around 1.9 m. It had a 4 m long uniform central (trunk) section and semicircular round heads which extended for a further 0.9 m each. The water depth was 0.3 m and the bed sediment $d_{50} = 0.24$ mm. The 3-D nature of the tests allowed interaction between the head and trunk sections of the breakwater complementing the 3-D head (Fredsøe and Sumer, 1997) and 2-D trunk (Sumer and Fredsøe, 2000) tests. The tests were for the live-bed condition, and sediment was transported both as bedload and partially as near-bed suspended load.

A total of approximately 30,000 (average period) waves was run for each test case.

The tests investigated how oblique-wave incidence (Fig. 7.43, bottom panel) and longshore currents affect the distribution of scour and deposition. The results clearly show the 3-D effects present, even for the normal incidence case (Fig. 7.43, top panel).

The normal incidence wave condition produced scour depths similar to the 3-D head-scour experiments by Fredsøe and Sumer (1997), described in Section 7.2.2, Fig. 7.44 - circle with cross (A). For the oblique wave case - square with cross - the scour depth at the head is considerably reduced (Ci) and is only marginally increased with the addition of the shore-parallel current (Bi) - diamond with cross. However, the scour depth at the junction between the head and trunk sections is seen to be greater than at the head, but similar for the oblique wave and oblique wave-current (Cii and Bii). For the normal incidence waves and oblique waves deposition dominates at the trunk section of the breakwater (in contrast to the 2-D trunk-section results in Fig. 7.17) due to the longshore gradients in sediment transport, resulting in reduced scouring in that region. However, the influence of the current becomes clear, translating the deposition pattern further downstream along the breakwater and contributing to the formation of a scour trough along the toe of the breakwater. In this case the scour depth scales more to the normal incidence situation (Biii).

A 0.5 m ($0.26B$) wide scour protection layer of stones (4 - 8 mm sieve

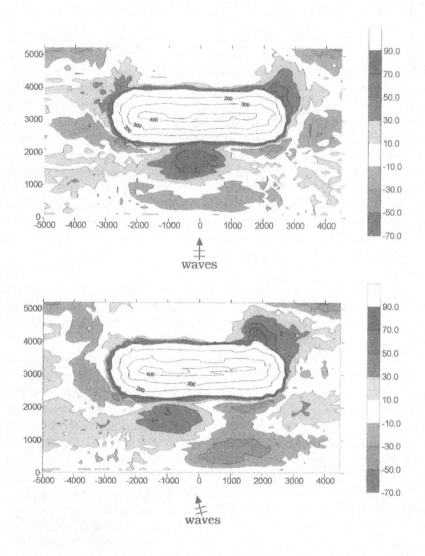

Figure 7.43: Bed topography after 30.000 waves around the rubble-mound breakwater. (a) Normal incidence waves. (b) Oblique incidence waves (the angle = 20^0). x, y in mm, and bed elevation bands in mm. Sutherland et al. (1999).

7.2. SCOUR AROUND THE HEAD OF A BREAKWATER

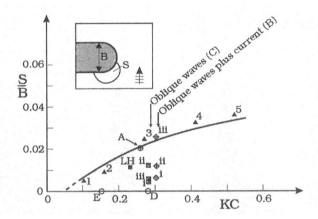

Figure 7.44: Maximum depth of scour in front of the breakwater. Comparison. A to E: Sutherland et al. (1999) (see text for details). ▲: Fredsøe and Sumer (1997). ■: Field data, Lillycrop and Hughes (1993). The latter two sets of data from Fig. 7.40.

diameter) was placed all the way around the toe of the structure on top of the flat bed prior to Test D of Sutherland et al. (1999, 2000). The scour protection layer did prevent any significant scour around the toe of the breakwater, either in front or behind the breakwater. Some scouring was evident adjacent to the non-protected bed. This may have been partly due to sand and protection layer falling into a scour pit at the edge of the layer and partly due to sand being winnowed up through the scour protection layer and then being transported when at the top of the protection layer. Significant areas of the scour protection layer at the up-wave roundhead, more exposed, became covered in sand during the experiment. In general the bed protection material was stable except right at the edge of the most exposed areas. Its performance could have been improved with an underlayer or increased thickness to reduce scour/winnowing within the protection layer.

Test E of Sutherland et al. (1999, 2000) with more energetic (steeper-wave) conditions was run on the final Test D bathymetry. There were no significant changes to the scour near the protection layer or additional displacement of the protection layer. Small changes began to develop in the bed as areas of deposition were eroded and some areas of scour were infilled.

7.3 Scour at jetties

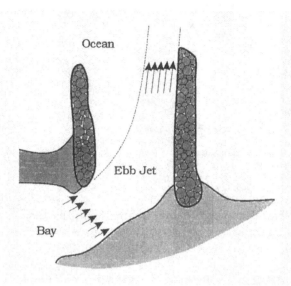

Figure 7.45: Ebb flow deflection. Hughes and Kamphuis (1996).

A jetty is a coastal inlet structure extending into the water to direct and confine river or tidal flow into a channel, and to prevent or reduce the shoaling of the channel by littoral material (Shore Protection Manual, 1977). Lillycrop and Hughes (1993) give numerous examples of scour hole formations at the tip and along the sides of jetties, and the resulting failures monitored in case studies in the U.S..

Scour hole formation near the channel-side toe and at/near the head of inlet jetty structures is a problem at many navigation inlets, Hughes and Kamphuis (1996). The experiments of the latter authors examined flood-flow and ebb-flow scour that develops at inlets under various combinations of tidal flow discharge, flow direction, and incoming wave action. It appears that one of the more important mechanisms of scour during ebb flow is strong ebb currents that exit the inner bay and impinge on the structure at an angle, Fig. 7.45. The deflection of the ebb flow results in increased velocities (due to the contraction of streamlines). These increased velocities scour the bottom over many ebb cycles until eventually flow velocities are reduced to non-scouring levels. Hughes and Kamphuis (1996) note that the scour process was further complicated by the influence of waves in the channel, entrainment at the flow shear interface, changes in the flow velocity over the ebb cycle, and the influence of a porous jetty structure. Hughes and Kamphuis (1996) also developed an easily-applied prediction capability for maximum flow velocity along with a simplified scour prediction method. Finally, note that waves breaking across the jetty head in the absence of currents may also cause scour (see Section 7.2.2).

7.4 References

1. Arneborg, L., Hansen, E.A. and Juhl, J. (1995 a): Numerical modelling of local scour at vertical structures. In: Final Proceedings of the project Monolithic (Vertical) Coastal Structures, Commission of the European Communities, Directorate General for Science, Research and Development, MAST contract No. MAS2-CT92-0042, Paper 3.2.

2. Arneborg, L., Hansen, E.A. and Juhl, J. (1995 b): Numerical modelling of local scour at partially reflective structures. In: Final Proceedings of the project Rubble-Mound Breakwater Failure Modes, Commission of the European Communities, Directorate General for Science, Research and Development, MAST contract No. MAS2-CT92-0047, vol. 2.

3. Bagnold, R.A. (1966): An approach to sediment transport problem from general physics. Prof. Paper No. 422-I, U.S. Geological Survey, U.S. Govt. Off., Washington, D.C.

4. Baquerizo, A. and Losada, M.A. (1998 a): Longitudinal current induced by oblique waves along coastal structures. Coastal Engineering, vol. 35, 211-230.

5. Baquerizo, A. and Losada, M.A. (1998 b): Sediment transport around a mound breakwater: The toe erosion problem. Proc. 26th International Conference on Coastal Engineering, Copenhagen, Denmark, vol. 2, 1721-1729.

6. Batchelor, G.K. (1965): The motion of small particles in turbulent flow. Proc. 2nd Australasian Conf. on Hydr. and Fluid Mech., Univ. of Auckland, Auckland, Australia, 019-041.

7. Bartels, A., Kloos, M. and Phelp, D. (2000): Failure and repair of the toe of an Accropode breakwater. In: Book of Abstracts, 26th International Coastal Engineering Conference, ASCE, Sydney, Australia, vol. 1, Paper No. 74.

8. Breusers, H.N.C. and Raudkivi, A.J. (1991): Scouring. A.A. Balkema, Rotterdam, viii + 143 p.

9. Carter, T.G., Liu, L.-F.P. and Mei, C.C. (1973): Mass transport by waves and offshore sand bedforms. J. Waterways, Harbors and Coastal Engineering, ASCE, vol. 99, No. WW2, 165-184.

10. Coastal Engineering Manual (2001): Scour and Scour Protection. Chapter VI-5-6, Engineer Manual EM 1110-2-1100, Headquarters, U.S. Army Corps of Engineers, Washington, D.C.

11. De Best, A., Bijker, E.W. and Wichers, J.E.W. (1971): Scouring of sand in front of a vertical breakwater. Proc. Conference on Port and Ocean Engineering under Arctic Conditions, The Norwegian Institute of Technology, Trondheim, Norway, vol. 2, 1077-1086.

12. Dean, R. and Dalrymple, R.A. (1984): Water Wave Mechanics for Engineers and Scientists. Prentice-Hall, Inc., xi + 353 p.

13. Engelund, F.A. and Fredsøe, J. (1976): A sediment transport model for straight alluvial channels. Nordic Hydrology, vol. 7, No. 5, 293-306.

14. Fredsøe, J. (1984): The turbulent boundary layer in combined wave-current motion. J. Hydraulic Engineering, ASCE, vol. 110, No. 8. 1103-1120.

15. Fredsøe, J. and Deigaard, R. (1992): Mechanics of Coastal Sediment Transport. World Scientific, Singapore, xviii + 369 p.

16. Fredsøe, J. and Sumer, B.M. (1997): Scour at the round head of a rubble-mound breakwater. Coastal Engineering, vol. 29, 231-262.

17. Fredsøe, J., Sumer, B.M. and Bundgaard, K. (2001): Scour at a riprap revetment in currents. In: Proc. 2nd IAHR Symposium on River and Estuarine Morphodynamics, 10-14. September, 2001, Obihiro, Japan.

7.4. REFERENCES

18. Gironella X. and Sánchez-Arcilla A. (2000): Hydrodynamic behaviour of submerged breakwaters. Some remarks based on experimental results. Proc. Coastal Structures '99, Santander, Spain, 7-9 June, 1999, vol. 2, 891-896, Balkema.

19. Gislason, K., Fredsøe, J., Mayer, S. and Sumer, B.M. (2000): The mathematical modelling of the scour in front of the toe of a rubble-mound breakwater. In: Book of Abstracts, 27th International Coastal Engineering Conference, ASCE, Sydney, Australia, vol. 1, Paper No. 130.

20. Gunbak, A.R. (1976): The stability of rubble mound breakwaters in relation to wave breaking and run-down characteristics and to the $\xi \sim \tan\alpha.T/\sqrt{H}$ number. Div. Port and Ocean Eng., The Norwegian Institute of Technology, Trondheim, Norway, Rep. 1- 1976.

21. Gunbak, A.R. (1979): Rubble mound breakwaters. Div. Port and Ocean Eng., The Norwegian Institute of Technology, Trondheim, Norway, Rep. 1-1979.

22. Gunbak, A.R., Gokce, T. and Guler, I. (1990): Erosion and protection of Samandag Breakwater. J. Coastal Research, Proc. of Skagen Symposium, 2-5. September, 1990, vol. 2, 753-771.

23. Hales, L.Z. (1980): Erosion Control of Scour During Construction. Technical Report HL-80-3, US Army Engineer Waterways Experiment Station, Coastal Engineering Research Center, Vicksburg, Mississippi.

24. Herbich, J. B. and Ko, S.C. (1969): Scour of sand beaches in front of seawalls. Proc. Eleventh Conference on Coastal Engineering, ASCE, London, England, September 1968, Chapter 40, 622-643.

25. Herbich, J.B., Schiller, R.E., Dunlap, W.A. and Watanabe, R.K. (1984): Seafloor Scour. Marcel and Dekker, Inc., New York and Basel.

26. Hsu, J.R.C. (1990): Short-crested waves. In: Handbook of Coastal and Ocean Engineering (ed. J.B. Herbich), vol. 1, 95-174, Gulf Publishing Company.

27. Hsu, J.R.C., Tsuchia, Y. and Silvester, R. (1979): Third-order approximation to short-crested waves. J. Fluid Mechanics, vol. 90, part 1, 179-196.

28. Hsu, J.R.C., Silvester, R. and Tsuchia, Y. (1980): Boundary-layer velocities and mass transport in short-crested waves. J. Fluid Mechanics, vol. 99, part 2, 321-342.

29. Hsu, J.R.C. and Silvester, R. (1989): Model test results of scour along breakwaters. J. Waterway, Port, Coastal, and Ocean Engineering, ASCE, vol. 115, No. 1, 66-85.

30. Hughes, S.A. (1992): Estimating wave-induced bottom velocities at a vertical wall. J. Waterway, Port, Coastal, and Ocean Engineering, ASCE, vol. 118, No. 2, 175-192.

31. Hughes, S.A. and Fowler, J.E. (1991): Wave-induced scour prediction at vertical walls, Proc. Coastal Sediments '91, Seattle, WA, ASCE, vol. 2, 1886-1900.

32. Hughes, S.A. and Fowler, J.E. (1995): Estimating wave-induced kinematics at sloping structures. J. Waterway, Port, Coastal, and Ocean Engineering, ASCE, vol. 121, No. 4, 209- 215.

33. Hughes, S.A. and Kamphuis, J.W. (1996): Scour at coastal inlet structures. Proc. 25th International Conference on Coastal Engineering, Orlando, Florida, ASCE, Chapter 175, vol. 2, 2258-2271.

34. Hughes, S.A. and Schwichtenberg, B.R. (1998): Current-induced scour along a breakwater at Ventura Harbor, CA - experimental study. Coastal Engineering, vol. 34, 1-22.

35. Irie, I. and Nadaoka, K. (1984): Laboratory reproduction of seabed scour in front of breakwaters. Proc. 19th International Conference on Coastal Engineering, Houston, TX, ASCE, Chapter 116, vol. 2, 1715-1731.

36. Klopman, G. and van der Meer, J.W.(1999): Random wave measurements in front of reflective structures, J. Waterway, Port, Coastal, and Ocean Engineering, ASCE, vol. 125, No. 1, 39- 45.

7.4. REFERENCES

37. Lillycrop, W.J. and Hughes, S.A. (1993): Scour hole problems experienced by the Corps of Engineers; Data presentation and summary. Miscellaneous papers. CERC-93-2, US Army Engineer Waterways Experiment Station, Coastal Engineering Research Center, Vicksburg, MS.

38. Liu, H.K., Chang, F.M. and Skinner, M.M. (1961): Effect of bridge constriction on scour and backwater. Civ. Eng. Section, Colorado State Univ., Fort Collins, Colorado, Report No. CER60HKL22, Prepared for Bureau of Public Roads under Contract CPR11-5480, February, 1961.

39. Longuet-Higgins, M.S. (1957): The mechanics of the boundary layer near the bottom in a progressive wave. Proc. 6th International Conference on Coastal Enigineering, Gainesville, Palm Beach and Miami Beach, Florida, December 1957, 184-193.

40. Losada, M.A. and Gimenez-Curto, L.A. (1981): Flow characteristics on rough, permeable slopes under wave action. Coastal Engineering, vol. 4, no. 3, 187-206.

41. Losada, I.J., Silva, R. and Losada, M.A. (1997): Effect of reflective vertical structures permeability on random wave kinematics. J. Waterway, Port, Coastal, and Ocean Engineering, ASCE, vol. 123, No. 6, 347-353.

42. Mantz, P.A. (1978): Bedforms produced by fine, cohesionless, granular and flakey sediments under subcritical water flows. Sedimentology, vol. 25, 83-103.

43. Mayer, S., Grapon, A. and Sørensen, L.S. (1998): A fractional step method for unsteady free-surface flow with applications to non-linear wave dynamics. Int. J. Numer. Meth. Fluids, vol. 28, 293-315.

44. Mei, C.C. (1989): The Applied Dynamics of Ocean Surface Waves. World Scientific, xx + 740 p.

45. Nezu, I. and Nakagawa, H. (1993): Turbulence in Open-Channel Flows. IAHR Monograph, A.A. Balkema Publisher, xii + 281 p.

46. Oumeraci, H. (1994 a): Review and analysis of vertical breakwater failures - lessons learned. Coastal Engineering, Special Issue on Vertical Breakwaters, vol. 22, 3-29.

47. Oumeraci, H. (1994 b): Scour in front of vertical breakwaters - Review of problems. Proc. International Workshop on Wave Barriers in Deep Water, Jan. 10-14, 1994, Port and Harbour Research Institute, Yokosuka, Japan, 281-307.

48. Sanchez-Arcilla, A., Gironella, X., Verges, D., Sierra, J.P., Pena, C. and Moreno, L. (2000): Submerged breakwater and "bars": From hydrodynamics to functional design. In: Proceedings of the 27th International Conference on Coastal Engineering (ICCE'00). Sydney, July 2000.

49. Sato, S. and Mitsunobu, N. (1990): A numerical model for sand transport under compound waves, Proc. 22nd Coastal Engineering Conference, 2-6- July 1990, Delft, The Netherlands, vol 3, 2658-2670.

50. Shields, A. (1936): Anwendug der Aehnlichkeitsmechanik und der Turbulenzforschung auf die Geschiebewegung. Mitteilungen Wasserbau und Schiffbau, Nr. 6.

51. Shore Protection Manual (1977): U.S. Army Coastal Engineering Research Center, vol. II, Department of the U.S. Army Corps of Engineers.

52. Silvester, R. (1990): Scour around breakwaters and submerged structures. In: Handbook of Coastal and Ocean Engineering (ed. J.B. Herbich), vol. 2, 959-996, Gulf Publishing Company.

53. Silvester R. and Hsu, J.R.C. (1989): Sines revisited. J. Waterway, Port, Coastal, and Ocean Engineering, ASCE, vol. 115, No.3, 327-344.

54. Silvester, R. and Hsu, J.R.C. (1997): Coastal Stabilization. World Scientific, xvi + 578 p.

55. Sumer, B.M. (1986): Recent developments on the mechanics of sediment suspension. In: Proc. Euromech 192: Transport of Suspended Solids in Open Channels, 11-15 June 1985, Neubiberg, Germany. A.A. Balkema Publisher, The Netherlands, 3-13.

56. Sumer, B.M. and Fredsøe, J. (1997): Scour at the head of a vertical-wall breakwater. Coastal Engineering, vol. 29, 201-230.

7.4. REFERENCES

57. Sumer, B.M. and Fredsøe, J. (2000): Experimental study of 2D scour and its protection at a rubble-mound breakwater. Coastal Engineering, vol.. 40, 59-87.

58. Sumer, B.M., Laursen, T. and Fredsøe, J. (1993 b): Wave boundary layers in a convergent tunnel. Coastal Engineering, vol. 20, 317-342.

59. Sumer, B.M., Whitehouse, R. and Tørum, A. (2001): Scour around coastal structures. A summary of recent research. Coastal Engineering.

60. Sutherland, J. and O'Donoghue, T. (1998): Wave phase shift at coastal structures. J. Waterway, Port, Coastal, and Ocean Engineering, ASCE, vol. 124, No. 2, 90- 98.

61. Sutherland, J., Chapman, B. and Whitehouse, R. (1999): SCARCOST Experiments in the UK Coastal Research Facility. Data on scour around a detached rubble mound breakwater. HR Wallingford Report TR 98, December 1999.

62. Sutherland, J., Whitehouse, R.J.S. and Chapman, B. (2000): Scour and deposition around a detached rubble mound breakwater. Coastal Structures'99, Santander, Spain, 7 - 10 June 1999. Editor I. Losada. Published by A.A. Balkema/Rotterdam/Brookfield.

63. US Army Corps of Engineers (1994): Coastal Groins and Nearshore Breakwaters. Technical Engineering and Design Guides as Adapted from the US Army Corps of Engineers, No. 6, ASCE, iii + 87 p.

64. Vergés, D. and Sánchez-Arcilla, A. (2000): Wave induced currents in the vicinity of coastal structures. Coastal Structures '99, Santander, Spain, 7-9 June, 1999, vol. 2, 883- 890, Balkema.

65. Xie, S.L. (1981): Scouring patterns in front of vertical breakwaters and their influence on the stability of the foundations of the breakwaters. Report, Department of Civil Engineering, Delft University of Technology, Delft, The Netherlands, September, 61 p.

66. Xie, S.L. (1985): Scouring patterns in front of vertical breakwaters. Acta Oceanologica Sinica, vol. 4, No. 1, 153-164.

Chapter 8
Scour at seawalls

Seawalls are structures placed parallel to the shoreline, to separate a land area from a water area. These structures are constructed basically to prevent landward retreat of the shoreline, therefore to prevent loss of the upland by wave action.

Fig. 8.1 displays various types of seawalls, adapted from Shore Protection Manual (1977). Although the majority of seawalls are vertical faced, or nearly vertical faced (Fowler, 1992) (Figs. 8.1 a and b), seawalls can be constructed with side slopes as well (rubble-mound seawalls) (Figs. 8.1 c and d).

Mechanism of scour in front of a seawall varies considerably. There are several 2-D and 3-D scour scenarios.

2-D scour scenarios (where the waves approach the coast at a right angle) may involve the following flow processes:

1. The waves may break before they reach the seawall (Fig. 8.2 a); or
2. They may break on the seawall itself (Fig. 8.2 b); or
3. They may reach the seawall without breaking and are reflected (Fig. 8.2 c); or
4. They may overtop the seawall (Fig. 8.2 d).

The latter two scenarios may occur during storm surges.

Possible 3-D scenarios, on the other hand, may include processes such as

1. The waves may be obliquely incident breaking waves; in this case, longshore currents may be generated; or

400   CHAPTER 8. SCOUR AT SEAWALLS

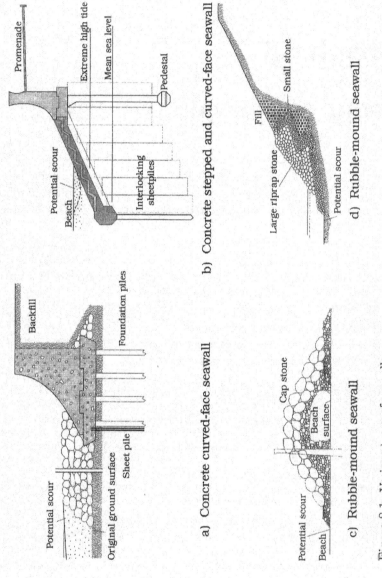

Figure 8.1: Various types of seawalls.

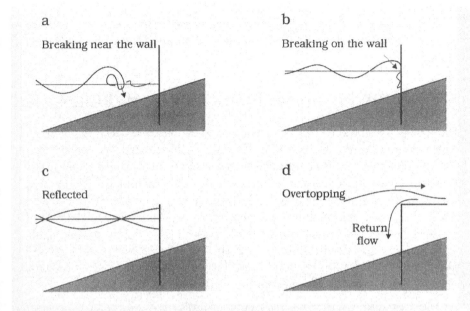

Figure 8.2: Flow processes leading to 2-D scour at seawalls.

2. The waves may be obliquely incident but nonbreaking waves; in this case, a steady streaming may be generated in the longshore direction; or

3. There may be also longshore currents generated by tidal motion.

Clearly, the scour mechanisms in all these cases will be different.

The toe scour at seawalls may result in most serious forms of damage. For instance, according to a CIRIA (1986) report, it accounted for over 12% of the case histories studied and indirectly responsible for up to a further 5% of cases, including collapses/breaching of seawalls and washing out of fill materials (see also Powell, 1987). Incidents have been reported worldwide (Powell, 1987, Kraus and McDougal, 1996, Powell and Whitehouse, 1998). Scour depths of from $O(1 \text{ m})$ to as much as $O(10 \text{ m})$ have been measured.

This chapter will review the existing work on scour at seawalls.

We shall first concentrate on scour induced by breaking waves, and then turn our attention to scour processes induced by nonbreaking waves. This will be followed by a brief discussion on scour generated by wave overtopping.

It may be noted that excellent reviews of the subject can be found in Kraus (1987, 1988), Powell (1987), Fowler (1992) and Kraus and McDougal (1996).

8.1 Scour by normally incident breaking waves

Unfortunately, our knowledge of scour due to breaking waves is very limited, and the mechanisms responsible for this kind of scour are not well understood.

The breaking process creates strong downward directed flows that erode the bed (Coastal Engineering Manual, 2001). For example, in the case of the plunging wave breaking (Fig. 8.2 a), the plunging breaker will penetrate down to the bed, and mobilize sediment at the toe. Likewise, in the case of the plunging wave breaking directly on the seawall (Fig. 8.2 b), the impinging wave will direct water down at the toe (in the form of a jet), and similarly mobilize the sediment at the toe (Coastal Engineering Manual, 2001). These processes will presumably lead to scour at the seawall.

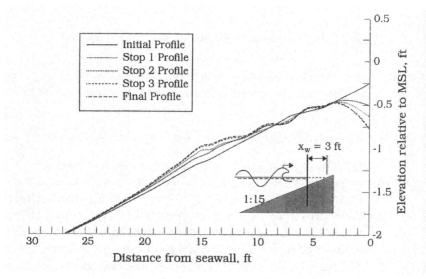

Figure 8.3: Typical bottom profile sequence. $H_0 = 0.68$ ft, $T = 1.97$ s, water depth = 3.8 ft, $h_w = 0.2$ ft. Fowler (1992).

Fig. 8.3 shows a sequence of bed profiles obtained in a laboratory exper-

8.1. SCOUR BY NORMALLY INCIDENT BREAKING WAVES 403

iment with breaking waves in the case of a vertical seawall (Fowler, 1992). The hole at the wall, generated during the scour process, is clear from the figure.

The following subsection will study the parameters governing the scour process.

Governing parameters

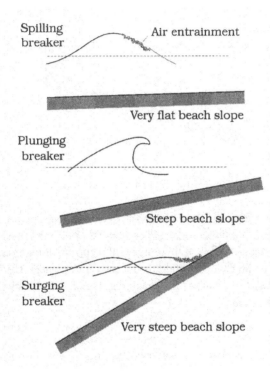

Figure 8.4: Breaker types.

Consider the scour process at a *vertical seawall*. The scour process may be governed by the following four effects:

1. The breaker type;

2. The presence of the seawall;

3. The sediment properties; and

4. The wave boundary layer over the seabed.

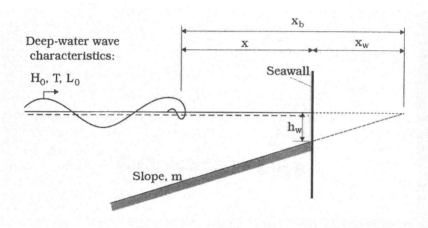

Figure 8.5: Definition sketch.

1. Breaker type. There are basically three kinds of breakers: spilling breakers, plunging breakers and surging breakers (Fig. 8.4) (see, for example, Fredsøe and Deigaard, 1992). Obviously, the type of breaker is an essential factor, influencing the scour. The parameters which govern the breaker type and its characteristics (such as the wave height, H_b, and water depth, h_b, at the breaking point) are

$$\frac{H_0}{L_0}, m \tag{8.1}$$

in which H_0 is the deep-water wave height, and L_0 is the deep-water wave length of the incident waves, and m is the beach slope (see, for example, Fredsøe and Deigaard, 1992, p. 86). This is when there is no wall present. When there is a wall (Fig. 8.5), the broken waves will be reflected from the wall. Therefore, an additional parameter, x/L_0, may be involved with regard to the breaker type and its characteristics. Here, x is the distance of the breaking point from the wall.

2. Presence of the seawall. This will obviously influence the scour process quite significantly (Fig. 8.5). First of all, the water depth at the wall, h_w, is a significant parameter. For a given value of h_w, on the other hand, the

8.1. SCOUR BY NORMALLY INCIDENT BREAKING WAVES

scour will change with x, the distance of the breaking point from the wall. These two quantities may be normalized to give the following nondimensional parameters: h_w/H_0, and x/L_0. Furthermore, there may be an additional parameter, $T_w\sqrt{gH_b}/h_w$, characterizing the penetration of the breaker down to the bed to mobilize the sediment there (Fredsøe and Sumer, 1997) (or alternatively $T_w\sqrt{gH_0}/h_w$, since there is a one-to-one correspondence between H_b and H_0, Fredsøe and Deigaard, 1992, p. 89) in which T_w is the wave period.

Hence, the nondimensional parameters responsible for the scour process regarding the presence of the seawall may be

$$\frac{h_w}{H_0} \text{ (or alternatively } \frac{h_w}{L_0}\text{)}, \quad \frac{x}{L_0}, \quad \frac{T_w\sqrt{gH_0}}{h_w} \qquad (8.2)$$

3. Sediment properties. These may be given in terms of the conventional Shields parameter and the fall-velocity-to-friction-velocity ratio

$$\theta, \quad \frac{w}{U_{fm}} \qquad (8.3)$$

in which θ is defined by

$$\theta = \frac{U_{fm}^2}{g(s-1)d} \qquad (8.4)$$

in which U_{fm} is the maximum value of a characteristic friction velocity.

4. Wave boundary layer. The flow in the boundary layer (and therefore, the sediment transport, and the scour process itself), may be influenced by the bed category of the boundary-layer flow; the governing parameters are

$$\frac{H_0}{d} \text{ (or alternatively } \frac{L_0}{d}\text{)}, \quad RE \qquad (8.5)$$

in which $RE = aU_m/\nu$, the boundary-layer Reynolds number (Eq. 7.10, Chapter 7).

The preceding analysis shows that the scour characteristics may be a function of the nine nondimensional parameters indicated in Eqs. 8.1-8.5.

It may be noted that the initial bed in the above analysis is assumed to be planar. However, if the seawall is placed on a developed coastal profile, then the resulting scour will obviously be a function of the initial bed profile as well. The latter, namely the coastal profile development (in the absence of any structures), has been studied fairly extensively in recent years (Fredsøe and Deigaard, 1992, Chapters 11 and 12), and is outside the scope of the present treatment.

Now, of the nine parameters given in Eqs. 8.1-8.5, RE and θ may be insignificant. RE may be insignificant because the bed under storm conditions acts as a rough boundary in most engineering problems. θ may be insignificant because it may influence the scour rather weakly if the bed is live, namely when $\theta > \theta_{cr}$.

Data regarding the functional dependence of the scour characteristics on the remainder of the parameters, namely

$$\frac{H_0}{L_0},\ m,\ \frac{h_w}{H_0}\text{(or alternatively }\frac{h_w}{L_0}\text{)},\ \frac{x}{L_0},\ \frac{T_w\sqrt{gH_0}}{h_w},\ \frac{w}{U_{fm}},\ \frac{H_0}{d}\text{(or alternatively }\frac{L_0}{d}\text{)} \tag{8.6}$$

are not extensive. The following subsection will present the existing experimental data for the scour depth, along with the results of a numerical investigation directed to identify the governing nondimensional variables responsible for the scour depth.

Scour depth

Fowler's (1992) experimental study. Fowler conducted a series of experiments where the seawall was simulated by a vertical wall. Three locations of the seawall were tested, namely at $x_w = 3$ ft (as in the test presented in Fig. 8.3), at $x_w = -3$ ft, and at $x_w = 0$. *Waves broke well seaward of the seawall, or immediately in front of the seawall in all the tests.* 18 irregular wave tests and 4 regular wave tests were conducted. The beach was initially planar, and had a slope of $m = 1 : 15$.

The maximum scour depth in Fowler's experiments occurred just at the seawall (as in Fig. 8.3) except for the tests where the seawall was located at $x_w = -3$ ft. In the latter tests, the maximum scour depth occurred a little distance away from the seawall, as anticipated.

Fig. 8.6 displays Fowler's (1992) data, the normalized scour depth versus the parameter h_w/L_0 (Eq. 8.2). Here, H_0 represents the significant wave height for the irregular waves.

Although Fig. 8.6 reveals a reasonable correlation between S/H_0 and h_w/L_0, the scatter is quite large. This may be due partly to the dependence of the scour on other parameters as well (Eq. 8.6). Nevertheless, the diagram does show that the scour depth increases with increasing h_w/L_0. As seen from the figure, the scour depth is rather small for negative values of h_w/L_0 (i.e., for the seawall locations at the onshore side of the intersection point

8.1. SCOUR BY NORMALLY INCIDENT BREAKING WAVES

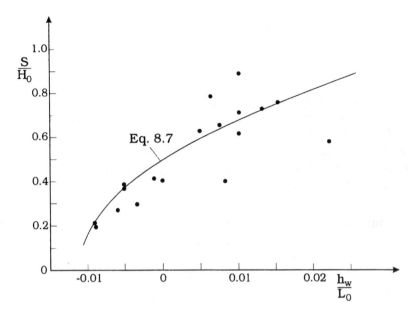

Figure 8.6: Maximum scour depth. Vertical wall breakwater. Initial beach slope = 1:15. Irregular-wave data only. H_0 is significant wave height for irregular waves. Fowler (1992).

between mean sea level and the beach, Fig. 8.5). This is expected; if the seawall is well away on the onshore side from this intersection point, there will even be no scour. As the seawall is moved seaward, however, (i.e., for the increasing values of h_w/L_0), the breaking will take place closer and closer to the seawall, therefore the scour depth will consequently increase. However, this is influenced not only by the parameter h_w/L_0, but also by the parameter x/L_0. Unfortunately, there are no data available revealing the role of x/L_0 alone.

The following empirical equation was proposed by Fowler (1992) (the solid line in Fig. 8.6):

$$\frac{S}{H_0} = (22.72\frac{h_w}{L_0} + 0.25)^{\frac{1}{2}} \tag{8.7}$$

in which L_0 is the deep-water wave length; in the case of irregular waves, $L_0 = \frac{g}{2\pi}T_p^2$ where T_p is the peak period. Fowler (1992) noted the ranges for application of this equation as follows

$$-0.011 < \frac{h_w}{L_0} < 0.045, \text{ and } 0.015 < \frac{H_0}{L_0} < 0.040 \qquad (8.8)$$

Fowler (1992) compared the preceding equation with the data from regular wave experiments of Barnett (1989) and Chesnutt and Schiller (1971) where H_0 was taken as the wave height for the regular waves. Although the scatter was even larger than that experienced in Fig. 8.6, the above equation generally seems to follow the trend indicated by the data.

Fowler's (1992) other findings are:

1. Significant wave height is the best irregular wave design parameter for matching results based on regular wave tests.

2. In the case of the regular wave tests, the scour depth increases by approximately 15%.

3. The data from Fowler's study and numerous field studies tend to support the most widely used rule of thumb, namely $S/H_0 \leq 1$.

Powell and Lowe's (1994) experimental study. An extensive series of tests have been conducted with coarse sediment (5 mm $< d <$ 30 mm) by Powell and Lowe (1994); see also Powell and Whitehouse (1998), and Whitehouse (1998). Irregular waves have been used in the study with an initially plane beach with a slope of 1:17.

Fig. 8.7 displays Powell and Lowe's results in the form of contour plots of the normalized scour depth, S/H_s, as a function of h_w/H_s and H_s/L_m in which H_s is the significant wave height, L_m is the mean wave length (both corresponding to the incident waves).

Fig. 8.7 clearly shows that the wave steepness is an important parameter, as implied by Eq. 8.1.

Also, the figure shows that there are two distinct regions: the erosion region, and the accretion region. Regarding the former, the figure shows that the scour depth at the seawall can reach values as high as $S/H_s = 1.5$. Although the data in Fig. 8.6 and that in Fig. 8.7 cannot be compared directly, the results seem to be consistent in the sense that the maximum scour depth does not exceed $S/H_s = O(1)$ (see item 3 above)

8.1. SCOUR BY NORMALLY INCIDENT BREAKING WAVES

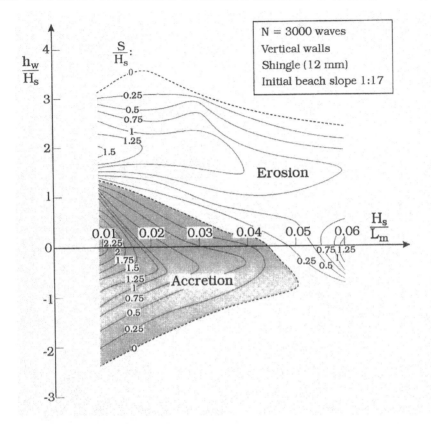

Figure 8.7: Contour plot of experimentally obtained scour depth in front of a vertical seawall. Powell and Lowe (1994).

New research by the same group (Carpenter and Powell, 1998; see also Powel and Whitehouse, 1998, and Whitehouse, 1998) has resulted in a similar diagram for sand with $d = 0.2$ mm (Fig. 8.8). In this latter work, the data have been obtained not from experiments but from numerical simulations with COSMOS-2D, a cross-shore process-based numerical model by HR Wallingford. (The accretion phase is not plotted on the grounds that the cross-shore models to predict the onshore movement of sediment on beaches only have limited capability, Whitehouse, 1998).

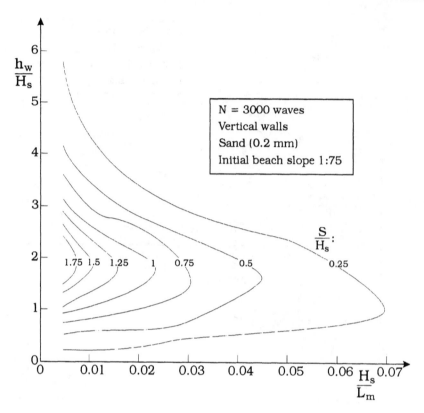

Figure 8.8: Contour plot of numerically obtained scour in front of a vertical seawall. Carpenter and Powell (1998).

To date, no detailed experimental information is available on the variation of the scour characteristics as a function of the other parameters given in Eq. 8.6, i.e., m, x/L_0, $T\sqrt{gH_0}/h_w$, θ, w/U_{fm}, H_0/d (or alternatively L_0/d), and RE. However, the numerical simulations of McDougal, Kraus and Ajiwibowo's (1996) give the functional dependency of the scour depth on some of these parameters. The following subsection will describe this study.

McDougal, Kraus and Ajiwibowo's (1996) numerical study. These authors carried out a numerical study of scour in front of a vertical seawall. Their numerical model basically comprises the following two elements: (1) a

8.1. SCOUR BY NORMALLY INCIDENT BREAKING WAVES   411

wave-transformation model, to predict the wave transformation (refraction, shoaling, breaking, and reflection from the seawall), and (2) a cross-shore sediment transport algorithm, to predict the beach profile. The model, widely known as SBEACH (Larson and Kraus, 1989), has been extended in McDougal et al.'s (1996) study to include the effect of reflection at the seawall as well. The model results have been compared with the results of a large-scale physical-model study. The latter tests have been conducted in the SUPERTANK Laboratory Data Collection Project (Kraus, Smith and Sollitt (1992), and described in McDougal et al. (1996).

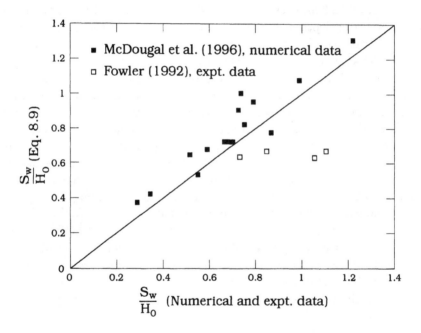

Figure 8.9: Scour depth approximation based on the deep-water wave height. McDougall et al. (1996).

An empirical expression has been determined from the obtained numerical scour-depth data. McDougal et al. (1996) note that the objective is not to develop a design equation, but rather to identify those variables which are most important in the scour process. The empirical equation developed in

McDougal et al.'s study is as follows

$$\frac{S}{H_0} = 0.41 m^{0.85} (\frac{L_0}{H_0})^{1/5} (\frac{h_w}{H_0})^{1/4} (\frac{H_0}{d})^{1/3} \qquad (8.9)$$

in which S is the scour depth at the seawall. McDougal et al. (1996) compared the preceding equation with the numerical results and experimental measurements from Fowler (1992), which are reproduced here in Fig. 8.9. (In Fig. 8.9, only monochromatic wave data with the seawall initially in the water are included).

Eqs. 8.7 and 8.9 have a similar dependency on the deep-water wave height and initial water depth at the wall. However, in Eq. 8.7 the dependency on the deep-water wave length (wave period) is opposite to that of Eq. 8.9.

A model similar to that of McDougal et al. (1996) has been developed by Rakha and Kamphuis (1997 b) to predict the scour in the vicinity of a seawall. The model essentially comprises four modules, namely the wave-transformation module (Rakha and Kamphuis, 1994), the wave-induced-current module (Rakha and Kamphuis, 1997 a), the sediment-transport module (Rakha and Kamphuis, 1997 b), and the morphology module (Rakha and Kamphuis, 1997 b). Rakha and Kamphuis (1997 b) tested their model against wave-flume and wave-basin tests (reported partly also in Kamphuis, Rakha and Jui, 1992), and predicted the beach evolution (including the scour in front of the seawall) well for the cases simulated.

Influence of sloping seawalls. In the case of sloping seawalls, there are no generally accepted methods for estimating maximum scour depth and other characteristics of the scour process.

Rules of thumb given by Coastal Engineering Manual (2001) regarding this issue are

1. Maximum scour at the toe of a sloping structure is expected to be somewhat smaller than that calculated for a vertical wall at the same location and under the same wave conditions. Hence, a conservative scour estimate is provided by the vertical-wall scour prediction equations.

2. Structures with larger porosity will experience smaller wave-induced scour.

8.1. SCOUR BY NORMALLY INCIDENT BREAKING WAVES

3. Scour depths are significantly increased when along-structure currents act concurrently with waves.

4. Obliquely incident waves may cause larger scour than normally incident waves because the short-crested waves increase in size along the structure. Also, oblique waves generate flows parallel to the structure.

Time scale

The time scale is expected to depend on the parameters responsible for the scour process itself (Eqs. 8.1-8.5), namely

$$T^* = T^*(\frac{H_0}{L_0}, m, \frac{h_w}{H_0}, \frac{x}{L_0}, \frac{T_w\sqrt{gH_0}}{h_w}, \theta, \frac{w}{U_{fm}}, \frac{H_0}{d}, RE) \qquad (8.10)$$

in which T^* is the normalized time scale defined by

$$T^* = \frac{(g(s-1)d^3)^{1/2}}{H_0^2}T \qquad (8.11)$$

in which T is the time scale of scour. No detailed study is yet available, investigating the time scale as a function of these parameters.

McDougal et al. (1996), in their numerical simulation of large-scale laboratory scour processes described in the preceding paragraphs, found that approximately 99% of the equilibrium scour occurs by 14.000 waves. This value corresponds to 39 hours of 10 sec-period waves. However, 50% of this scour occurred in the first 6 hours of the storm.

3-D effects

Rakha and Kamphuis (1997a), referred to earlier in conjunction with the 2-D scour at a seawall, developed a numerical model capable of predicting the **longshore current** caused by obliquely incident breaking waves for the case of a beach backed by a seawall. They tested their model against the wave-flume, and wave-basin, fixed bed flow measurements.

Rakha and Kamphuis (1997b) also developed a morphology model to predict the erosion in the vicinity of a seawall, as mentioned previously. The latter authors reported results from the numerical tests (Rakha and Kamphuis, 1997b, Figs. 7.15 and 7.16) where the scour is generated by obliquely

414 CHAPTER 8. SCOUR AT SEAWALLS

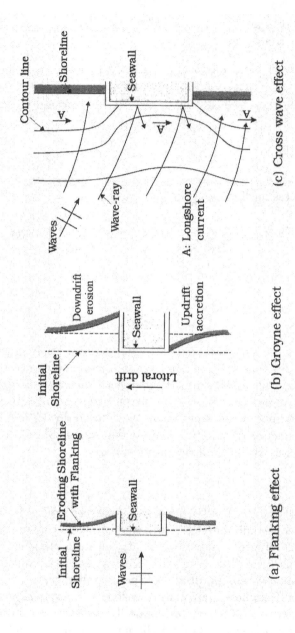

Figure 8.10: 3-D effects.

incident breaking waves (the angle of incidence was 10⁰), and compared the results with those from physical-model experiments (reported in Kamphuis et al., 1992). The scour depth at the seawall was found to be rather close to the value of the breaking significant wave height in these tests.

Other 3-D effects will be present when the seawall has a finite length, such as

1. the **flanking effect** (Fig. 8.10 a), McDougal, Sturtevant and Komar (1987) and Toue and Wang (1990); see the former reference for design equations regarding the scour depths and the plan-view extents of the scour holes;

2. the **groyne effect** (Fig. 8.10 b), Toue and Wang (1990); see Fredsøe and Deigaard (1992, p. 343) for the effect of a groyne on the coastline; and

3. the **cross-wave effect** (Fig. 8.10 c); Toue and Wang (1990).

8.2 Scour by normally incident nonbreaking waves

In the case when the waves are reflected, and there is no wave breaking, the situation will be rather similar to that described for vertical-wall breakwaters. Therefore, the design equations and design diagrams given for vertical-wall breakwaters (Eqs. 7.11-7.14) can be adopted for the present vertical seawall case, too.

The result in Fig. 8.6 for the breaking waves together with those obtained for nonbreaking reflected waves (the standing waves) depicted in Fig. 7.10 (Chapter 7) are plotted in Fig. 8.11. As seen, the scour depth first increases with h_w/L_0 in the range where the scour is caused by the breaking waves, and reaches values like $S/H_0 = O(1)$, and, subsequently, it begins to decrease with increasing h_w/L_0 in the range where the scour is caused by the nonbreaking, reflected waves.

Influence of sloping seawalls. Similar to the case of the vertical-wall breakwaters, the design equations and design diagrams given for rubble-mound breakwaters (Eqs. 7.18 and 7.19) can be used for the present case as well. *However, caution must be exercised in that the seawall case with*

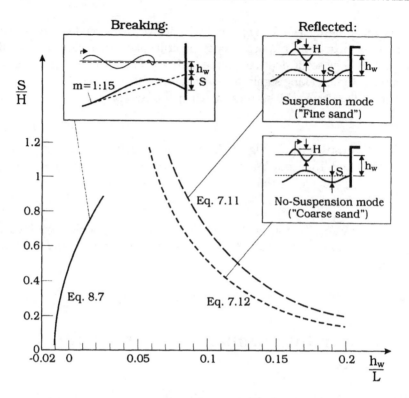

Figure 8.11: Scour depth for vertical seawalls. H and L are the deep-water wave height and wave length, respectively. Breaking-wave curve from Fig. 8.6 and reflected-wave curve from Fig. 7.10.

rubble-mound material may experience larger scour than the breakwater case because of larger reflection coefficients.

Kim, Park, O'Connor, Lee, Hwang and Kim (1999) report a case study in conjunction with the rubble-mound Sooyung seawall in Korea. The front slope of the seawall was 1:1.5, the estimated maximum offshore wave height was 7.6 m, the wave period was 9.9 s and the water depth was 8.5 m with a tidal range of 0.5 m. Kim et al. (1999) report a scour hole formation, 2 m in depth and 20 m in width, along the length of the seawall. The reported scour depth is somewhat larger than expected (Eqs. 7.18 and 7.19), supporting the argument in the preceding paragraph.

8.3. SCOUR INDUCED BY WAVE OVERTOPPING

Influence of irregular waves. Likewise, the influence of irregular waves on the end results can be taken as in the case of breakwaters in the previous chapter (Sections 7.1.1 and 7.1.2).

Influence of obliquely incident waves. The information given in Chapter 7 (Section 7.1.1) in conjunction with scour at the trunk section of vertical-wall breakwaters may also be applicable to the present case as well.

8.3 Scour induced by wave overtopping

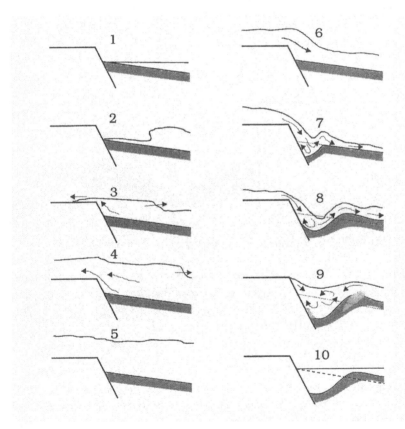

Figure 8.12: Sketch of wave deformation and scouring processes. Nishimura et al. (1978).

There is not much information on scour induced by wave overtopping.

Nishimura, Watanabe and Horikawa (1978) studied scour at seawalls caused by an incident tsunami. They reasoned that coastal structures on the Pacific coasts of Japan are continuously exposed to tsunamis, noting that damage to such structures is often caused by scouring at their toe.

Kadib (1963) studied scour induced by overtopping short-period waves.

In the Nishimura et al. (1978) study, the incident tsunami was simulated with a single, solitary wave in a wave flume. The following variables were changed in the study: the face slope of the seawall, the slope of the backland, the beach slope, the water depth at the seawall, the height of the seawall, the wave height, and the sediment properties.

Fig. 8.12 shows a sequence of sketches illustrating the scour process generated by the overtopping incident tsunami.

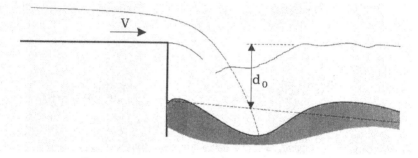

Figure 8.13: "Water jet" of the return flow from the backland.

Nishimura et al.'s study indicated that there are two major parameters governing the scour: (1) the rate of return flow from the backland, which enters into the main body of the water as a "jet" flow; and (2) the depth of the water receiving the return flow (Fig. 8.13). The latter is obviously important with regard to the penetration of the jet down to the bed.

Some of the results from Nishimura et al.'s (1978) study are

1. Scour depth is decreased with decreasing wave height, and with increasing crown elevation; however, the area of serious scouring is displaced towards the seawall in this case.

2. Scour increases, and more importantly it occurs precisely at the toe, when the face slope is mild.

8.4. SCOUR PROTECTION

3. Scour decreases markedly, when the water depth at the seawall increases.

4. When waves are applied repeatedly, much less scouring is induced by each successive wave.

Unfortunately, no prediction method has been given in Nishimura et al.'s (1978) paper. However, given the key mechanism of the scour process, namely the jet flow of the return flow, one may make assessments of scour. For this, the following procedure may be followed: (1) Assess the entrainment velocity and the corresponding thickness of the jet into the water body; (2) Assess the water depth at the instant when the jet enters the main body of the water; and (3) Based on the quantities obtained in the previous items, estimate the scour characteristics (the scour depth and the scour width), using the information on scour generated by submerged vertical jets (Rajaratnam, 1981; see a review on this topic in Breusers and Raudkivi, 1991).

8.4 Scour protection

Toe details should ensure that protection is provided to a depth below that of the predicted scour. This is obviously to prevent undermining of the structure. If scour appears to be a problem, then modifications to the structure to reduce scour may be necessary, such as adopting a milder slope angle. A detailed design manual regarding the seawalls taking into account all design considerations has been given in McConnell (1998). Incidentally, in a parallel publication, design considerations have been compiled for river and channel revetments (Escaramelia, 1998), similar structures but normally exposed to currents parallel to the structure.

8.5 References

1. Barnett, M.R. (1989): Laboratory study of the effects of a vertical seawall on beach profile response. Report 87/005, Coastal and Oceanographic Engineering Department, University of Florida, Gainesville, FL.

2. Breusers, H.N.C. and Raudkivi, A.J. (1991): Scouring. A.A. Balkema, Rotterdam, viii + 143 p.

3. Carpenter, K. and Powell, K.A. (1998): Toe Scour at Vertical Seawalls. Mechanism and Prediction Methods. HR Wallingford SR 506.

4. Chesnutt, C.B. and Schiller, R.E. (1971): Scour of simulated Gulf coast sand beaches due to wave action in front of seawalls and dune barriers. COE Report No. 139, TAMU-SG-71-207, Texas A&M University, College Station, TX.

5. CIRIA (1986): Sea walls: Survey of Performance and Design Practice. Technical Note 125.

6. Coastal Engineering Manual (2001): Scour and Scour Protection. Chapter VI-5-6, Engineer Manual EM 1110-2-1100, Headquarters, U.S. Army Corps of Engineers, Washington, D.C.

7. Escaramelia, M. (1998): River and Channel Revetments. A Design Manual. Thomas Telford, London, U.K., xx + 245 p.

8. Fowler, J.E. (1992): Scour Problems and Methods for Prediction of Maximum Scour at Vertical Seawalls. Department of the Army, Waterways Experiment Station, Corps of Engineers, Vicksburg, Mississippi, U.S., Technical Report CERC-92-16.

9. Fredsøe, J. and Deigaard, R. (1992): Mechanics of Coastal Sediment Transport. World Scientific, Singapore, xviii + 369 p.

10. Fredsøe, J. and Sumer, B.M. (1997): Scour at the round head of a rubble-mound breakwater. Coastal Engineering, vol. 29, 231-262.

11. Kadib, A.L. (1963): Beach profile as affected by vertical walls. Technical Memorandum No. 134, Beach Erosion Board, U.S. Army Corps of Engineers.

12. Kamphuis, J.W., Rakha, K.A. and Jui, J. (1992): Hydraulic model experiments on seawalls. Proc. 23rd Coastal Engineering Conference, Venice, Italy, ASCE, vol. 2, 1272-1284.

13. Kim H., Park W., O'Connor B.A., Lee T.H., Hwang K.N. and Kim T.H. (1999): Scour at Sooyung Seawall caused by wave reflection and liquefaction", Coastal Eng. and Marina Developments, Ed. C. A. Brebbia and P. Anagnostopoulos, WIT Press, Southampton, U.K., 411 - 424.

8.5. REFERENCES

14. Kraus, N.C. (1987): The effects of seawalls on the beach: A literature review. Proc. Coastal Sediments '87 (ed. N. Kraus), New Orleans, LA, American Society of Civil Engineers, 945-960.

15. Kraus, N.C. (1988): The effect of seawalls on the beach: An extended literature review. In: N.C. Kraus and O. Pilkey (eds.). The Effects of Seawalls on the Beach, J. Coastal Research, Special Issue 4, 1-29.

16. Kraus, N.C., Smith, J.M. and Sollitt, C.K. (1992): SUPERTANK Laboratory Data Collection Project. Proc. 23rd Coastal Engineering Conference, Venice, Italy, ASCE, vol. 2, 2191-2204.

17. Kraus, N.C. and McDougal, W.G. (1996): The effects of seawalls on the beach: Part I, An updated literature review. J. Coastal Research, vol. 12 Nr. 3, 691-701.

18. Larson, M. and Kraus, N.C. (1989): Numerical Model for Simulating Storm-Induced Beach Change, Report I: Theory and Model Foundation. Technical Report CERC-89-9, U.S. Army Engineering Waterways Experimental Station, Coastal Engineering Research Center, Vicksburg, Miss.

19. McConnell, K. (1998): Revetment Systems Against Wave Attack. A Design Manual. Supervising Ed. William Allsop. Thomas Telford, London, U.K., xviii + 162 p.

20. McDougal, W.G, Sturtevant and Komar, P.D. (1987): Laboratory and field investigations of the impact of shoreline stabilization structures on adjacent properties. Proceedings Coastal Sediments '87 (ed. N. Kraus), New Orleans, LA, American Society of Civil Engineers, 961-973.

21. McDougal, W.G., Kraus, N. and Ajiwibowo, H. (1996): The effects of seawalls on the beach: Part 2, Numerical modelling of SUPERTANK seawall tests. J. Coastal Research, vol. 12, Nr. 3, 702-713.

22. Nishimura, H., Watanabe, A. and Horikawa, K. (1978): Scouring at the toe of a seawall due to tsunamis. Proc. 16th Coastal Engineering Conference, Hamburg, Germany, ASCE, 2540-2547.

23. Powell, K.A. (1987): Toe Scour at Seawalls Subject to Wave Action. Report No SR 119, HR Wallingford, U.K.

24. Powell, K.A. and Lowe, J.P. (1994): The scouring of sediments at the toe of seawalls. Proc. Hornafjordor Int. Coastal Symposium (ed. G. Viggosson), Iceland, 20-24. June, 1994.

25. Powell, K.A. and Whitehouse, R. (1998): The occurrence and prediction of scour at coastal and estuarine structures. Proc. 33rd MAFF Conference of River and Coastal Engineers, 1-2 July 1998, Keele University, U.K., 3.3.1-3.3.13.

26. Rajaratnam, N. (1981): Erosion by plane turbulent jets. J. Hydraulic Research, vol. 19, No. 4, 339-358.

27. Rakha, K.A. and Kamphuis, J.W. (1994): Wave transformation in the vicinity of a seawall. In: Symp. Waves - Physical and Numerical Modelling, Vancouver, B.C. IAHR, 1011-1020.

28. Rakha, K.A. and Kamphuis, J.W. (1997 a): Wave-induced currents in the vicinity of a seawall. Coastal Engineering, vol. 30, 23-52.

29. Rakha, K.A. and Kamphuis, J.W. (1997 b): A morphology model for an eroding beach backed by a seawall. Coastal Engineering, vol. 30, 53-75.

30. Shore Protection Manual (1977): U.S. Army Coastal Engineering Research Center, vol. II, Department of the Army, Corps of Engineers.

31. Toue, T. and Wang, H. (1990): Three dimensional effects of seawall on the adjacent beach. Proc. 22nd Coastal Engineering Conference, ASCE, 2-6. July, 1990, Delft, The Netherlands, 2782-2795.

32. Whitehouse, R. (1998): Scour at Marine Structures. Thomas Telford, London, U.K., xix +198 p.

Chapter 9

Ship-propeller scour

The bed and banks of harbour basins and navigation channels, particularly the quay walls, are exposed to the lee-wake (hereafter termed the wash to follow the terminology used in the literature) generated from a ship's propeller or from the bow thruster or stern thruster of a ship (Fig. 9.1).

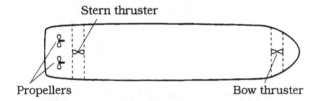

Figure 9.1: Ship's propellers, bow and stern thrusters.

The quay structures may be "closed" (Fig. 9.2 a), or they may be "open" and supported with piles (Fig. 9.2 b). Fig. 9.3 illustrates the deflection of the wash toward the bed and the quay wall in the case of roll-on-roll-off ferries (ro-ro) with a closed quay structure, while Fig. 9.4 illustrates the deflection of the wash in the case of a bow thruster.

The velocities experienced in the wash, at the quay wall, and at the bed may be very high, in the order of magnitude of several metres per second. The continued mooring and unmooring of the vessel may therefore result in an accumulative scour, and therefore lead to structural instability.

A number of case studies have been reported, cataloging the problems of propeller-wash-induced damage at quay structures (Hamill, Johnston and

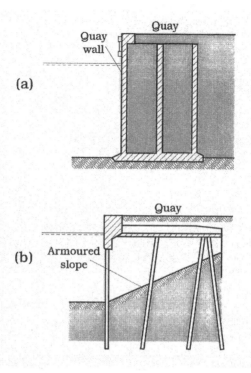

Figure 9.2: (a) Closed and (b) Open quays.

Stewart, 1999). Bergh and Cederwall (1981) reported a survey of harbours in Sweden, indicating that 16 out of 18 ports have suffered propeller-induced scour. Ryan, Hamill and Hughes (1999) refer to a similar survey of ports in the United Kingdom carried out by Quarrain (1994); this latter survey showed that 42% of all British ports have been affected by propeller-induced scour.

The objective of this chapter is to give a brief account of scour induced by ship propellers. Clearly, the knowledge of propeller-induced scour is essential to make adequate provision in the design of quay foundation levels so that the occurrence of quay instability during port operations can be reduced. The chapter will also give a short account of scour protection measures in relation to quay structures.

9.1. SCOUR DUE TO UNCONFINED PROPELLER WASH

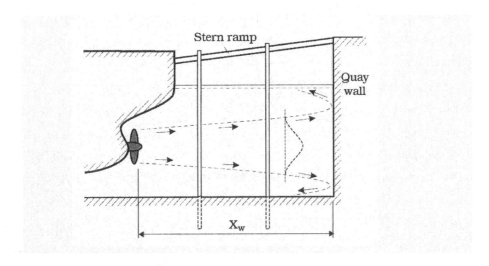

Figure 9.3: Jet flow deflection in the case of ro-ro ship with stern ramp.

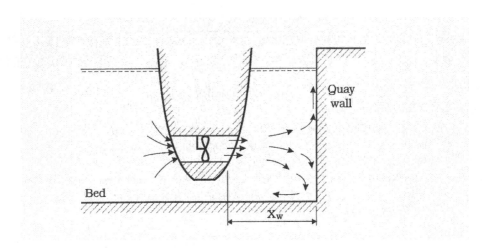

Figure 9.4: Jet flow deflection in the case of bow thrusters.

9.1 Scour due to unconfined propeller wash

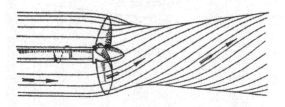

Figure 9.5: Rotation in a propeller jet.

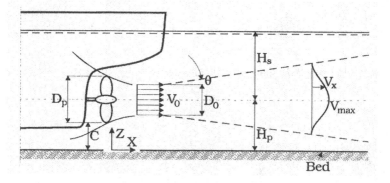

Figure 9.6: Definition sketch.

A ship's propeller produces a jet with a swirl component, Fig. 9.5. Now, consider such a jet (the wash of the propeller) in a water body with no quay walls, Fig. 9.6. This jet is termed the *unconfined propeller wash*.

An unconfined propeller wash induces scour on the bed, when the bed is erodible (Fig. 9.7). Let S be the maximum equilibrium scour depth (Fig. 9.7). This quantity can be written

$$S = f(V_0, D_p, d_{50}, C, \rho, g, \Delta\rho, \nu) \tag{9.1}$$

in which V_0 = the efflux velocity , D_p = the propeller diameter, C = the clearance distance between the propeller tip and the seabed, d_{50} = the sediment grain size, ρ = the water density, $\Delta\rho = \rho_s - \rho$ = the difference between

9.1. SCOUR DUE TO UNCONFINED PROPELLER WASH

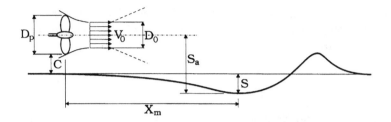

Figure 9.7: Scour profile at the centreline plane.

the sediment-grain density and the water density, ν = the kinematic viscosity and g = the acceleration due to gravity. Note that the efflux velocity may be calculated from

$$V_0 = nD_p\sqrt{C_t} \qquad (9.2)$$

in which n = the number of propeller revolutions per second and C_t = the propeller thrust coefficient (Blaauw and Kaa, 1978).

On dimensional considerations, the maximum equilibrium scour depth normalized by the propeller diameter may, from Eq. 9.1, be expressed as

$$\frac{S}{D_p} = f_1\left(\frac{V_0}{\sqrt{gd_{50}\frac{\Delta\rho}{\rho}}}, \frac{V_0 D_p}{\nu}, \frac{D_p}{d_{50}}, \frac{C}{d_{50}}\right) \qquad (9.3)$$

in which the first term in function f_1 is termed the densimetric Froude number

$$F_0 = \frac{V_0}{\sqrt{gd_{50}\frac{\Delta\rho}{\rho}}} \qquad (9.4)$$

(This quantity plays a role similar to that of the Shields parameter). The second term in f_1 is the Reynolds number of the jet

$$\text{Re}_j = \frac{V_0 D_p}{\nu} \qquad (9.5)$$

The preceding formulation is due to Hamill et al. (1999). The latter authors argue that the effect of viscosity can be neglected for large Reynolds numbers, referring to the work of Rajaratnam (1981) where it was shown that the effect of the Reynolds number for a plain wall jet on scour can be neglected when the Reynolds number is larger than 10^4. Hence, the normalized maximum equilibrium scour depth will be

$$\frac{S}{D_p} = f_1(F_0, \frac{D_p}{d_{50}}, \frac{C}{d_{50}}) \qquad (9.6)$$

Hamill (1988) and Hamill et al. (1999) studied the time variation of the maximum scour depth. From their experiments, they obtained the following empirical expression for the maximum scour depth, S_t, as a function of time and as a function of the previously mentioned parameters, namely F_0, D_p/d_{50} and C/d_{50}:

$$S_t = k\Omega[\ln(t)]^\Gamma \qquad (9.7)$$

in which k is a constant taking a value of 38.97, Γ is

$$\Gamma = F_0^{-0.53}(\frac{D_p}{d_{50}})^{-0.48}(\frac{C}{d_{50}})^{0.94} \qquad (9.8)$$

and Ω is

$$\Omega = \Gamma^{-6.38} \qquad (9.9)$$

In Eq. 9.7, S_t is in millimeters and t in seconds.

Apparently, Eq. 9.7 can not give the maximum equilibrium scour depth because $S_t \to \infty$, as $t \to \infty$. However, Hamill et al. (1999) recommend that different time intervals be attempted, and when the difference is smaller than a previously chosen accuracy value, the equilibrium scour depth (termed by Hamill et al. the asymptotic value) can be approximated. Hamill et al. (1999) also note that the preceding equations only apply for depths of seabed below the propeller in the range $0.5D_p < C < 2.5D_p$.

Hamill et al. (1999) also give an empirical equation for the distance of the position of the maximum scour from the propeller, X_m (Fig. 9.7):

$$\frac{X_m}{C} = F_0^{0.94} \qquad (9.10)$$

They note that X_m corresponds to the equilibrium scour profile.

More information about the research results regarding the scour due to unconfined propeller wash can be found in Hamill (1988) and Hamill et al. (1999).

Example 21 *Effect of rudder*

9.1. SCOUR DUE TO UNCONFINED PROPELLER WASH 429

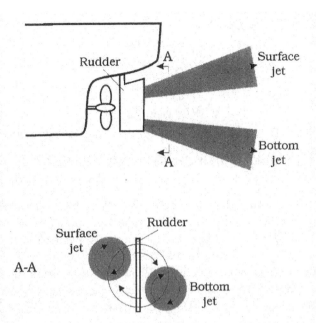

Figure 9.8: Rudder splits the propeller jet into two jets.

The rudder used for steering is located in front of the propeller, as illustrated in Fig. 9.8. Observations show that the presence of the rudder splits the wash into two streams, one directed upwards to the surface and the other directed downwards towards the bottom (termed the surface jet and the bottom jet, respectively, see Section A-A in Fig. 9.8). The influence of the rudder on the propeller wash has been included in the studies of Verhey (1983), Robakiewicz (1987) and Fuehrer, Pohl and Romisch (1987), and more recently of Hamill and McGarvey (1996) and Hamill, McGarvey and Mackinnon (1998). Hamill and McGarvey (1997) have further reported the results of an experimental investigation where the influence of the rudder on the scouring process was studied. The following paragraphs will present the highlights of the study by Hamill and McGarvey (1997).

1. The maximum scour depth increases with the rudder present (at the zero angle, $\alpha = 0$, Fig. 9.9), by approximately 25%.

2. In the experiments, the rudder angle was changed in the range $-35^0 <$

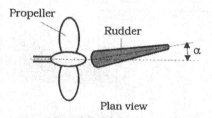

Figure 9.9: Definition sketch for the rudder angle.

$\alpha < +35^0$, noting that 35 degrees was found to be the practical limit within which a rudder is an effective steering device. By changing the rudder angle, the rudder was turned into the bottom jet or into the surface jet. (In their experiments, the bottom jet was achieved by negative rudder angles, and the surface jet by positive rudder angles). The authors found that the maximum scour depth produced by the bottom-jet mode is larger than that produced by the surface-jet mode, as anticipated.

3. The location of the maximum scour depth corresponds quite well to the location of the maximum velocity (Fig. 9.10). The authors recommend that the equations developed to locate the locus of the maximum velocities in Hamill and McGarvey (1996) may be used to locate the region within which the maximum scour depth will occur (consult also Hamill et al., 1998).

4. The maximum equilibrium scour depth in the presence of the rudder may be calculated from the following empirical equation

$$\frac{S_r}{S} = 0.75 - 0.07F_0 + 0.02(\frac{D_p}{d_{50}}) - 0.15(1+\alpha) \qquad (9.11)$$

in which $S =$ the maximum equilibrium scour depth without a rudder.

5. The authors note that the final eroded depth could be increased by up to a factor of 3 with the rudder present.

9.2. SCOUR DUE TO CONFINED PROPELLER WASH

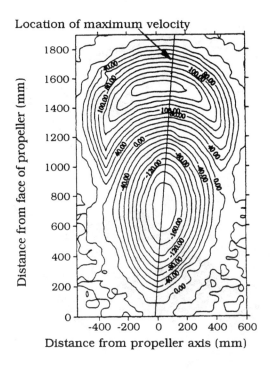

Figure 9.10: Location of maximum scour and that of maximum velocity. Hamill and McGarvey (1997).

9.2 Scour due to confined propeller wash

The presence of a quay wall (Figs. 9.3 and 9.4) will interfere with the development of the propeller wash. This will obviously result in a significant alteration to the scoured bed profile. Hamill et al. (1999) also studied the effect of the confinement on scour. Figs. 9.11 a and b (from Hamill et al., 1999) show this effect when the quay wall is placed at a location between the position of maximum scour and the deposited crest in the unconfined case (Fig. 9.11 b). For this configuration, the effect is to cause a deepening of the scour hole to a maximum at the wall, on the centreline of the profile (Fig. 9.11 b). The scour hole is also widened at the wall with the extra-scoured material being deposited in two crests on either side of the wash centreline (Fig. 9.11 b). Note that, in Hamill et al.'s (1999) experiments, no rudder

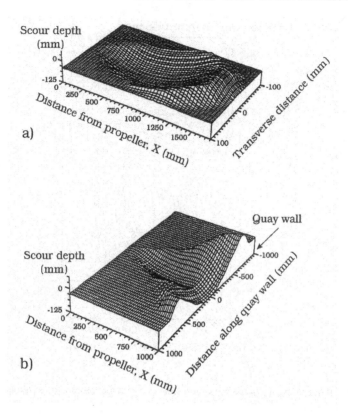

Figure 9.11: (a) Scour hole, unconfined case. (b) Scour hole, confined case. Hamill et al. (1999).

was present.

(Incidentally we may mention that rudders are not necessary for manoeuvering in the case where four thrusters - two each fore and aft - are fitted symmetrically and in-nozzles are installed; in this case, the thrusters can turn 360 degrees and have reversible CP (controllable pitch) propellers with variable velocity, and therefore the rudders are not needed for manoeuvering (Pankchik, Gravesen and Thomsen, 1994)).

Figs. 9.12 a-f depict the equilibrium scour profiles obtained in Hamill et al.'s (1999) experiments. The scour profile corresponding to the unconfined case (the dashed line) is included in the figure as a reference line. In the

9.2. SCOUR DUE TO CONFINED PROPELLER WASH

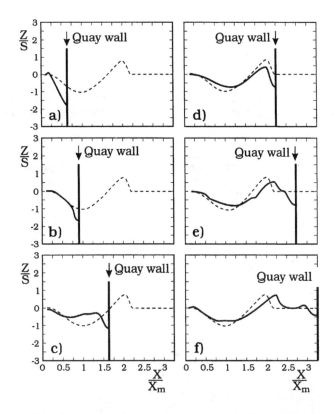

Figure 9.12: Equilibrium scour profiles. Solid line: with quay wall. Dashed line: without quay wall, plotted as a reference line. Hamill et al. (1999).

figure, S = the maximum equilibrium scour depth in the unconfined-scour case (Eq. 9.6), while X_m = the distance of the position of the maximum scour from the propeller for the same case (Eq. 9.10). As seen from Fig. 9.12 a-b, the maximum scour depth is increased quite considerably when the wall is close to the propeller (X_w, the distance of the wall from the face of the propeller, is $X_w = 0.636X_m$ in Fig. 9.12 a, and $X_w = 0.909X_m$ in Fig. 9.12 b). However, it is seen from Figs. 9.12 c-f that the influence of the wall decreases with increasing X_w, as expected.

Hamill et al. (1999), from their experiments, obtained the following empirical relationship between the maximum equilibrium scour depth, S_w, and the wall distance, X_w, as follows

$$\frac{S_w - S}{S_a} = 1.18(\frac{X_w}{X_m})^{-0.2} - 1 \qquad (9.12)$$

in which S_a = the maximum equilibrium scour depth in the unconfined scour case, measured from the shaft axis of the propeller (Fig. 9.7). This expression is plotted in Fig. 9.13.

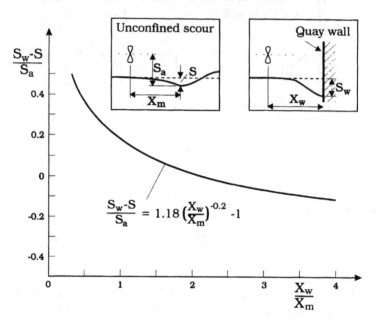

Figure 9.13: Maximum equilibrium scour depth for the confined case. Hamill et al. (1999).

Example 22 *Effect of rudder in the confined case*

The preceding results are applicable to the case of a ro-ro ship with stern ramp but without the rudder (Fig. 9.3), or they are applicable to the case of bow trusters (Fig. 9.4).

In the latter case, in addition to the scour induced by the bow trusters, scour will also be induced by the ship's propeller (Fig. 9.14). This scour is

9.2. SCOUR DUE TO CONFINED PROPELLER WASH

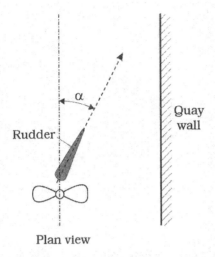

Figure 9.14: Confined case with rudder.

certainly influenced by the presence of the rudder and also by the presence of the quay wall (Fig. 9.14). Ryan et al. (1999) have studied this aspect of the problem. The following paragraphs give the principal findings of this study.

1. The maximum scour depth lies along a particular line, which is approximately equal to the rudder angle. This may be observed from Fig. 9.15 where the rudder angle was 17 degrees.

2. The scour profiles plotted along the line of maximum scour depth practically coincide, irrespective of the distance of the quay wall from the propeller Y_w (in the range of Y_w tested in the experiments, $Y_w/D_p = 1.5 - 3$). (See Fig. 9.15 for the definition of Y_w).

3. However, the scour profiles at the toe of the quay wall vary, depending on the distance of the wall to the propeller and the rudder angle (Fig. 9.16).

4. The maximum equilibrium scour depth at the toe of the quay wall is given by the following empirical equation in terms of the rudder angle

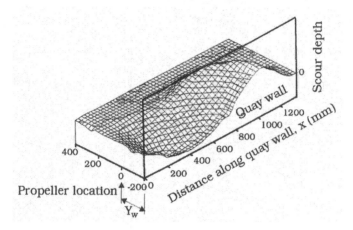

Figure 9.15: Equilibrium scour hole with $Y_w/D_p = 1.52$, and $\alpha = 17^0$. Ryan et al. (1999).

and the distance of the wall from the propeller (plus in terms of the previously described parameters F_0, D_p/d_{50} and C/d_{50}) as follows

$$\frac{S_r}{D_p} = 2.3 \times 10^{-4} (\frac{C}{d_{50}})^{0.581} (\frac{D_p}{d_{50}})^{0.427} (\frac{Y_w}{D_p})^{-0.052} (1+\alpha)^{-0.772} F_0^{4.403} \quad (9.13)$$

in which S_r is the maximum equilibrium scour depth, measured from the initial undisturbed bed level.

9.3 Scour protection

Counter measures for scouring are essential to ensure safe operations of ports. There are several methods for scour protection such as stone protection, grouted stone layers, gabions/mattresses, bag concrete, concrete slabs, concrete slab carpets, underwater concreting and deflectors/systems for dissipation of jet energy.

The design strategy for a stone cover requires the stability of the top layer, as described in conjunction with the scour protection for pipelines (Chapter 2, Section 2.6). For the stability of the top layer, the Shields criterion may be used; namely, the Shields parameter calculated for the stones must be smaller than θ_{cr}, the critical value of the Shields parameter corresponding to the

9.3. SCOUR PROTECTION

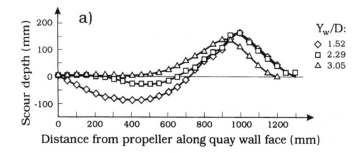

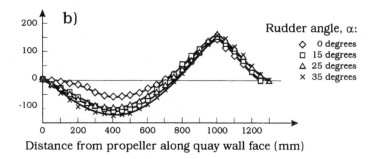

Figure 9.16: Equilibrium scour profiles at the toe of quay wall. (a): $\alpha = 25^0$. (b): $Y_w = 200$ mm. Ryan et al. (1999).

initiation of motion at the top layer of the stone protection. To calculate the Shields parameter, the maximum jet velocity at the bed is needed. Fuehrer and Romisch (1977) give this velocity as a function of the clearance of the propeller from the bed (Fig. 9.17) for three "standard" situations illustrated in Fig. 9.18. In Fig. 9.17, U_0 is the efflux velocity (Fig. 9.7) which may be calculated from Eq. 9.2.

Example 23 *Scour protection in Ferry Terminal of Elsinore, Denmark (Pankchik et al., 1994).*

Danish State Railways (DSB) constructed the Ferry Terminal of Elsinore for the ferry route to Helsingborg, Sweden. The ferry terminal was opened in November 1991. It is placed outside the existing harbour and comprises two

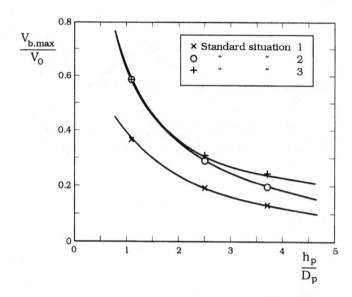

Figure 9.17: Maximum bottom velocity (see the next figure for 1, 2, 3). Fuehrer and Romisch (1977).

berths, Berth A and Berth B (Fig. 9.19). Berth A is a railway-ferry berth while Berth B is a car-ferry berth, and there is a 135 m long pier between the two (Fig. 9.19).

The water depth in the area at the terminal reaches up to 30 m. To minimize the initial expenditures, the terminal was constructed without a breakwater. Because of the very large water depths at the berths, the bottom has been raised to the level -14.5 m by undersea sand filling in layers of about 5 m between submerged stone dikes (Fig. 9.20).

The ferries are equipped with two sets of thrusters - two each fore and aft. No rudders were necessary for the arrangement of the thrusters as they can turn 360 degrees and have reversible CP (controllable pitch) propellers with variable velocity. The nozzle velocities can reach values as high as 8-9 m/s. The velocities generated at the bed by the wash were found to be 3-3.5 m/s. The stability calculations for the top layer of the scour protection required a stone size of 0.5-0.6 m.

Physical model tests were also carried out for scour protection (DHI, 1990). These tests showed that the scour protection was to fulfil two criteria:

9.3. SCOUR PROTECTION

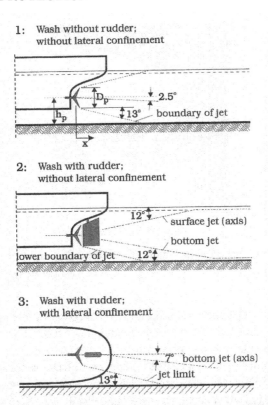

Figure 9.18: Standard situations in conjunction with the previous figure. Fuehrer and Romisch (1977).

(1) The horizontal area exposed to the direct thruster impact requires protection; and (2) Likewise, the outer end and the upper slope of the sandfill enclosed by quarry run (Fig. 9.20) also require protection.

On the basis of the stability calculations and the physical model experiments, the scour protection was designed, as sketched in Fig. 9.21. Basically, for the horizontal area, stones with mean size of 0.4 m (130 kg) were selected.

The ferry berth is entered by a ferry every 40 minutes (day and night, all year round).

One year after the opening of the terminal (November 1992), the scour protection was inspected by the divers. (1) In the area just below the outer-

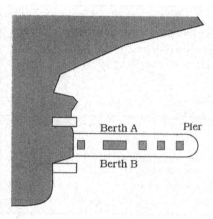

Figure 9.19: Ferry Terminal of Elsinor, Denmark. Adapted from Pankchik et al. (1994).

most thrusters in the berth, there had been some movements of the 130 kg stones. (2) Two holes of 3-5 m² penetrating the armour layer were found in the scour protection, one in the vicinity of the sheet-pile wall and the other near the centreline of the berth. (3) The rest of the scour protection was fully intact including the area under the innermost thrusters in the berth. (Note that, in this latter location, the stone weight was 450 kg, because the bottom was raised from -13.2 m to -9.7 m).

As a repair, an area with an extent of 20 m × 15 m (including the damaged area) was covered with stones of an average weight of 450 kg in two layers. A diver inspection one year after the repair (at the end of October 1993) showed that the scour protection including the repair was fully intact.

As mentioned at the beginning of this chapter, quays may be "open" and supported with piles (Fig. 9.2 b). A guidelines for the design of armoured slopes under such quay structures (including the results of a survey (carried out by the PIANC working group who prepared this guidelines), design approach, types of slope protection, practical aspects, design guidelines and execution) has been published by PIANC (1997) and can be consulted for the design of armoured slopes in the case of open quays.

9.3. SCOUR PROTECTION

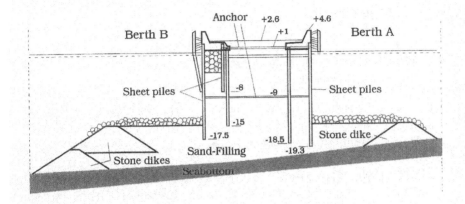

Figure 9.20: Cross-section of the pier in Ferry Terminal of Elsinor, Denmark. Pankchik et al. (1994).

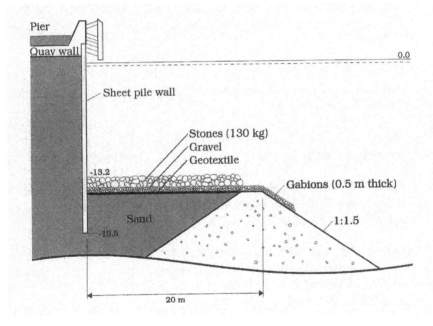

Figure 9.21: Details of scour protection. Ferry Terminal of Elsinor, Denmark. Pankchik et al. (1994).

9.4 References

1. Bergh, H. and Cederwall, K. (1981): Propeller Erosion in Harbours. TRITA-VBI-107, Hydraulics Laboratory, Royal Institute of Technology, Stockholm, Sweden.

2. Blaauw, H.G. and Kaa, E.J. (1978): Erosion of Bottom and Sloping Banks Caused by the Screw Race of the Manoeuvering Ships. Publ. No. 202, Delft Hydraulics Laboratory, Delft, the Netherlands.

3. DHI (1990): Ferry Terminal of Elsinore, Hydraulic Investigations. A Report prepared by the Danish Hydraulic Institute (DHI) for the joint venture Cowi/Carl Bro/DSB. (in Danish).

4. Fuehrer, M. and Romisch, K. (1977): Effects of modern ship traffic on inland- and ocean-waterways and their structures. Proc. 24th International Navigation Congress, Permanent International Association of Navigational Congresses (PIANC), Leningrad, 1997. Section I, Subject 3, 79-94.

5. Fuehrer, M., Pohl, H. and Romisch, K. (1987): Propeller Jet Erosion and Stability Criteria for Bottom Protection of Various Construction. PIANC Bulletin No. 58.

6. Hamill, G.A. (1988): The Scouring Action of the Propeller Jet Produced by a Slowly Manoeuvreing Ship. PIANC Bulletin No. 62.

7. Hamill, G.A. and McGarvey, J.A. (1996): Designing for propeller action in harbours. Proc. 25th International Conference on Coastal Engineering, Orlando, Florida, USA, vol. 4, 4451-4463.

8. Hamill G.A. and McGarvey, J.A. (1997): The influence of a ship's rudder on the scouring action of a propeller wash. Proc. Seventh International Offshore and Polar Eng. Conf. (ISOPE), Honolulu, USA, May 25-30, 1997, vol. IV, 754-757.

9. Hamill, G.A., Johnston, H.T. and Stewart, D.P. (1999): Propeller wash scour near quay walls. J. Waterway, Port, Coastal, and Ocean Engineering, ASCE, vol. 125, No. 4, 170-175.

9.4. REFERENCES

10. Hamill, G.A., McGarvey, J.A. and Mackinnon, P.A. (1998): A method for estimating the bed velocities produced by a ship's propeller wash influenced by a rudder. Proc. 26th International Conference on Coastal Engineering, Copenhagen, Denmark, vol. 3, 3624-3633.

11. Pankchik, B., Gravesen, H. and Thomsen, J. (1994): Scour protection in new Ferry Terminal of Elsinore (DK). Proc. 28th International Navigation Congress, PIANC, Seville, Spain, 23-28. May, 1994, Special Issue, Bulletin No. 83/84, 15-19.

12. PIANC (1997): Guidelines for Design of Armoured Slopes under Open Piled Quay Walls. Report of Working Group 22, Supplement to Bulletin No. 96, PIANC.

13. Quarrain, R.M.M. (1994): Influence of the Sea Bed and the Bed Geometry on the Hydrodynamics of the Wash from a Ship's Propeller. A thesis submitted to the Queen's University of Belfast for the Degree of Doctor of Philosophy.

14. Rajaratnam, N. (1981): Erosion by plain turbulent jets. J. Hydraulic Res., vol. 19, No. 4, 339-358.

15. Robakiewicz, W. (1987): Bottom Erosion as an Effect of Propeller Action near the Harbour Quays. PIANC Bulletin No. 58.

16. Ryan, D., Hamill, G.A. and Hughes, D.A.B. (1999): Designing for protection against propeller scour in harbours. Proc. Fifth International Conference on Coastal and Port Engineering in Developing Countries, Cape Town, South Africa, 19-23. April, 1999, 242-253.

17. Verhey, H.J. (1983): The Stability of Bottom and Banks Subjected to the Velocities in the Propeller Jet Behind Ships. DH Publication No. 303.

Chapter 10

Impact of liquefaction

In the geotechnical-engineering terminology, liquefaction stands for the state of the soil where the effective stresses between the individual grains in the bed vanish, and therefore the water-sediment mixture as a whole acts like a fluid. Under this condition, the soil fails, thus precipitating failure of the supported structure. Some such failures have been catastrophic. With the soil liquefied, buried pipelines may float to the surface of the seabed; large individual blocks (like those used for scour protection) may penetrate into the seabed; sea mines may enter into the seabed and eventually disappear.

Several examples of liquefaction have been reported in the literature in conjunction with marine engineering. Christian, Taylor, Yen and Erali (1974) and Herbich, Schiller, Dunlap and Watanabe (1984) report incidents where sections of pipelines floated to the surface of the soil during storms; Dunlap, Bryant, Williams and Suheyda (1979) report storm-induced pore pressures in soft, clayey sediments in the Mississippi Delta where sinking of several of the measuring instruments up to 6-14 ft was noted; Miyamoto, Yoshinaga, Soga, Shimizu, Kawamata, and Sato (1989) report the subsidence of offshore breakwaters composed of concrete blocks at Niagata Coast, Japan (see also Goda, 1994), to mention but a few.

This chapter will review the subject in conjunction with the above applications.

As will be seen in the following section, the wave-induced stresses in the soil, the pore pressure and the ground water flow are essential "ingredients" of the liquefaction processes. For this reason, we shall, after a brief introduction regarding the physics of liquefaction, first concentrate on the Biot consolidation equations, the equations that govern the previously mentioned quantities. Then we shall turn our attention to the liquefaction processes under waves. Subsequently, we shall study the self-burial (sinking)/floatation of pipelines, and the penetration of marine objects (such as armour blocks) in the seabed.

10.1 Physics of liquefaction

Liquefaction is generated mainly by two different mechanisms, namely

1. by the buildup of pore pressure (residual liquefaction); and

2. by the upward vertical pressure gradient in the soil during the passage of a wave trough (momentary liquefaction).

Each mechanism is now considered individually.

10.1.1 Liquefaction induced by the buildup of pore pressure. Residual liquefaction

This mechanism can best be described by reference to a progressive wave over a horizontal seabed (Fig. 10.1).

The seabed in the case of a progressive wave will undergo a periodic pressure variation, as sketched in Fig. 10.1 b. Owing to the increased bed pressure under the wave crest, and the opposite effect under the wave trough, the soil will be compressed under the wave crest, and expanded under the wave trough. Therefore the water-soil interface will be (nearly) 180^0 out of phase with the water surface elevation Fig. 10.1 c. This will result in the generation of shear stresses in the soil, as illustrated in Fig. 10.1 c. These shear stresses will vary periodically in time, as the wave continues. (Normal stresses will also be generated in the soil. However, for the time being, we put aside these latter stresses for the sake of simplicity).

Now, if the grains are initially loosely packed, the previously mentioned periodic shear stresses and their associated shear deformations in the soil will gradually rearrange the soil grains at the expense of the pore volume of the soil. The latter effect will "press" the water in the pores, and presumably lead to a buildup of pore-water pressure in the case of an undrained soil (i.e., in the case of silt, for example; see Appendix II, Section 10.8 at the end of

10.1. PHYSICS OF LIQUEFACTION

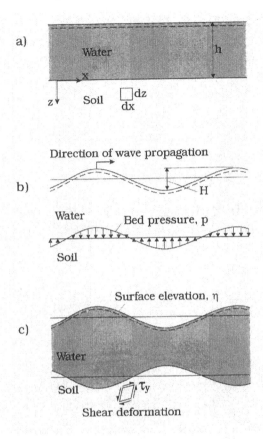

Figure 10.1: Elastic deformation of the soil under a progressive wave.

this chapter). As the wave action continues, the pore-water pressure will continue to accumulate.

During this progressive buildup, the pore-water pressure may reach such levels that it may exceed the value of the overburden pressure. In this latter situation, the soil grains will become unbound and completely free, and the soil will begin to act like a liquid. This process is called the *residual liquefaction*.

It may be noted that the liquefaction due to the buildup of pressure in the coastal environment occurs not only by the action of waves but also by other effects such as

1. earthquakes;

2. shocks (the shock effects may be caused by a sudden failure of a slope, or blasting effects. Chaney and Fang (1991) give a comprehensive review of the case histories experienced in coastal areas); and

3. rocking motions that structures may execute under cyclic loadings (rocking motion of vertical-wall breakwaters under waves, for example).

10.1.2 Liquefaction induced by the upward-directed pressure gradient. Momentary liquefaction

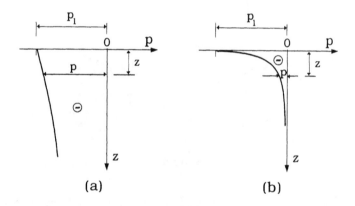

Figure 10.2: Typical distributions of pore pressure (in excess of hydrostatic pressure) during the passage of wave trough. (a) Saturated soil. (b) Unsaturated soil.

The second mechanism generating soil liquefaction is related to the phase-resolved component of the waves. This kind of liquefaction occurs during the passage of the wave trough. Under the wave trough, the pore pressure (in excess of the hydrostatic pressure) has a negative sign (Fig. 10.1 b). Therefore the pressure distribution across the soil depth will be as sketched in Fig. 10.2 a. This figure describes the pressure distribution in the case of a completely saturated soil, while Fig. 10.2 b describes that in the case of

10.1. PHYSICS OF LIQUEFACTION

an unsaturated soil. In the latter case, the soil contains some air/gas, and therefore the pore pressure is "dissipated" at a very fast rate with the depth, as sketched in Fig. 10.2 b.

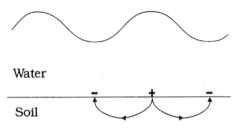

Figure 10.3: Seepage flow under a progressive wave.

Now, in the case of the completely saturated soil, the pressure gradient is not tremendously large (Fig. 10.2 a). However, in the case of the unsaturated soil (Fig. 10.2 b), the pressure gradient can be very large, particularly at small values of z, meaning that quite a substantial amount of lift can be generated at the top layer of the soil during the passage of the wave trough. If this lift exceeds the submerged weight of the soil, the soil will fail, and as a result, it will be liquefied. This type of liquefaction is termed the *momentary liquefaction*. (The liquefaction here occurs over a short period of time during the passage of the wave trough; for the rest of the wave period, the soil will be in the no-liquefaction regime). It may be noted that the seepage flow under the wave trough (Fig. 10.3) may help enhance the momentary liquefaction due to the upward drag acting on the individual grains.

The large pressure gradient in the case of the unsaturated soil (Fig. 10.2 b) is caused by the air/gas content of the soil. It may be mentioned that only a very small amount of gas (less than 1%) will cause a very large dissipation. Gas bubbles may form in the offshore environment where the methane is generated around nuclei of bacteria locally within a soft, consolidating soil. Sills, Wheeler, Thomas and Gardner (1991) report that the gas bubbles produced in this way are considerably larger than the fine-grained soil particles, and the resulting soil structure consists of large bubble "cavities" within a matrix of saturated soil. Incidentally, the latter authors developed a laboratory technique to mimic as closely as possible the process of bubble formation in the offshore environment. Fig. 10.32 a, adapted from Sills et al., illustrates a schematic representation of a gassy soil.

450   CHAPTER 10. IMPACT OF LIQUEFACTION

As seen from the preceding discussion, the wave-induced shear stresses in the soil, the pore pressure and the ground-water flow are essential components of the liquefaction processes. Basically these quantities are governed by the Biot consolidation equations. The following section will describe these equations and their solutions.

10.2 Biot equations and their solutions

10.2.1 Biot equations

It is a common practice that the soil stresses induced by waves are calculated, using the classical elasticity theory in which the soil is assumed to be a poro-elastic medium (e.g., Terzaghi, 1948, p.265 and Biot, 1941). The voids of the elastic skeleton are filled with water. A good example of such a model is a rubber sponge saturated with water (Biot, 1941). In the following paragraphs, the governing equations related to a poro-elastic soil will be derived.

1) Equilibrium conditions for a stress field

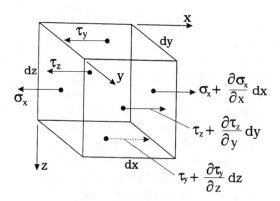

Figure 10.4: Stresses in the x-direction on the surface of a small soil element.

The equilibrium conditions for a stress field shown in Fig. 10.4 (where only the stress components in the x−direction are shown, to keep the figure

10.2. BIOT EQUATIONS AND THEIR SOLUTIONS

relatively simple) are as follows

$$\frac{\partial \sigma_x}{\partial x} + \frac{\partial \tau_z}{\partial y} + \frac{\partial \tau_y}{\partial z} = 0 \tag{10.1}$$

$$\frac{\partial \tau_z}{\partial x} + \frac{\partial \sigma_y}{\partial y} + \frac{\partial \tau_x}{\partial z} = 0 \tag{10.2}$$

$$\frac{\partial \tau_y}{\partial x} + \frac{\partial \tau_x}{\partial y} + \frac{\partial \sigma_z}{\partial z} = 0 \tag{10.3}$$

in which the normal stresses are denoted by the symbol σ, and the shear stresses by τ. (Note that we adopt here Biot's (1941) original notation; for example, σ_x is the normal stress in the $x-$ direction, while τ_y is the shear stress in the $x-$ direction acting on the side perpendicular to $z-$ direction, or τ_z is the shear stress in the $x-$ direction acting on the side perpendicular to $y-$ direction).

2) The stress-strain relationships

The stress-strain relationships for an isotropic, elastic material are given as follows (Hooke's law)

$$e_x = \frac{\sigma_x}{E} - \frac{\nu}{E}(\sigma_y + \sigma_z) \tag{10.4}$$

$$e_y = \frac{\sigma_y}{E} - \frac{\nu}{E}(\sigma_x + \sigma_z) \tag{10.5}$$

$$e_z = \frac{\sigma_z}{E} - \frac{\nu}{E}(\sigma_x + \sigma_y) \tag{10.6}$$

$$\gamma_x = \frac{\tau_x}{G} \tag{10.7}$$

$$\gamma_y = \frac{\tau_y}{G} \tag{10.8}$$

$$\gamma_z = \frac{\tau_z}{G} \tag{10.9}$$

in which E is Young's modulus, and G the shear modulus. The quantity ν is Poisson's ratio which links Young's modulus to the shear modulus by:

$$G = \frac{E}{2(1+\nu)} \tag{10.10}$$

The quantities e_x, e_y, e_z, represent the strains (namely, the linear deformations per unit length), while γ_x, γ_y, γ_z represent the shear (angular) deformations in the $x-$, $y-$ and $z-$directions, respectively:

$$e_x = \frac{\partial u}{\partial x}, \qquad e_y = \frac{\partial v}{\partial y}, \qquad e_z = \frac{\partial w}{\partial z} \qquad (10.11)$$

$$\gamma_x = \frac{\partial w}{\partial y} + \frac{\partial v}{\partial z}, \qquad \gamma_y = \frac{\partial u}{\partial z} + \frac{\partial w}{\partial x}, \qquad \gamma_z = \frac{\partial v}{\partial x} + \frac{\partial u}{\partial y} \qquad (10.12)$$

in which u, v, w are the $x-$, $y-$ and $z-$components of the soil displacement, respectively.

Now, for later use, solving σ_x, σ_y, σ_z, and , τ_x, τ_y, τ_z from Eqs. 10.4-10.6, we get

$$\sigma_x = 2G(e_x + \frac{\nu \epsilon}{1 - 2\nu}) \qquad (10.13)$$

$$\sigma_y = 2G(e_y + \frac{\nu \epsilon}{1 - 2\nu}) \qquad (10.14)$$

$$\sigma_z = 2G(e_z + \frac{\nu \epsilon}{1 - 2\nu}) \qquad (10.15)$$

and

$$\tau_x = G\gamma_x, \qquad \tau_y = G\gamma_y, \qquad \tau_z = G\gamma_z \qquad (10.16)$$

in which ϵ is termed the volume expansion, and given by

$$\epsilon = \frac{\partial u}{\partial x} + \frac{\partial v}{\partial y} + \frac{\partial w}{\partial z} \qquad (10.17)$$

3) The stress-strain relationships in the case of a poro-elastic soil

In the case of a poro-elastic soil where the voids are filled with water (Fig. 10.32 a), the normal stresses will be apportioned by the soil skeleton and the pore water. For example, σ_x, the normal stress in the $x-$direction, will be

$$\sigma_x = \text{Normal stress carried by soil} + \text{Normal stress carried by water} \qquad (10.18)$$

The first part is (from Eq. 10.13):

$$\text{Normal stress carried by soil} = 2G(e_x + \frac{\nu \epsilon}{1 - 2\nu}) \qquad (10.19)$$

10.2. BIOT EQUATIONS AND THEIR SOLUTIONS

while the second part is:

$$\text{Normal stress carried by water} = -p \tag{10.20}$$

in which p is the pore water pressure. (Note that tensile stresses have positive signs while pressures have negative signs for convenience).

Hence from Eqs. 10.18, 10.19 and 10.20,

$$\sigma_x = 2G(e_x + \frac{\nu\epsilon}{1-2\nu}) - p \tag{10.21}$$

This is the stress-strain relationship for the x-direction for a poro-elastic soil. Similarly, the other normal stresses (Eqs. 10.14-10.16) will be

$$\sigma_y = 2G(e_y + \frac{\nu\epsilon}{1-2\nu}) - p \tag{10.22}$$

$$\sigma_z = 2G(e_z + \frac{\nu\epsilon}{1-2\nu}) - p \tag{10.23}$$

in which ϵ is given by Eq. 10.17. The shear stresses, on the other hand, will remain unchanged, since they are carried only by the soil (Eqs. 10.16):

$$\tau_x = G\gamma_x \tag{10.24}$$

$$\tau_y = G\gamma_y \tag{10.25}$$

$$\tau_z = G\gamma_z \tag{10.26}$$

The soil part of the stress (i.e., Eq. 10.19, and similar equations for the other two directions, namely $y-$, and $z-$ directions) is termed the **effective stress**:

$$\sigma'_x = 2G(e_x + \frac{\nu\epsilon}{1-2\nu}) \tag{10.27}$$

$$\sigma'_y = 2G(e_y + \frac{\nu\epsilon}{1-2\nu}) \tag{10.28}$$

$$\sigma'_z = 2G(e_z + \frac{\nu\epsilon}{1-2\nu}) \tag{10.29}$$

4) The equations of equilibrium for a poro-elastic soil

Inserting the stress-strain relationships in Eqs. 10.21-10.26 into Eqs. 10.1-10.3, the following equations are obtained

$$G\nabla^2 u + \frac{G}{1-2\nu}\frac{\partial\epsilon}{\partial x} = \frac{\partial p}{\partial x} \tag{10.30}$$

$$GV^2v + \frac{G}{1-2\nu}\frac{\partial \epsilon}{\partial y} = \frac{\partial p}{\partial y} \qquad (10.31)$$

$$GV^2w + \frac{G}{1-2\nu}\frac{\partial \epsilon}{\partial z} = \frac{\partial p}{\partial z} \qquad (10.32)$$

(Note that the above equations do not contain the inertia terms. It can be shown that the inertia effect can be ignored in most engineering problems, Cheng and Liu (1986). Mei and Foda (1981) can be consulted for the full version of the previous equations including the inertia effect).

5) *Darcy's law*

The variation in the pore-water pressure will drive a flow in the pores. The flow velocities are related to the pressure gradient through Darcy's law in the following way

$$V_x = -\frac{k}{\gamma}\frac{\partial p}{\partial x}, \qquad V_y = -\frac{k}{\gamma}\frac{\partial p}{\partial y}, \qquad V_z = -\frac{k}{\gamma}\frac{\partial p}{\partial z} \qquad (10.33)$$

in which V_x, V_y and V_z are the velocity components in the $x-$, $y-$ and $z-$directions, respectively, k is the coefficient of permeability of the soil, and γ is the specific weight of water.

6) *Continuity equation for the pore water*

Finally, from the conservation of mass of pore water, the following equation is obtained

$$\frac{\partial}{\partial t}(\epsilon + \frac{n}{K'}p) + \frac{\partial V_x}{\partial x} + \frac{\partial V_y}{\partial y} + \frac{\partial V_z}{\partial z} = 0 \qquad (10.34)$$

in which n is the porosity of the soil (see Appendix I, Section 10.7 at the end of this chapter for the relationships among various soil quantities), and K' is the *apparent* bulk modulus of elasticity of water (Biot, 1941).

Regarding the first term in the preceding equation, there are two contributions: First, $\frac{\partial}{\partial t}(\epsilon)$, which represents the increase in the volume of water due to the expansion of the soil skeleton, and second, $\frac{\partial}{\partial t}(\frac{n}{K'}p)$, which represents the increase in the volume of water due to the compressibility of water itself (including the effect of gas/air content in water).

10.2. BIOT EQUATIONS AND THEIR SOLUTIONS

The quantity K' is related to the *true* bulk modulus of elasticity of water, K, by (Verruijt, 1969)

$$\frac{1}{K'} = \frac{1}{K} + \frac{1 - S_r}{p_0} \qquad (10.35)$$

in which S_r is the degree of saturation, and p_0 is the absolute (not excess) pore-water pressure and can be taken equal to the initial value of pressure. When the pore water is gas/air free, S_r will be unity, therefore, in this case, K' will be equal to the true bulk modulus of elasticity of water, K.

Now, inserting Eqs. 10.33 in Eq. 10.34, one gets

$$\frac{k}{\gamma}\nabla^2 p = \frac{n}{K'}\frac{\partial p}{\partial t} + \frac{\partial \epsilon}{\partial t} \qquad (10.36)$$

This equation is known as the storage equation, while the entire set of equations (namely, Eqs. 10.30-10.32, and 10.36) are known as the **Biot consolidation equations** (Biot, 1941).

The Biot consolidation equations are to be solved to get the four unknown quantities, namely u, v, w and p. Once the solution is obtained, then the stresses in the soil can be found from Eqs. 10.21-10.26.

10.2.2 Stresses in soil under a progressive wave

The case of infinitely large soil depth

We now consider a soil (with an infinitely large depth) subject to a progressive wave (Fig. 10.1 b). As discussed in the preceding paragraphs, the waves induce a pressure distribution on the bed as sketched in Fig. 10.1 b, and this will, in turn, cause an elastic deformation in the soil (Fig. 10.1 c), resulting in the generation of shear stress, τ_y, and pore-water pressure, p, in the soil (Fig. 10.1 c). Our objective is to describe τ_y and p. As will be seen in the following sections, we need these quantities to describe the liquefaction processes, i.e.,

1. the process of residual liquefaction and

2. the process of momentary liquefaction.

The governing equations are the Biot consolidation equations, i.e.,

1. the $x-$ and $z-$components of the equations of equilibrium, Eqs. 10.30 and 10.32, and

2. the storage equation, Eq. 10.36 (the 2-D case):

$$G\nabla^2 u + \frac{G}{1-2\nu}\frac{\partial \epsilon}{\partial x} = \frac{\partial p}{\partial x} \qquad (10.37)$$

$$G\nabla^2 w + \frac{G}{1-2\nu}\frac{\partial \epsilon}{\partial z} = \frac{\partial p}{\partial z} \qquad (10.38)$$

$$\frac{k}{\gamma}\nabla^2 p = \frac{n}{K'}\frac{\partial p}{\partial t} + \frac{\partial \epsilon}{\partial t} \qquad (10.39)$$

There are three unknowns: (1) the two components of the soil displacement u and w, and (2) the pore-water pressure p. Once the soil displacements u and w are obtained, then the stresses τ_y, σ'_x and σ'_z in the soil can be found from Eqs. 10.25, 10.27 and 10.29, respectively.

The solution to these equations is to be sought under the following three boundary conditions.

1) Pressure at the bed surface:

At the bed surface, the excess pore pressure generated by a small amplitude, linear, progressive wave (Fig. 10.1 b)

$$\eta = \frac{H}{2}\exp\left[i(\lambda x + \omega t)\right] \qquad (10.40)$$

is given by

$$z = 0: \qquad p = p_b \exp\left[i(\lambda x + \omega t)\right] \qquad (10.41)$$

in which λ is the wave number

$$\lambda = \frac{2\pi}{L} \qquad (10.42)$$

L is the wave length, ω is the angular frequency of the waves

$$\omega = \frac{2\pi}{T} \qquad (10.43)$$

10.2. BIOT EQUATIONS AND THEIR SOLUTIONS

T is the wave period, i the imaginary unit, $i = \sqrt{-1}$, and p_b, the maximum value (the amplitude) of the pressure exerted on the bed by the progressive wave (Fig. 10.1 b), given by

$$p_b = \gamma \frac{H}{2} \frac{1}{\cosh(\lambda h)} \tag{10.44}$$

(Eq. A.17, Appendix A. See also e.g., Dean and Dalrymple, 1984, p. 89).

2) Stresses at the bed surface:

At the bed surface, the vertical effective stress (the vertical stress carried by the soil) must be zero (from Eq. 10.29):

$$z = 0: \quad \sigma'_z = 2G\left[\frac{\partial w}{\partial z} + \frac{\nu}{1-2\nu}(\frac{\partial u}{\partial x} + \frac{\partial w}{\partial z})\right] = 0 \tag{10.45}$$

and the shear stress is (from Eqs. 10.25 and 10.12):

$$z = 0: \quad \tau_y = G(\frac{\partial u}{\partial z} + \frac{\partial w}{\partial x}) = \tau_0 (= \rho U_f^2) \tag{10.46}$$

in which τ_0 is the bed shear stress due to the wave boundary layer, U_f is the corresponding friction velocity. However, normally, $\tau_0 \ll \tau_y$. Hence the right-hand-side of Eq. 10.46 can be put equal to zero.

3) The boundary conditions at large depths:

At large depths, obviously no soil displacement will be experienced, and also no pore-water pressure will develop. Therefore

$$z \to \infty: \quad u, w \text{ and } p \to 0 \tag{10.47}$$

Solution:

The solution to the present set of equations (Eqs. 10.37-10.39) under the preceding boundary conditions (Eqs. 10.41, 10.45, 10.46 and 10.47) was obtained by Yamamoto, Koning, Sellmeijer and van Hijum (1978) for the general case where the soil is not completely saturated, i.e., $S_r < 1$.

In the solution, the ratio G/K' emerges as a key parameter. In the case of a *completely saturated soil*, however, it can be shown that this parameter

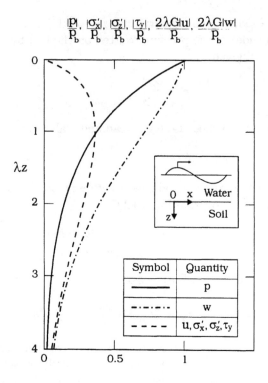

Figure 10.5: Distribution of the amplitudes of pore pressure, effective stresses and displacements for a soil (with infinitely large depth) exposed to a progressive wave. Yamamoto et al. (1978).

becomes practically zero for most soils except for dense sand. For completely saturated soil, $K' = K = 1.9 \times 10^6$ kN/m^2 and $G = 4.8 \times 10^2$ (silt and clay)-4.8×10^5 (dense sand) kN/m^2 (Yamamoto et al., 1978), and therefore G/K' will be extremely small.

In the limit when $G/K' \to 0$, Yamamoto et al.'s (1978) solution reduces to

$$u = -i\lambda z \exp(-\lambda z)(p_b/2\lambda G) \exp\left[i(\lambda x + \omega t)\right] \tag{10.48}$$

$$w = [\exp(-\lambda z) + \lambda z \exp(-\lambda z)]\,(p_b/2\lambda G) \exp\left[i(\lambda x + \omega t)\right] \tag{10.49}$$

$$p = p_b \exp(-\lambda z) \exp\left[i(\lambda x + \omega t)\right] \tag{10.50}$$

10.2. BIOT EQUATIONS AND THEIR SOLUTIONS

The stresses in the soil can then be calculated by inserting these solutions into Eqs. 10.27, 10.29 and 10.25. The effective stress in the x-direction will be

$$\sigma'_x = 2G(e_x + \frac{\nu\epsilon}{1-2\nu}) = p_b \lambda z \exp(-\lambda z) \exp[i(\lambda x + \omega t)] \qquad (10.51)$$

that in the z-direction

$$\sigma'_z = 2G(e_z + \frac{\nu\epsilon}{1-2\nu}) = -p_b \lambda z \exp(-\lambda z) \exp[i(\lambda x + \omega t)] \qquad (10.52)$$

and the shear stress τ_y,

$$\tau_y = -ip_b \lambda z \exp(-\lambda z) \exp[i(\lambda x + \omega t)] \qquad (10.53)$$

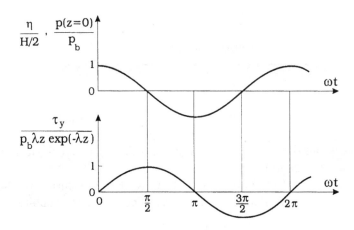

Figure 10.6: The surface elevation, the bed pressure and the shear stress in the soil for a soil with infinitely large depth.

Fig. 10.5 illustrates how the maximum values of the various quantities vary with respect to the depth. As seen, the quantities asymptotically go to zero for larger depths, as dictated by Eq. 10.47. Also, while the pore pressure is dissipated in a monotonous manner with the depth, the soil stresses appear to first increase, attain their maximum values, and then decrease with z.

Furthermore, from Eqs. 10.40, 10.41 and 10.53, it is seen that the shear stress is zero below the wave crest and trough, while it attains its maximum value below the zero-crossing points of the surface elevation (Fig. 10.6), a result anticipated from the physical arguments given in Section 10.1.1.

The case of finite soil depth

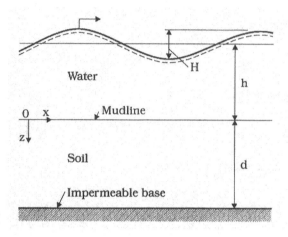

Figure 10.7: Definition sketch. Finite soil depth.

We now consider a soil with a finite depth. The soil is subject to a progressive wave, as in the previous case.

Hsu and Jeng (1994) have developed an analytical solution to the Biot equations (Eqs. 10.37-10.39) for this case. The boundary conditions (Fig. 10.7) are:

$$z = 0: \quad p = p_b \cos(\lambda x - \omega t) \tag{10.54}$$

$$z = 0: \quad \sigma'_z = 0 \tag{10.55}$$

$$z = 0: \quad \tau_y = 0 \tag{10.56}$$

$$z = d: \quad u \text{ and } w = 0 \tag{10.57}$$

$$z = d: \quad \frac{\partial p}{\partial z} = 0 \tag{10.58}$$

The last two conditions imply zero displacements of the soil and no vertical flow at the impermeable base, respectively.

10.2. BIOT EQUATIONS AND THEIR SOLUTIONS

As discussed in the previous section, in the case of the *completely saturated soil*, the parameter G/K' is extremely small. For this case ($G/K' \to 0$), Hsu and Jeng's (1994, p. 793) solution gives the soil displacements as

$$u = \frac{ip_b}{2G\lambda} U(z) e^{i(\lambda x - \omega t)} \qquad (10.59)$$

$$w = \frac{p_b}{2G\lambda} W(z) e^{i(\lambda x - \omega t)} \qquad (10.60)$$

the pore pressure

$$p = \frac{p_b}{(1-2\nu)} P(z) e^{i(\lambda x - \omega t)} \qquad (10.61)$$

and the effective normal and shear stresses

$$\sigma'_x = -p_b \Xi_x(z) e^{i(\lambda x - \omega t)} \qquad (10.62)$$

$$\sigma'_z = p_b \Xi_z(z) e^{i(\lambda x - \omega t)} \qquad (10.63)$$

$$\tau_y = ip_b \Upsilon(z) e^{i(\lambda x - \omega t)} \qquad (10.64)$$

in which the functions $U(z), W(z), P(z), \Xi_x(z), \Xi_z(z)$ and $\Upsilon(z)$ are given as

$$U(z) = (C_1 - C_2 \lambda z) e^{-\lambda z} + (C_3 - C_4 \lambda z) e^{\lambda z} + \qquad (10.65)$$
$$+ \lambda^2 C_5 e^{-\delta z} + \lambda^2 C_6 e^{\delta z}$$

$$W(z) = [C_1 - (1+\lambda z)C_2] e^{-\lambda z} - [C_3 + (1-\lambda z)C_4] e^{\lambda z} + \qquad (10.66)$$
$$+ \lambda \delta (C_5 e^{-\delta z} - C_6 e^{\delta z})$$

$$P(z) = (1-2\nu)(C_2 e^{-\lambda z} - C_4 e^{\lambda z}) + \qquad (10.67)$$
$$+ (1-\nu)(\delta^2 - \lambda^2)(C_5 e^{-\delta z} + C_6 e^{\delta z})$$

$$\Xi_x(z) = (C_1 - C_2 \lambda z) e^{-\lambda z} + (C_3 - C_4 \lambda z) e^{\lambda z} + \qquad (10.68)$$
$$+ \left[\lambda^2 - \frac{(\delta^2 - \lambda^2)\nu}{1-2\nu} \right] (C_5 e^{-\delta z} + C_6 e^{\delta z})$$

$$\Xi_z(z) = (C_1 - C_2 \lambda z) e^{-\lambda z} + (C_3 - C_4 \lambda z) e^{\lambda z} + \qquad (10.69)$$
$$+ \frac{1}{1-2\nu} \left[\delta^2(1-\nu) - \lambda^2 \nu \right] (C_5 e^{-\delta z} + C_6 e^{\delta z})$$

$$\Upsilon(z) = (C_1 - C_2 \lambda z) e^{-\lambda z} - (C_3 - C_4 \lambda z) e^{\lambda z} + \qquad (10.70)$$
$$+ \lambda \delta (C_5 e^{-\delta z} - C_6 e^{\delta z})$$

in which δ is given by

$$\delta^2 = \lambda^2 - \frac{i\omega\gamma(1-2\nu)}{2k(1-\nu)G} \qquad (10.71)$$

and i is the imaginary unit. The coefficients C_1 to C_6 are given in Appendix III, Section 10.9 at the end of this chapter.

It can easily be shown that the above solution converges to the solution of Yamamoto et al. (1978) (given in the previous section, Eqs. 10.48-10.53) for large soil depths. Incidentally, Cheng, Sumer and Fredsøe (2001) numerically solved the Biot equations subject to the boundary conditions given in Eqs. 10.54-10.58, and found that their numerical solution and Hsu and Jeng's analytical solution are in good agreement.

Finally, it may be noted that Hsu and Jeng (1994) also have developed analytical solutions for the case of standing waves and for the case of short-crested waves with an unsaturated and anisotropic soil of finite depth.

In the next section, we shall return to Hsu and Jeng's solution in conjunction with the shear stress in the soil given in Eq. 10.64.

The present subject, namely soil stresses under waves, has been investigated quite extensively over the years. Table 10.1 presents a partial review of these studies (the review has been adapted mainly from McDougal, Tsai, Liu and Clukey, 1989).

10.2. BIOT EQUATIONS AND THEIR SOLUTIONS

Table 10.1. A partial list of the past work regarding stresses and pore pressure in soil under waves.

Author	Soil	Pore water	Note
Putnam (1949)	Rigid skeleton	Incompressible	See (1)
Reid and Kajiura (1957)	,,	,,	,,
Hunt (1959)	,,	,,	,,
Murray (1965)	,,	,,	,,
Sleath (1970)	,,	,,	,,
Moshagen and Tørum (1975)	Rigid skeleton	Compressible	See (2)
Gade (1958)	See (3)	-	-
Mallard and Dalrymple (1977)	,,	-	-
Dawson (1978)	,,	-	-
Dalrymple and Liu (1978)	,,	-	-
MacPherson (1980)	,,	-	-
Hsiao and Shemdin (1980)	,,	-	-
Dawson (1981)	,,	-	-
Yamamoto (1977)	See (4)	Incomp./Comp.	See (5)
Madsen (1978)	,,	,,	,,
Yamamoto (1978, 1981a, 1981b)	,,	,,	,,
Yamamoto et al. (1978)	,,	,,	,,
Yamamoto and Suzuki (1980)	,,	,,	,,
Mei and Foda (1981)	,,	,,	,,
Dalrymple and Liu (1982)	,,	,,	,,
McDougal and Sollitt (1984)	,,	,,	,,
McDougal et al. (1989)	,,	,,	,,
Hsu and Jeng (1994)	,,	,,	,,
Cheng et al. (2001)	,,	,,	,,

(1) The governing equation is the Laplace equation satisfied by the pore pressure.

(2) An additional term (time-derivative of the pore pressure) appears in the Laplace equation satisfied by the pore pressure.

(3) Soil behaves as a viscous liquid, as an elastic soil or as a combined viscoelastic medium.

(4) Soil behaves as a linearly elastic medium.

(5) The Biot consolidation equations are the governing equations.

10.3 Residual liquefaction

10.3.1 Peacock and Seed's (1968) experiment

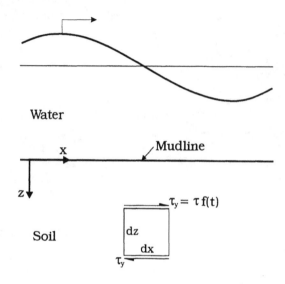

Figure 10.8: The shear stress in the soil. $f(t)$: Periodic function of time.

As described earlier (Section 10.1), under a progressive wave, a soil element of the seabed will undergo a cyclic shear stress variation, Fig. 10.8.

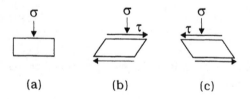

Figure 10.9: Idealized stress condition for element of soil below ground surface during an earthquake. Peacock and Seed (1968).

A similar process also occurs in the case of an earthquake. A soil element below a horizontal ground surface during the earthquake undergoes a cyclic

10.3. RESIDUAL LIQUEFACTION

shear stress variation, as sketched in Fig. 10.9. This cyclic shear stress variation will lead to a progressive buildup of pore pressure, in precisely the same fashion as in waves (Section 10.1), which may lead to the liquefaction of the soil.

In order to simulate the pore-pressure accumulation during an earthquake, Peacock and Seed (1968) conducted laboratory experiments under cyclic stress conditions causing liquefaction of saturated sand in *undrained* simple shear tests. The following paragraphs will summarize the highlights of this important work. (As will be seen in the next sub-section, Section 10.3.2, the end result of Peacock and Seed's work will be one of the key elements of the theory to be developed to describe the buildup of pore pressure under a progressive wave).

The equipment used for these tests essentially consisted of a simple shear box and an arrangement for applying a horizontal, cyclic, shear-stress load to the soil (Fig. 10.10).

The soil sample was *consolidated* under an initial confining pressure. This initial confining pressure is obviously apportioned by the soil and the water. The soil portion of the initial confining pressure, i.e., the effective stress, was $\sigma'_0 = 5$ kg/cm^2, while the water portion, i.e., the initial pore pressure, was p_0 (the initial pore pressure) $= 1$ kg/cm^2 (Fig. 10.11) in Peacock and Seed's (1968) experiment.

The vertical load remained constant during the application of the cyclic shear stress in the experiments.

Fig. 10.11 displays the results of a typical test, reproduced from Peacock and Seed's paper. The top diagram shows the time series of the pore pressure, the middle diagram the time series of the shear strain, and the bottom diagram the time series of the applied shear stress.

As seen clearly from Fig. 10.11, the application of the cyclic shear stress on the soil sample generates an excess pore pressure, p, in the soil, and this pressure progressively builds up, as the cyclic loading continues. The action of the cyclic shear stress on the soil can be explained in the same way as in the case of the waves, Section 10.1.1.

The process of the buildup of pore pressure will come to an end when the accumulated pore pressure reaches the level of the initial effective stress, σ'_0. When this point is reached, i.e., when

$$p = \sigma'_0 \qquad (10.72)$$

the total load will be carried by the water alone, and the effective stress will

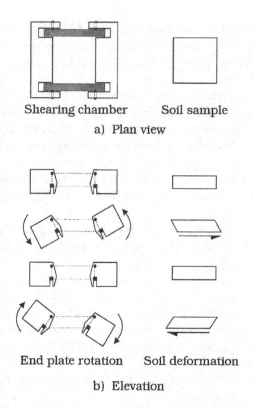

Figure 10.10: Idealized stress condition for element of soil below ground surface during an earthquake. Peacock and Seed (1968).

become zero, and therefore the soil will fail, the *soil liquefaction*. As seen, in the test presented in Fig. 10.11, this point is reached after 24 cycles; we shall return to this point later in the section.

(Fig. 10.11 b clearly shows that the failure sets in precisely at the same instant as the liquefaction occurs (Fig. 10.11 a). The fact that practically no significant shear strain/deformation occurs until this moment is reached (Fig. 10.11 b) indicates that the soil failure is due only to the liquefaction alone, but not due to a combination of liquefaction and shear failure).

The description in the preceding paragraphs implies that, for the generation of the buildup of pressure, the drainage of the pore water from the soil

10.3. RESIDUAL LIQUEFACTION

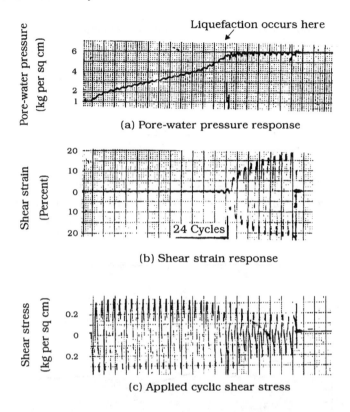

Figure 10.11: Time series of pore pressure, shear strain and shear stress in Peacock and Seed's (1968) experiment.

must be zero (undrained soil) or very small. If the water can "escape" from the soil relatively quickly, the pore pressure will be relieved, therefore no significant buildup of pressure will develop. This implies that, in practice, the pore-pressure accumulation occurs normally in soils with low permeability (such as silt).

Likewise, the pore-pressure accumulation develops only when the frequency of the cyclic loading is sufficiently high. If the frequency is low, the accumulated pore pressure will dissipate as rapidly as it develops; therefore no significant pore pressure accumulation will take place.

An important quantity in the analysis of liquefaction is the number of cycles to cause liquefaction, N_ℓ. This quantity is mainly dependent on the following parameters

$$N_\ell = f(\tau, \sigma_0', D_r) \tag{10.73}$$

in which τ is the amplitude of the oscillating shear stress acting on the soil, and D_r is the relative density of the soil

$$D_r = \frac{e_{max} - e}{e_{max} - e_{min}} \tag{10.74}$$

in which e = the void ratio, e_{max} and e_{min} = the maximum void ratio (in the loosest condition) and minimum void ratio (in the densest condition), respectively, obtained in the way as described, for example, in Peacock and Seed (1968).

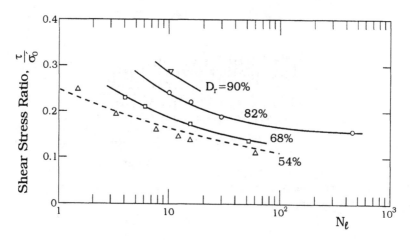

Figure 10.12: Ratio of the amplitude of the shear stress in the soil to the initial effective stress versus the number of cycles to cause liquefaction. Alba et al. (1976).

Peacock and Seed (1968) and Alba, Seed and Chan (1976) carried out tests for different values of the relative density, and plotted the data N_ℓ versus the normalized shear stress τ/σ_0'. Fig. 10.12 displays this diagram (reproduced from Alba, Seed and Chan, 1976). It is seen that the number of cycles to cause liquefaction increases tremendously with decreasing shear stress, and with increasing relative density, as expected.

10.3. RESIDUAL LIQUEFACTION

The variation in Fig. 10.12 can be represented by the following empirical equation

$$N_\ell = (\frac{1}{\alpha}\frac{\tau}{\sigma_0'})^{1/\beta} \quad (10.75)$$

in which α and β are two empirical constants in which $\alpha = \alpha(D_r)$ and $\beta = \beta(D_r)$ (McDougal et al., 1989). For $D_r = 0.54$, for example, $\alpha = 0.246$ and $\beta = -0.165$, the dashed line in Fig. 10.12.

10.3.2 Equation governing the buildup of pore pressure

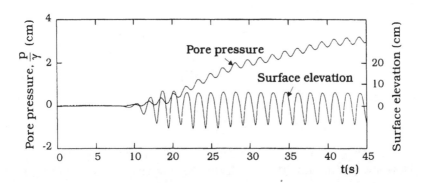

Figure 10.13: Time series of pore pressure and surface elevation in Sumer et al.'s (1999) experiment. $h = 42$ cm, $H = 10$ cm, $z = 16.5$ cm, $d = 17$ cm, $T = 1.6$ s.

Now, we return to the case where the soil is exposed to a progressive wave, Fig. 10.8.

As described earlier (Section 10.1.1), the pore pressure will begin to build up, as the waves progress, similar to Peacock and Seed's oscillating shear stress tests (Fig. 10.11 a). Fig. 10.13 displays the time series of the pore pressure and the surface elevation obtained in a wave flume with a silt bottom, exposed to a progressive wave (Sumer, Fredsøe, Christensen and Lind, 1999). The figure illustrates clearly that the pore pressure begins to accumulate upon the introduction of the waves.

The purpose of this subsection is to derive the equation governing this process.

1) Period-averaged pore pressure

First, consider the pore pressure. This quantity is governed by the Biot consolidation equations (Eqs. 10.30, 10.31, 10.32 and 10.36). Considering that the present process is a 2-D process (independent of the y-direction), and furthermore that the variations with respect to x are negligible (Fig. 10.8), Eq. 10.32 for the present case will be

$$G \frac{2-2\nu}{1-2\nu} \frac{\partial^2 w}{\partial z^2} = \frac{\partial p}{\partial z} \qquad (10.76)$$

and Eq. 10.36

$$\frac{k}{\gamma} \frac{\partial^2 p}{\partial z^2} = \frac{n}{K'} \frac{\partial p}{\partial t} + \frac{\partial^2 w}{\partial z \partial t} \qquad (10.77)$$

Differentiating Eq. 10.76 with respect to t, and Eq. 10.77 with respect to z, one obtains

$$c_v \frac{\partial^3 p}{\partial z^3} = \frac{\partial^2 p}{\partial z \partial t} \qquad (10.78)$$

in which c_v is

$$c_v = \frac{Gk}{\gamma} \frac{2-2\nu}{(1-2\nu) + (2-2\nu)\frac{nG}{K'}} \qquad (10.79)$$

This quantity is termed the *coefficient of consolidation*.

Now, integrating Eq. 10.78 with respect to z gives

$$\frac{\partial p}{\partial t} = c_v \frac{\partial^2 p}{\partial z^2} + c \qquad (10.80)$$

in which c is an integration constant.

In the context of the present analysis, we are not interested in the phase-resolved values of p but rather the period-averaged value, $\bar{p}$ (Fig. 10.14). "Moving" averaging therefore gives

$$\frac{\partial \bar{p}}{\partial t} = c_v \frac{\partial^2 \bar{p}}{\partial z^2} + f \qquad (10.81)$$

in which $\bar{p}$ is the period-averaged pore pressure, namely

$$\bar{p} = \frac{1}{T} \int_t^{t+T} p \, dt \qquad (10.82)$$

and T is the wave period.

10.3. RESIDUAL LIQUEFACTION

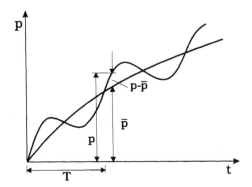

Figure 10.14: Definition sketch. Two components of pore pressure: the residual component, $\bar{p}$, and the oscillating component, p-$\bar{p}$.

In Eq. 10.81, apparently f is a source term, and represents the total amount of pore pressure generated per unit time and per unit volume of soil (including the pores). The quantity f, in general, may be considered to be a function of both space, z, and time, t (McDougal et al., 1989).

As seen, the period-averaged pore pressure satisfies the diffusion equation (Eq. 10.81). This is the *governing equation for the buildup of pore pressure*. Eq. 10.81 implies that the pore pressure is generated through the source term f, and it spreads in the soil according to a diffusion process where c_v, the coefficient of consolidation, plays the role of the familiar diffusion coefficient.

Next, we shall study the source term, f.

2) The source term f

From Section 10.3.1, the generation of the pore pressure is achieved through the action of the cyclic shear stress in the soil. This will, at each point in the soil, lead to a pressure generation similar to that in Fig. 10.11 a. This pressure can, in its simplest form, be represented by a linear variation with time, Fig. 10.15. (This is obviously valid up to the point of liquefaction). From Fig. 10.15:

$$\text{Pressure generated} = \sigma'_0 \frac{N}{N_\ell} \qquad (10.83)$$

in which N is the number of cycles. This is the pressure generated over the time period of NT. From Eq. 10.83, the pressure generated per unit time

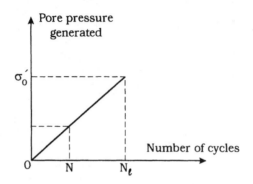

Figure 10.15: The accumulated pore pressure increases linearly with the number of cycles. The simple, linear model.

and volume (i.e., the quantity f) will then be

$$f = \frac{\sigma'_0 \frac{N}{N_\ell}}{NT} \tag{10.84}$$

or

$$f = \frac{\sigma'_0}{N_\ell T} \tag{10.85}$$

Here, σ'_0 is the initial effective stress, i.e., the overburden-pressure value, and can be taken as

$$\sigma'_0 = \frac{1}{3}(\gamma' z + k_0 \gamma' z + k_0 \gamma' z) = \gamma' z \frac{1 + 2k_0}{3} \tag{10.86}$$

where z is the distance from the mudline downwards (Fig. 10.8), k_0 is the coefficient of lateral earth pressure, and γ' is the submerged specific weight of the soil.

The quantity N_ℓ in Eq. 10.83 is the number of cycles to cause liquefaction, and may be taken as that obtained from Peacock and Seed's (1968) cyclic shear-stress tests discussed in Section 10.3.1 (Eq. 10.75), namely

$$N_\ell = (\frac{1}{\alpha} \frac{\tau}{\sigma'_0})^{1/\beta} \tag{10.87}$$

where τ is, in the present case, the amplitude of the shear stress in the soil under the "forcing" progressive wave.

10.3. RESIDUAL LIQUEFACTION

(Finally, it may be noted that nonlinear forms of the relationship in Eq. 10.83 have been suggested; see Seed and Booker, 1976, Rahman, Seed and Booker, 1977, Sekiguchi, Kita and Okamoto, 1995).

Great many works have been devoted to the theoretical and experimental investigation of buildup of pore pressure induced by waves.

Seed and Rahman (1978) were the first to adopt the model in Eq. 10.81 to describe the buildup of pore pressure under a progressive wave. Spierenburg (1987) and McDougal et al. (1989) have subsequently adopted similar approaches. The works by Barends and Calle (1985) and de Groot, Lindenberg and Meijers (1991) have considered similar theoretical descriptions of the process of pressure accumulation in the soil.

Clukey, Kulhawy and Liu (1985) report the results of an experimental investigation of the buildup of pore pressure induced by progressive waves in a laboratory flume. Tzang, Hunt and Foda (1992), Tzang (1998) and Sumer et al. (1999) present similar laboratory results. The sediment in the latter investigations was silt.

de Wit and Kranenburg (1992) report the results of an experimental study where the soil was a cohesive material; particular attention was concentrated on the liquefaction and erosion of China clay due to waves and current. Their results have shown that similar buildup of pore pressure is experienced in this kind of soil. Further results of the research undertaken by the same group have been reported in the publications by de Wit, Kranenburg and Battjes (1994), de Wit (1995), van Kessel, Kranenburg and Battjes (1996), de Wit and Kranenburg (1997) and van Kessel and Kranenburg (1998).

Foda and Tzang (1994) give an account of what they call the resonant fluidization, a process where massive liquefaction failure due to buildup of pore pressure occurs by waves. The latter authors link this process to a strong channeling of the seepage flow within the silt bed. (See also the review article by Foda, 1995).

Sekiguchi et al.'s (1995) study focuses on the generation of the pore pressure (characterized in the present description with the term f, Eq. 10.81). Their poro-elastoplastic formulation enables the researchers to obtain closed-form solutions for the accumulated pore pressure under cycling loading.

Finally, Tzang (1998) studies the oscillating component of the pore pressure in silt where there is a buildup of pore pressure (Fig. 10.13). Tzang's results show that the oscillating component behaves like that of a sandy soil.

However, it may be noted that the accumulated pressure in Tzang's experiments was relatively small, mostly $O(0.1)$ times the overburden pressure (or even less) (Tzang, 1998, Table 4).

10.3.3 Solution to the equation of buildup of pore pressure

The case of infinitely large soil depth

As seen from Eqs. 10.85 and 10.87, to be able to calculate the source term f (Eq. 10.85) we need N_ℓ; and for the latter we need τ, the amplitude of the cyclic shear stress in the soil (Fig. 10.8). The stresses in the soil (with an infinitely large soil depth) exposed to a progressive wave have been studied in Section 10.2.2, and closed solutions have been presented. From the latter, the amplitude of the shear stress, τ, is obtained as (Eq. 10.53)

$$\tau = p_b \lambda z \exp(-\lambda z) \tag{10.88}$$

in which p_b is the bed-pressure, given in Eq. 10.44.

(Note that the shear stress given in Eq. 10.53 is obtained in the case of a soil in which no buildup of pressure takes place (Section 10.2.2), and therefore it may not be entirely correct for the present case where there is a pore-pressure accumulation. No study is yet available, investigating the soil stresses in the presence of pore pressure accumulation).

Now, the solution to Eq. 10.81 (with f given in Eq. 10.85, σ'_0 in Eq. 10.86, N_ℓ in Eq. 10.87, and τ in Eq. 10.88) is to be sought under the following initial and boundary conditions.

1. At the initial instant, there is no accumulated pore pressure:

$$t = 0: \quad \bar{p} = 0 \tag{10.89}$$

2. At the mudline, the pore pressure continuously dissipates:

$$z = 0: \quad \bar{p} = 0 \tag{10.90}$$

10.3. RESIDUAL LIQUEFACTION

The solution to Eq. 10.81 can be obtained by the method of Fourier sine transform (Sneddon, 1957, p. 128 and p. 302):

$$\overline{p}(z,t) = \frac{2}{\pi} \int_{t'=0}^{t} dt' \int_{\xi=0}^{\infty} \exp\left[-c_v \xi^2 (t-t')\right] \sin(\xi z) \left[\int_{z'=0}^{\infty} f(z') \sin(\xi z') dz'\right] d\xi \quad (10.91)$$

where $f(z')$ is to be calculated from Eqs. 10.85-10.88.

The case of finite soil depth

In this case (Fig. 10.7), τ, the amplitude of the shear stress in the soil is obtained from Eqs. 10.64 and 10.70 as

$$\tau = |\tau_y| = p_b \left\{ (C_1 - C_2 \lambda z) e^{-\lambda z} - (C_3 - C_4 \lambda z) e^{\lambda z} + \lambda \delta (C_5 e^{-\delta z} - C_6 e^{\delta z}) \right\} \quad (10.92)$$

in which the coefficients C_i are depicted in Appendix III, Section 10.9 at the end of this chapter.

(Similar to the previous case, the shear stress given, this time, in Eq. 10.92 is obtained in the case of a soil in which no buildup of pressure takes place (Section 10.2.2), and hence it may not be entirely correct for the present case where there is a pore-pressure accumulation. As pointed out in the previous sub-section, no study is yet available, investigating the soil stresses in the presence of pore pressure accumulation).

Regarding the initial and boundary conditions, those employed for the case of the infinite soil depth are also valid for this case (Eqs. 10.89 and 10.90), plus the pressure "flux" at the impermeable base should be zero:

$$z = d: \quad \frac{\partial \overline{p}}{\partial z} = 0 \quad (10.93)$$

The solution to the governing equation (Eq. 10.81) under these initial and boundary conditions is given by Sumer and Cheng (1999) as:

$$\overline{p}(z,t) = \frac{2}{\pi} \sum_{m=1}^{\infty} \sin\left[(m - \frac{1}{2})\frac{\pi}{d}z\right] \quad (10.94)$$

$$\times \int_{t'=0}^{tc_v(\frac{\pi}{d})^2} dt'$$

$$\times \int_{\xi=0}^{\pi} \exp\left[-\frac{1}{4}(2m-1)^2 \left[c_v(\frac{\pi}{d})^2 t - t'\right]\right] \sin\left[(m - \frac{1}{2})\xi\right] g(\xi) d\xi$$

in which $g(\xi)$ is

$$g(\xi) = \frac{1}{c_v}\left(\frac{d}{\pi}\right)^2 \frac{1}{T}\gamma'\frac{1+2k_0}{3}\xi\frac{d}{\pi}\left[\frac{1}{\alpha}\frac{3}{1+2k_0}\frac{\tau}{\gamma'\xi(d/\pi)}\right]^{-1/\beta} \qquad (10.95)$$

in which τ is given by Eq. 10.92.

It may be noted that Cheng et al. (2001) solved Eq. 10.81 numerically (where the analytical expression given by Hsu and Jeng (1994), Eqs. 10.64 and 10.70, was used). They obtained a good agreement between the numerical results and the analytical solution given in Eq. 10.94. They emphasized that a small error in the soil shear stress can lead to a large error in the accumulated pore pressure.

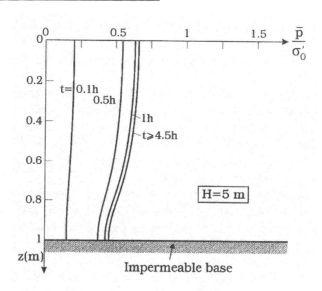

Figure 10.16: Time development of the accumulated pore pressure. Liquefaction does not occur; $\overline{p}$ never exceeds σ'_0.

Example 24 *Assessment of liquefaction potential*

The solutions obtained in the preceding paragraphs can be used to make assessment of liquefaction potential. The following numerical example will demonstrate this.

10.3. RESIDUAL LIQUEFACTION

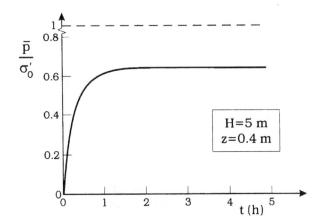

Figure 10.17: Time development of the accumulated pore pressure at $z = 0.4$ m. Liquefaction does not occur; $\bar{p}/\sigma'_0$ never exceeds unity.

The soil properties are given as follows: The soil depth, $d = 1$ m, the submerged specific weight, $\gamma' = 10.8$ kN/m^3, the shear modulus, $G = 926$ kN/m^2, the coefficient of permeability, $k = 1 \times 10^{-6}$ m/s, the porosity, $n = 0.333$, the degree of saturation, $S_r = 1$, the coefficient of lateral earth pressure, $k_0 = 0.4$, Poisson's ratio, $\nu = 0.35$, and the empirical constants in the Seed equation (Eq. 10.75), $\alpha = 0.246$ and $\beta = -0.165$.

The water properties, on the other hand, are: The specific weight of water, $\gamma = 9.81$ kN/m^3, and the bulk modulus of elasticity of water, $K = 1.9 \times 10^6$ kN/m^2.

The soil is exposed to a progressive wave with the following properties: The wave height, $H = 5$ m, the period $T = 13.7$ s, and the water depth, $h = 19$ m.

The question is whether the soil will be liquefied under the given wave climate.

Using Eq. 10.79, the coefficient of consolidation is found as $c_v = 4.1 \times 10^{-4}$ m^2/s. From the linear wave theory (Appendix A), the wave length is $L = 174$ m, and the wave number $\lambda = 2\pi/L = 0.036$ m^{-1}. The pore pressure is calculated from Eqs. 10.94 and 10.95. The results are given in Figs. 10.16 and 10.17.

As seen from the figures, the pore pressure never reaches the value of the overburden pressure, σ'_0, i.e., $\bar{p}/\sigma'_0$ is always

$$\bar{p}/\sigma'_0 < 1$$

Hence, liquefaction will not occur under this wave climate.

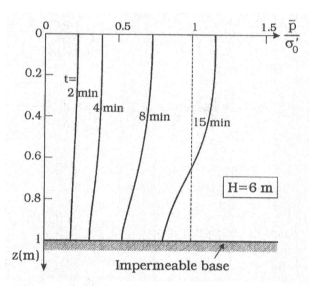

Figure 10.18: Time development of the accumulated pore pressure. Liquefaction occurs; $\bar{p}$ exceeds σ'_0 within less than 15 minutes.

Now, Fig. 10.18 displays the corresponding pressure distributions when the wave height is $H = 6$ m. The pore pressure in this case reaches the initial overburden pressure value within less than 15 minutes. Given the fact that 15 minutes is a reasonably short period of time for a given sea state (which might last as long as O(4-5 hours) or even more), it may be concluded that there is a liquefaction potential for this second wave climate. Also note that liquefaction first starts at the surface of the soil and spreads downwards (Fig. 10.18).

Example 25 *Laboratory observations of liquefaction and no-liquefaction regimes*

Fig. 10.19 presents the time series of pore pressure for two cases recorded in the laboratory study of Sumer et al. (1999): the top time series is obtained

10.3. RESIDUAL LIQUEFACTION

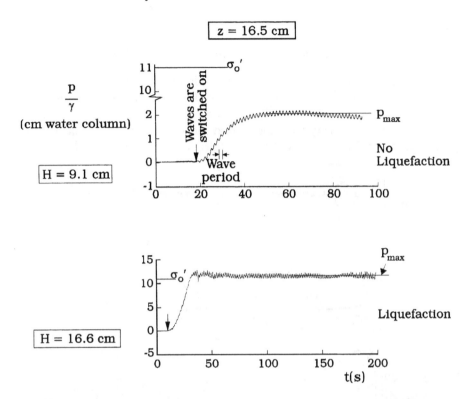

Figure 10.19: Two regimes: No-liquefaction Regime and Liquefaction Regime.

for a wave height of $H = 9.1$ cm while the bottom one is obtained for a wave height of $H = 16.6$ cm, both at the depth $z = 16.5$ cm. The soil in Sumer et al.'s experiments was silt with $d_{50} = 0.045$ mm, and exposed to a progressive wave with $T = 1.6$ s with a water depth of $h = 0.42$ m. The soil depth was $d = 17$ cm.

As seen from the figure, $\bar{p}$ does not reach the value of the overburden pressure (p_{max} remains smaller than σ'_0) for the wave height $H = 9.1$ cm, meaning that no liquefaction occurs for this case (the *no-liquefaction regime*) (cf. Fig. 10.17), while it does reach σ'_0, and slightly exceeds it for the wave height $H = 16.6$ cm, meaning that liquefaction occurs for the latter case (the *liquefaction regime*). Sumer et al. (1999) found that, for the liquefaction regime to occur, the wave height should exceed a critical value which was

10.2 cm.

Example 26 *Behaviour of the accumulated pressure for large times*

For large times, the water will begin to drain from the soil due to the continuous buildup of pore pressure. This will obviously lead to a pressure relief in the soil. The pressure will therefore begin to fall. The time scale of the latter process, namely the time needed for a substantial decrease in the pore pressure, is $O(z^2/c_v)$ (Lambe and Whitman, 1969, Chapter 27).

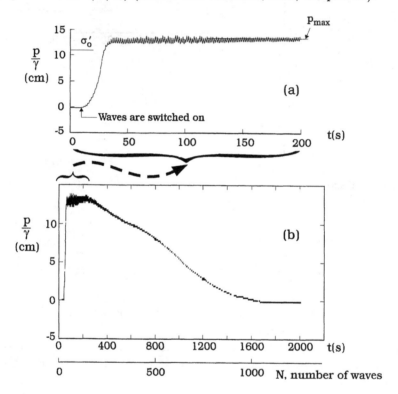

Figure 10.20: Behaviour of the accumulated pore pressure for large times. $h/L = 0.145$, $d/L = 0.059$, $H/d = 0.75$, $c_v T/d^2 = 7.1 \times 10^{-4}$, $z = 16.5$ cm, $H = 12.8$ cm. Sumer et al. (1999).

Fig. 10.20 displays two time series of the pore pressure measured in Sumer et al.'s (1999) experiments with $H = 12.8$ cm. (The test conditions

10.3. RESIDUAL LIQUEFACTION

are the same as in the previous example). Fig. 10.20 a is a close-up picture (corresponding to the first 200 s) of the time series given in Fig. 10.20 b. Compare Fig. 10.20 a with Fig. 10.11 a.

Fig. 10.20 b clearly shows that the accumulated pore pressure after reaching an equilibrium value, namely $p_{\max}$ (Fig. 10.20 a), begins to fall, and eventually tends to the zero-pressure level (that it started with), as the waves continue - the complete dissipation of the accumulated pore pressure.

Sumer et al.'s (1999) experiments indicated that this dissipation of the pressure first begins at the largest depth, and it gradually spreads to the smaller depths. This is because the pore water begins to drain (and therefore, the pressure begins to dissipate) at the larger depths where the pressure is highest.

Furthermore, Sumer et al.'s experiments showed that the same behaviour is experienced in the no-liquefaction regime, too, in which the accumulated pore pressure, after reaching a steady-state value, $p_{\max}$ (which was smaller than the overburden pressure value), began to fall in the same fashion as in Fig. 10.20 b.

Note that the buildup of pressure, followed by the stage where the pressure is progressively dissipated (similar to that in Fig. 10.20 b), may be observed in other contexts such as in the familiar consolidation process (Lambe and Whitman, 1969, p. 392), and in the process where the pore-water pressure builds up due to an earthquake (as reported by Tanaka (1996, Fig. 12) in conjunction with the 1995 Kobe earthquake).

Example 27 *Effect of irregular waves*

Sumer et al.'s (1999) experiments showed that the process of buildup of pressure in irregular waves occurs in much the same way as in the case of the regular waves.

These authors addressed the question of how to define the wave parameters in the case of irregular waves with regard to the process of pore pressure accumulation. Several combinations of the wave heights (H_s, $H_s/\sqrt{2}$, etc.) and the wave periods (T_p, T_z, etc.) were tested to see which combination would give the best comparison with the regular wave results, when the results were plotted in terms of the wave height versus the number of waves to cause liquefaction. Here, H_s = the significant wave height, T_p = the peak period, and T_z = the mean zero upcrossing period. The results showed that the combination $H_s/\sqrt{2}$ and T_z gave the best agreement. Incidentally, the quantity $H_s/\sqrt{2}$ can be interpreted as the equivalent wave height of the irregular waves, since $H_s/\sqrt{2} \simeq (4\sigma_\eta)/\sqrt{2} = 2(\sqrt{2}\sigma_\eta) = 2\,a_\eta$, in which $a_\eta =$ the amplitude of the surface elevation η when the waves are sinusoidal.

Example 28 *Effect of history of wave exposure on the pressure buildup*

To observe the effect of the history of wave exposure, Sumer et al. (1999) carried out the following systematic test where the soil was exposed more than once to a progressive wave with $H = 16.6$ cm, and $T = 1.6$ s. The following procedure was used.

1. Expose the soil to the waves for 20 minutes. Monitor the pore pressure.

2. Stop the waves. Wait for 10 minutes.

3. Switch on the waves again, and expose the soil to the same waves for another 20 minutes. Monitor the pore pressure.

4. Repeat Steps 1-3 a number of times.

From these experiments it was found that $p_{\max}$, the maximum accumulated pore pressure (Fig. 10.20 a), was reduced tremendously when the soil was exposed to the waves for the second time. While $p_{\max}/\gamma$ at $z = 16.5$ cm was about 11 cm for the first exposure, it was only 0.6 cm when the soil was exposed to the same waves for the second time, an order of magnitude reduction. The pressure $p_{\max}$ was virtually nil, when the soil was exposed to the same waves for the third time. This behaviour is explained in the following way.

The soil grains (when they are exposed to the waves for the first time) rearrange through the process of the buildup and dissipation of the pore pressure (e.g., Fig. 10.20 b). When the soil is exposed to the waves for the second time, there will not be too much "room" for the grains to rearrange (the grains now being much more densely packed after the first "round" of buildup and dissipation of the pore pressure), and hence the pore pressure will accumulate only slightly. The work by van Kessel and Kraneneburg (1998) can be consulted for a somewhat detailed discussion of this issue in relation to the seabed liquefaction.

From the preceding paragraphs it may be concluded that, *in a real-life situation, unless the soil is "softened" (for example, when it is used as a backfill material), it will not be liquefied due to the wave action because of its long time history of wave exposure.*

10.4 Momentary liquefaction

10.4.1 General description

As mentioned previously (Section 10.1), liquefaction may also develop momentarily during the passage of the wave trough (without the buildup of pore pressure). The present section will describe this process.

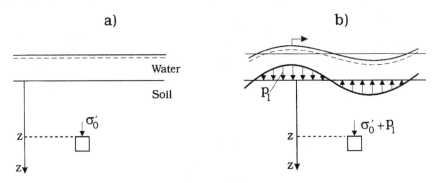

Figure 10.21: Definition sketch.

The mechanism of the momentary liquefaction is described in Section 10.1.

Now, consider the depth z in the soil (Fig. 10.21). Let σ'_0 be the overburden pressure (i.e., the initial effective stress) at this depth, i.e., the portion of the total normal stress carried by the soil (Eq. 10.86) before the waves are introduced. When the waves are introduced (Fig. 10.21 b), the normal stress at z at time $t = 0$ will be

$$\sigma'_0 + p_1 \tag{10.96}$$

where p_1 is the pressure induced by the waves on the bed (Eq. 10.41):

$$p_1 = p_b \exp[i(\lambda x + \omega t)] \tag{10.97}$$

in which p_b is the maximum value of the bed pressure given in Eq. 10.44. This normal stress, namely $\sigma'_0 + p_1$, will be apportioned by the soil and the pore water at time t as follows:

$$\sigma'_0 + p_1 = \sigma' + p \tag{10.98}$$

in which σ' represents the soil part of the normal stress (i.e., the effective stress) at time t and at depth z, and p is the water part of the normal stress at the same time instant at depth z.

Solving σ',

$$\sigma' = \sigma'_0 - (p - p_1) \qquad (10.99)$$

Now, liquefaction occurs when σ' becomes zero, namely

$$\sigma'_0 - (p - p_1) = 0 \qquad (10.100)$$

This is simply because the soil in this case will carry no load, and therefore it will fail.

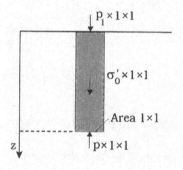

Figure 10.22: Forces on the soil element the size $1 \times 1 \times z$.

(It can readily be shown that the second term on the left-hand side of the above equation, namely $(p - p_1)$, represents the lift force on the soil column illustrated in Fig. 10.22; the soil fails when this lift force becomes equal to the submerged soil weight, i.e., σ'_0).

10.4.2 The case of completely saturated soil

In the case of a completely saturated soil with an infinitely large depth, the pore pressure p has been obtained from the solution of the Biot equations in Section 10.2.2, and is given in Eq. 10.50 with p_b given in Eq. 10.44. Inserting Eq. 10.50, and Eqs. 10.86 and 10.97 in Eq. 10.99, the effective stress is found to be

10.4. MOMENTARY LIQUEFACTION

$$\frac{\sigma'}{\gamma H} = \frac{\gamma'}{\gamma}\frac{1+2k_0}{3}\frac{z}{H} - \frac{0.5}{\cosh(\lambda h)}\left[1 - \exp(-\lambda H \frac{z}{H})\right]\exp\left[i(\lambda x + \omega t)\right]$$
(10.101)

or in terms of wave steepness H/L

$$\frac{\sigma'}{\gamma H} = \frac{\gamma'}{\gamma}\frac{1+2k_0}{3}\frac{z}{H} - \frac{0.5}{\cosh(\lambda h)}\left[1 - \exp(-2\pi \frac{H}{L}\frac{z}{H})\right]\exp\left[i(\lambda x + \omega t)\right]$$

The soil will be most susceptible to liquefaction when the wave height is largest. The largest wave height for a given wave length may be given approximately as (Isaacson, 1979)

$$(\frac{H}{L})_{max} = 0.14 \tanh(\lambda h)$$
(10.102)

(The waves higher than that in the preceding equation will break). Hence, σ' for $(H/L)_{max}$ will be

$$\frac{\sigma'}{\gamma H} = \frac{\gamma'}{\gamma}\frac{1+2k_0}{3}\frac{z}{H} - \frac{0.5}{\cosh(\lambda h)}\left[1 - \exp(-0.88 \tanh(\lambda h)\frac{z}{H})\right]\exp\left[i(\lambda x + \omega t)\right]$$
(10.103)

For $\gamma'/\gamma = 0.8$ and $k_0 = 0.4$, typical values, it can be seen that σ' would normally remain positive. This means that *normally the momentary liquefaction will not occur in a completely saturated soil*.

10.4.3 The case of unsaturated soil

In the case of an unsaturated soil with an infinitely large depth, the pressure in the soil has been calculated by various authors; e.g., Yamamoto et al. (1978) (see Section 10.2.2) and Mei and Foda (1981).

Sakai et al. (1992) adopted Mei and Foda's solution, namely

$$p = p_b \frac{1}{1+m}\exp(-\lambda z)\cos(\lambda x - \omega t) + p_b \frac{m}{1+m}\exp(\frac{-z}{\sqrt{2\delta}})\cos(\lambda x - \omega t + \frac{z}{\sqrt{2\delta}})$$
(10.104)

in which

$$m = \frac{n}{(1-2\nu)}\frac{G}{K'}$$
(10.105)

$$\delta = \left[\frac{(k/\gamma)G}{\omega}\right]^{1/2} \left[n\frac{G}{K'} + \frac{1-2\nu}{2(1-\nu)}\right]^{-1/2} \qquad (10.106)$$

and, making the shallow water, long wave approximations, namely

$$\cosh(\lambda h) \simeq 1, \quad \frac{2\pi}{\lambda} \equiv L \simeq T\sqrt{gh} \qquad (10.107)$$

they examined whether or not σ' in Eq. 10.99 can become zero (the momentary liquefaction). Fig. 10.23 depicts Sakai et al.'s results (corrected by Law, 1993), giving the depth of the momentary liquefaction.

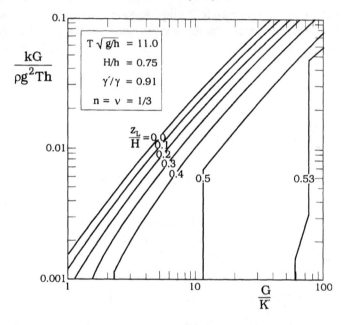

Figure 10.23: Depth of momentary liquefaction. Sakai et al. (1992) and corrected by Law (1993).

As implied by Sakai et al.'s analysis, the momentary liquefaction does occur for the values of the parameters indicated in the box in the figure, and it can penetrate to depths as large as $0.5H$. Sakai et al. notes that it is difficult to produce the momentary liquefaction in a small-size wave flume using a normal sand bed, because the value of $kG/(\rho g^2 Th)$ is two orders of magnitude larger than that in actual sandy beaches.

10.5. SINKING/FLOATATION OF PIPELINES

Gratiot and Mory (2000) extended the analysis of Sakai et al. (1992) to include the effect of the water depth variations. In the analysis, the long wave approximation was not made. Basically, Gratiot and Mory's work is a sensitivity study of the depth of liquefaction with respect to the degree of saturation, the porosity, the permeability, the shear modulus, the Poisson ratio of the soil, the water depth, the wave height and the wave period.

Finally, it may be noted that, in Sakai et al.'s analysis, the initial effective stress is taken not as given in Eq. 10.86, but as $\sigma'_0 = \gamma' z$, ignoring the effect of the lateral earth pressure. However, an intercomparison study of Jeng (1997) implied that the initial effective stress, when taken as in Eq. 10.86, gives a more realistic result for the liquefaction criterion in Eq. 10.100. See Zen and Yamazaki (1993) and Jeng (1997) for further discussion.

10.5 Sinking/floatation of pipelines

A pipeline buried in the seabed may float to the surface of the seabed when the soil supporting the pipeline may fail due to liquefaction. There are reported incidents in the literature of pipeline floatation. Christian et al. (1974) report that a 10-ft-diameter steel pipeline in Lake Ontario (with a backfill of 7 ft deep over the top of the pipe) has failed several times, apparently because of liquefaction, where sections of the pipeline floated to the surface of the soil during storms. Likewise, Herbich et al. (1984) report that a 10-ft-diameter pipeline, under construction, was found on the surface after a rather severe storm. The pipeline floatation has been the subject of extensive research. Early research which dates back as early as 1949 (see ASCE Pipeline Floatation Research Council, 1966) involved the agitation of the soil in laboratory tests by different means (different from waves). Silvis (1990) reports the results of a study where liquefaction potential has been assessed for Zeepipe, a 810 km long and 1.25 m diameter gas pipeline laid from Sleipner in the North Sea to Zeebrugge at the Belgian coast. de Groot and Meijers (1992) present a study undertaken in connection with a 36" gas pipeline off the Dutch coast; the main concern was to study the risk of the floatation of the pipeline placed in a trench which was filled with sand from a nearby source. Siddhartan and Norris (1993) identify all of the important processes associated with floatation mechanisms and provide methods and guidelines to quantify these factors with due consideration given to the buildup of pore pressure. Damgaard and Palmer (2001) make assessment of liquefaction potential for a pipeline seated on a seabed. The latter authors present a diagram indicating various regimes of the pipeline and seabed interaction starting with the threshold of the sediment motion at the bed, and continuing with the "threshold" of the pipeline stability, the onset of liquefaction due to buildup of pore pressure, and the onset of momentary liquefaction. They point out that the marginal pipeline stability can under realistic field conditions be accompanied by seabed liquefaction, which, in turn, is likely to cause sinking of the pipeline.

Likewise, a pipeline laid on the seabed may sink along the stretch where the soil may fail due to liquefaction. This will result in the self-burial of the pipeline.

In the previously mentioned instabilities (sinking/floatation of pipelines), the liquefaction-failure of the soil may be due to the buildup of pore pressure (the residual liquefaction), or it may be due to the presence of upward pressure gradient during the passage of the wave trough (the momentary liquefaction), the processes described in greater detail in the preceding sections. In the following sections, we shall first look at the self-burial (sinking)/floatation of pipelines in the case of the residual liquefaction, and then we shall give a brief account of the pipeline sinking in the case of the momentary liquefaction.

10.5.1 The case of residual liquefaction

Sumer et al. (1999) have studied the sinking/floatation of pipelines in liquefied soils in the laboratory where the liquefaction was due to the buildup of pore pressure. The following account is mainly based on this latter work.

General description of the process

Fig. 10.24 displays two kinds of time series; the top ones are for the pore pressure at several depths in the soil (Fig. 10.24 a), while the bottom one (Fig. 10.24 b) is the displacement of a buried pipeline, recorded simultaneously with the pressure time series. The specific gravity of the pipe in this test,

$$s_p = \frac{\gamma_p}{\gamma} \qquad (10.108)$$

was $s_p = 1.0$. The pipe diameter was $D = 4$ cm. The pore pressures were

10.5. SINKING/FLOATATION OF PIPELINES

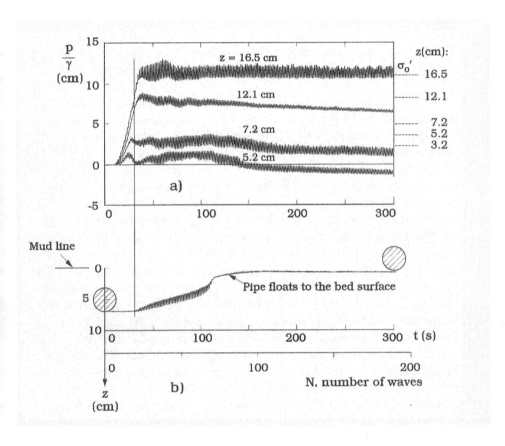

Figure 10.24: Time series of pipe displacement (flotation) in the vertical and pore pressure recorded simultaneously. $h/L = 0.145$, $d/L = 0.059$, $H/d = 1$, $c_v T/d^2 = 7.1 \times 10^{-4}$, $H = 16.6$ cm. Pipe specific gravity, $s_p = 1$. Sumer et al. (1999).

measured away from the pipe, representing the far-field values of the pore pressure.

Fig. 10.25 gives similar time series in the case of a relatively heavier pipe. with $s_p = 3.1$. Fig. 10.26 illustrates the close-up picture of the initial stage of the process in Fig. 10.25.

The following observations can be made from Figs. 10.24-10.26:

1. A pipe buried in a soil, which is exposed to waves, may float to the surface of the soil, or it may sink in the seabed, depending on its specific weight.

2. The sinking of the pipe stops well away from the impermeable base (Fig. 10.25). The depth at which the sinking stops apparently depends on several quantities such as the initial position of the pipe, its specific gravity, and the soil depth, as will be discussed later in the section.

3. A close examination of Figs. 10.24 and 10.26 indicates that the pipe begins to float/sink in the soil before the pore pressure reaches the value of the overburden pressure, σ_0'. However, for the onset of floatation/sinking, the accumulated pore pressure apparently needs to reach a substantial value. For example, in the case of the sinking (Fig. 10.26), p reaches a value of about 70 % of σ_0' for $z = 16.5$ cm, about 60 % of σ_0' for $z = 12.8$ cm, and about 40 % of σ_0' for $z = 7.2$ cm in the test presented in Fig. 10.26. The reason why the pipe begins to sink before the pore pressure reaches the value of σ_0', i.e., before the soil is fully liquefied, may be due partly to the considerable reduction of the soil strength caused by a high degree of pore-pressure buildup in the soil, and partly to the additional accumulation of the pore pressure in the neighbourhood of the pipe, which may presumably lead to the soil liquefaction earlier around the pipe than in the undisturbed-flow situation. Sumer et al. (1999) note, however, that, in six tests out of the total fourteen in their study, the pressure at $z = 16.5$ cm and 12.8 cm reached the overburden pressure value at the time when the pipe began to sink, and that the latter behaviour did not reveal any correlation with the parameters such as the wave height, the initial pipe position, the specific gravity of the pipe, and the soil depth.

Similar experiments were carried out by the authors with various wave heights, both in the liquefaction regime and in the no-liquefaction regime. No displacement of the pipe was observed in the case of the no-liquefaction regime. In one no-liquefaction test ($H = 9.1$ cm, $D = 4$ cm, and e_0 (the burial depth from the mudline to the pipe bottom) $= 7$ cm), the pipe replaced with a much heavier pipe ($s_p = 8.9$) with the same outcome, namely no sinking.

10.5. SINKING/FLOATATION OF PIPELINES

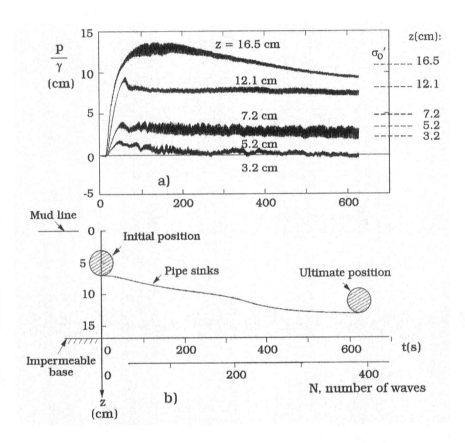

Figure 10.25: Time series of pipe displacement (sinking) in the vertical and pore pressure recorded simultaneously. $h/L = 0.145$, $d/L = 0.059$, $H/d = 1$, $c_v T/d^2 = 7.1 \times 10^{-4}$, $H = 16.6$ cm. Pipe specific gravity, $s_p = 3.1$. Sumer et al. (1999).

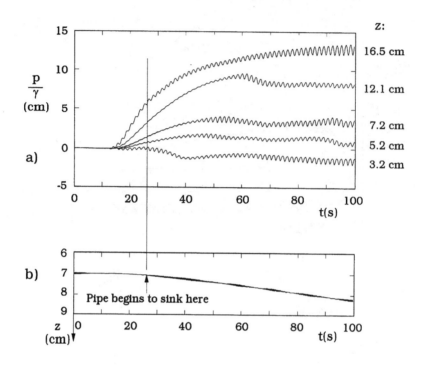

Figure 10.26: Close-up of the time series depicted in the previous figure. Sumer et al. (1999).

The depth of sinking

From dimensional considerations, the depth of sinking normalized by the pipe diameter, e/D (see Fig. 10.27 for definition sketch), depends on the following quantities:

1. The pipe properties

$$\frac{e_0}{D}, s_p, \frac{d}{D} \tag{10.109}$$

in which D is the pipe diameter, e_0 is the initial pipe position, and s_p is the specific gravity of the pipe. (The third parameter in the preceding equation, d/D, appears through the effect of the presence of the pipe itself on the pore-pressure accumulation).

10.5. SINKING/FLOATATION OF PIPELINES

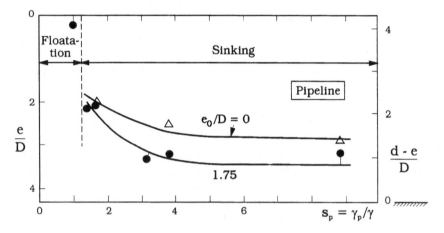

Figure 10.27: Ultimate sinking/flotation depth of pipeline. $h/L = 0.145$, $d/L = 0.059$, $H/d = 1$, $c_v T/d^2 = 7.1 \times 10^{-4}$, $H = 16.6$ cm. Sumer et al. (1999).

2. The quantities which govern the pore-pressure accumulation

$$\nu, k_0, \gamma', \alpha, \beta, D_r, G, S_r, n, \frac{d}{L}, \frac{h}{L}, \frac{H}{d}, \frac{c_v T}{d^2} \qquad (10.110)$$

Of these parameters, the first three parameters are of particular interest. Each parameter is now considered individually.

Effect of initial position, e_0/D. Fig. 10.27 displays the results of the tests carried out with two different initial pipe positions; in one, the pipe was initially sitting on the bed ($e_0/D = 0$), and in the other, it was completely buried in the soil with $e_0/D = 1.75$. The ultimate sinking position of the pipe is plotted against the specific gravity of the pipe.

Regarding the case of the bottom-seated pipe ($e_0/D = 0$), scour will occur initially, as described in Chapter 2. Sumer et al.'s (1999) analysis indicated that the pipe would sink to a depth of $O(1\text{ cm})$ (or alternatively, $e/D = O(0.3)$) due to scour before it sank due to liquefaction. Comparing the estimated sinking $e/D = O(0.3)$ with the data in Fig. 10.27, namely $e/D = 2 - 2.5$, clearly indicates that the sinking in the tests of Sumer et al. (1999) (Fig. 10.27) is mainly governed by liquefaction.

Finally, Fig. 10.27 clearly shows that a pipe initially sitting on the bed sinks to a relatively shallower depth than a pipe initially buried in the soil. This may be due to the fact that a bed-seated pipe, during its sinking process, will help drain more water, presumably resulting in relatively smaller accumulation of the pore pressure around the pipe, leading to a relatively smaller sinking depth.

Effect of pipe's specific gravity, s_p. Fig. 10.27 shows that the critical value of s_p beyond which the pipe sinks appears to be around 1.3 (the vertical dashed line). This value is apparently somewhat smaller than the generally accepted value for the specific gravity of a completely liquefied soil, namely ~1.8 (see, for example, de Groot and Meijers, 1992). This may be due to an increase in the porosity as a result of liquefaction in the tests in Fig. 10.27. Also, note that the specific gravity values of liquefied soils as small as 1.3-1.4 have been reported in the literature (ASCE Pipeline Floatation Research Council, 1966, Table 3, Figs. 7 and 11).

Secondly, Fig. 10.27 shows that the depth of sinking first increases with increasing s_p, and subsequently it attains an asymptotic value for $s_p > 4-5$. The pipe with small s_p values (such as $s_p = 1.36$ and 1.7 in Fig. 10.27) stops in its sinking motion at relatively shallower depths. This suggests that the specific gravity of the liquefied soil increases rather rapidly with increasing depth, and the pipe stops at these shallower depths mainly because the specific gravity of the soil surrounding the pipe becomes equal to that of the pipe for these small values of s_p.

Fig. 10.27 further shows that, for large s_p values ($s_p > 4-5$), the pipe's sinking motion terminates at a distance from the impermeable base. By contrast, Sumer et al.'s sphere experiments indicated that the sinking of the sphere for such large values of s_p ($s_p > 4-5$) terminated at the impermeable base (see Section 10.6). This suggests that the ultimate depth of sinking of an object is governed not only by the parameters given in Eq. 10.109, but also by the volume and, perhaps, by the shape of the sinking object. We shall return to this issue later in the chapter.

10.5. SINKING/FLOATATION OF PIPELINES

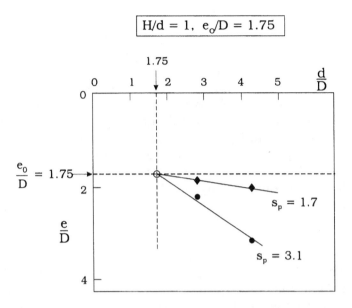

Figure 10.28: Ultimate sinking depth of pipeline. Sumer et al. (1999).

Effect of soil depth, d/D. Fig. 10.28 presents the results where the effect of the soil depth has been investigated. As seen, the soil depth is also an influencing parameter. The larger the soil depth, the larger the sinking depth. However, it should be expected that e/D tends to an asymptotic value, as $d/D \to \infty$. The experimental constraints in the tests did not permit this asymptotic value to be predicted. However, the data in Fig. 10.28 suggests that this value should be larger than $O(4)$ for $s_p = 3.1$.

It is interesting to note that, in addition to the parameters in Eqs. 10.109, the effect of the wave height (see Eq. 10.110) has also been investigated, to observe if sinking would be different in a liquefied soil with different wave heights for a given water depth. The results showed that this effect is practically nil. This is because once the soil is liquefied, the pipe will not be able to differentiate whether the wave height is large or small (while the accumulated pore pressure increases with increasing wave height for the no-liquefaction regime, it remains constant, at a value slightly larger than σ'_0, for the liquefaction regime).

Sinking/floatation velocity

Figs. 10.24 and 10.25 show that the pipe's motion (sinking or floatation) reaches a steady state (in which the pipe moves with practically a constant velocity) shortly after the onset of motion. This continues until the pipe comes close to the impermeable base in the case of the sinking, and to the surface of the soil in the case of the floatation. In this steady-state pipe motion, obviously the gravity force is balanced by the resistance force. The resistance force may have several components such as (1) the viscous force, (2) the force due to the accumulated-pressure gradient in the vertical direction, and (3) the force necessary to displace the soil in the immediate neighbourhood of the pipe. Obviously the process of resistance force in a liquefied soil may be different from that in an ordinary fluid. Nevertheless, making an analogy to the steady motion of a body in a fluid, the resistance force (per unit pipe length) can, to a first approximation, be written as $\frac{1}{2}\rho_\ell C_D D w^2$ in which C_D is the drag coefficient, ρ_ℓ the density of the liquefied soil (i.e., the density of the mixture of soil and water in the liquefied state) and w the sinking/floatation velocity. Therefore, the force balance for the steady-state pipe motion will be

$$(\gamma_p - \gamma_\ell)\frac{\pi D^2}{4} = \frac{1}{2}\rho_\ell C_D D w^2 \qquad (10.111)$$

Here, γ_ℓ = the specific weight of the liquefied soil, $\gamma_\ell = \rho_\ell g$. It is expected that C_D is governed primarily by the Reynolds number, defined by

$$Re_p = \frac{wD}{\nu_\ell} \qquad (10.112)$$

in which ν_ℓ = the kinematic viscosity of the liquefied soil. Clearly, ν_ℓ is different from the water viscosity. ν_ℓ can be expressed as $\nu_\ell/\nu = f$, in which f is a function of concentration (Happel and Brenner 1973, p. 453).

From Eq. 10.111, the drag coefficient is

$$C_D = \frac{\pi}{2}\frac{g(s_p - s_\ell)D}{w^2 s_\ell} \qquad (10.113)$$

in which $s_p(=\gamma_p/\gamma)$, $s_\ell(=\gamma_\ell/\gamma)$ are the specific gravities of the pipe and the liquefied soil, respectively. In Fig. 10.29 is plotted the drag coefficient data against Re. In this plot, (1) s_ℓ is taken as 1.3 (although s_ℓ may be expected

10.5. SINKING/FLOATATION OF PIPELINES

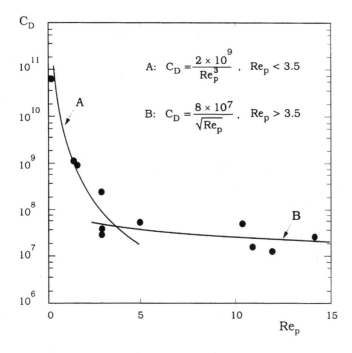

Figure 10.29: Drag coefficient for a pipeline in liquefied soil. Sumer et al. (1999).

to change with the depth, Sumer et al. (1999) argue that this refinement in the calculations may be inconsistent with the accuracy of the tests achievable in these kinds of experiments); and (2) the Reynolds number is taken as

$$Re_p = \frac{wD}{\nu} \tag{10.114}$$

to avoid uncertainties and also unnecessary complication related to the kinematic viscosity ν_ℓ. This is not a problem, however, since ν_ℓ in a real life situation would not be radically different from that in the experiments in Fig. 10.29. Hence, ν_ℓ would not be a scaling parameter.

First of all, Fig. 10.29 shows that C_D decreases with increasing Re_p, a behaviour quite similar to that in the case of a fluid. However, the drag coefficient in the present case is 7-9 orders of magnitude larger than the ordinary fluid drag (Schlichting, 1979, p. 17). The variations of C_D with Re_p

in Fig. 10.29 can be represented by the following empirical expressions

$$C_D = \frac{2 \times 10^9}{Re_p^3} \text{ for } Re_p < 3.5 \qquad (10.115)$$

and

$$C_D = \frac{8 \times 10^7}{\sqrt{Re_p}} \text{ for } Re_p > 3.5 \qquad (10.116)$$

These equations can be used to estimate the sinking (or floatation) velocity w of a pipe in a liquefied soil. The latter information may prove useful in making assessments about the time scale of the sinking/floatation process. Caution must be exercised, however, when the preceding equation is extrapolated to large Re_p numbers.

Finally, it may be noted that a backward calculation where C_D is assumed to be given by the Lamb-Oseen relation (see, for example Sumer and Fredsøe, 1997, p.219) (the creeping flow)

$$C_D = \frac{8\pi}{Re_p \ln(7.4/Re_p)} \qquad (10.117)$$

with $Re_p = wD/\nu_\ell$, gave the kinematic viscosity of the liquefied soil as $\nu_\ell \simeq 400$ m^2/s, a value which is 8-9 orders of magnitude larger than the water value, i.e., 0.01 cm^2/s at 20^0. However, again, it must be emphasized that the process of flow of liquefied soil around the pipe may be different from that in the case of the ordinary fluid (involving other mechanisms as well), as already stated in the preceding paragraphs.

Remarks on practical applications

The soil used in the tests in Figs. 10.24-10.29 was prepared by the authors themselves to achieve repeatability throughout the experiments. Hence, when a test was performed, the soil was exposed to waves for the first time, meaning that it acted as a "soft" soil. Whereas the soil in an actual field situation has a long history of wave action, and hence it is a stiff soil, unless it is composed of mechanical backfill, as in the case where a pipeline is laid in a trench, and the trench is then backfilled. Likewise, an earthquake can reduce the soil strength; although the soil in this case will eventually consolidate, and gain its strength, this may take some time (from several hours to several days, Tanaka (1996)), and therefore during this time the soil stiffness

10.5. SINKING/FLOATATION OF PIPELINES

will be relatively small, and hence the soil may be prone to wave-induced liquefaction.

As far as the drag coefficient is concerned (Eqs. 10.115 and 10.116), the empirical expressions given in these equations should be used with caution for large Reynolds numbers.

Finally, we should note that the soil response observed in a laboratory model may not be properly extrapolated to field conditions; such a laboratory model should be treated as an individual prototype in itself. Centrifuge wave testing on a soil bed has been tried to extend its potential for exploring wave-induced instability of sediments. Sassa and Sekiguchi (1999) can be consulted for further details regarding this latter issue.

10.5.2 The case of momentary liquefaction

No study is yet available, investigating the sinking/floatation of pipelines under momentary-liquefaction regime. This may be due to the fact that it is difficult to produce the momentary liquefaction in a small-size wave flume using a normal sand bed, as pointed out by Sakai et al. (1992) (see Section 10.4.3).

However, Maeno, Magda and Nago (1999) have studied the floatation of a pipeline in the laboratory where the effect of a progressive wave was simulated by an oscillating water table. A model pipeline buried in the soil floated to the surface of the soil, apparently due to the lift force on the pipeline induced by the oscillating water table. The lift force on the pipeline (and therefore the upward displacement of the pipe) was found to be associated with the "trough" half period of the motion of the water table (Maeno et al., 1999, Fig. 4). (The soil used in the experiments was sand with $d_{50} = 0.25$ mm). The paper by Maeno et al. (1999) furthermore presents a stability analysis for the floatation of the pipeline in which the pressure and the soil shear stresses are calculated by the numerical model of Magda (1997) (see also Magda, Maeno and Nago, 1998). It may be noted that, in the study of Maeno et al. (1999) (also see Sakai et al., 1992), the degree of saturation plays a significant role; even a slight reduction in the degree of saturation (a reduction from the value of unity) may lead to the momentary liquefaction (Sakai et al., 1992), or to a considerable uplift on the pipe (Maeno et al., 1999).

10.6 Sinking of armour blocks in liquefied soil

10.6.1 The case of residual liquefaction

Sumer et al. also (1999) studied the sinking of armour blocks. The test set-up was the same as that in the study of pipeline floatation/sinking described in the previous section. The armour blocks were simulated by spherical bodies. To see the effect of the shape of the object, the experiments were also carried out for cube-shaped objects as well. The study indicated that the process of sinking of these 3D objects is basically the same as in the case of pipelines, described in the previous section. The following paragraphs will summarize the highlights of this latter study.

Sphere-shaped body

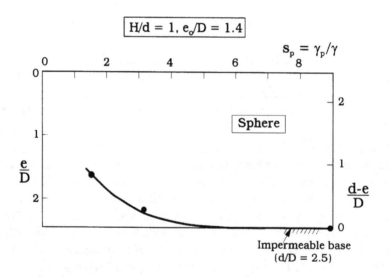

Figure 10.30: Ultimate sinking depth of sphere. $h/L = 0.145$, $d/L = 0.059$, $H/d = 1$, $c_v T/d^2 = 7.1 \times 10^{-4}$, $H = 16.6$ cm. Sumer et al. (1999).

Fig. 10.30 shows the sinking depth as a function of the specific gravity of the sphere for an initially buried sphere where $e_0/D = 1.4$. The variation of e/D with the specific gravity s_p is quite similar to that given for the pipeline case (Fig. 10.27); e/D increases with s_p, and subsequently attains

10.6. SINKING OF ARMOUR BLOCKS IN LIQUEFIED SOIL

an asymptotic value for $s_p > 4-5$. However, in contrast to the case of the pipeline, this asymptotic value in the present case is 2.5, namely the sinking of the sphere stops precisely at the impermeable base.

To see the influence of the initial position of the sphere on the sinking, two tests were made with the sphere with $s_p = 3.2$ and $D = 6.9$ cm: one with a bottom-seated sphere ($e_0/D = 0$), and the other with a buried sphere with $e_0/D = 1.4$. The sinking depth was measured to be 15 cm (or $e/D = 2.2$) in both cases, again in contrast to the pipeline case where the initially bottom-seated pipe ($e_0/D = 0$) sank to a somewhat shallower depth. The latter has, in the preceding paragraphs, been linked with the fact that the sinking process of an initially bottom-seated pipe helps drain more water from the soil, resulting in relatively smaller buildup of pore-water pressure around the pipe, presumably leading to a shallower e/D. Considering the much smaller volume of the sphere, the previously mentioned effect will not be very strong, hence the depth of the sinking will be practically uninfluenced by the initial position of the sphere.

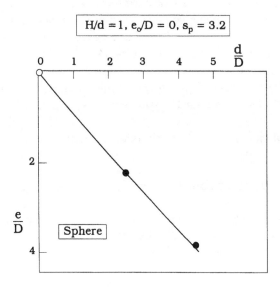

Figure 10.31: Ultimate sinking depth of sphere. $h/L = 0.145$. Sumer et al. (1999).

Fig. 10.31 illustrates the effect of d/D on the depth of sinking (cf. Fig. 10.28, the pipeline case). As stated also in the pipeline case, e/D should go to an asymptotic value as $d/D \to \infty$. This value could not be captured in Sumer et al.'s (1999) work due to the experimental constraints. As in the case of the pipeline, the data in the figure suggests that this asymptotic value is larger than $O(4)$ for $s_p = 3.2$.

The sinking motion of the sphere terminates at the impermeable base when $d/D = 4.25$ (Fig. 10.31), in contrast to the pipeline case where the sinking of the pipeline stops at a relatively large distance from the impermeable base. This may be attributed to the much smaller volume of the object (the volume of the sphere is a factor of 5 smaller than that of the pipe); the latter implies that the volume of the soil which needs to be displaced by the sinking motion of the sphere is much smaller in the case of the sphere, hence the sphere meets much smaller resistance as it sinks through the liquefied soil, presumably reaching much larger depths.

Cube-shaped body

Two tests in Sumer et al.'s study were carried out with a cube the size $D = 5.5$ cm, its volume being equal to that of the sphere with $D = 6.9$ cm, to observe the influence of the actual shape of the sinking object. The cube was initially buried at a depth of about $e_0/D = 1.4$. In one test ($s_p = 3.2$), e was measured to be $e = 16$ cm (cf., $e = 15$ cm for the sphere with $s_p = 3.2$ for the same initial burial depth e_0/D), while in the other test ($s_p = 8.9$), $e = 16.5$ cm (cf., $e = 17$ cm for the sphere with $s_p = 8.9$, again, for the same initial burial depth). As seen, given the volume of the sinking object, the actual shape of the object apparently does not have any dramatic influence on e.

Sinking velocity of sphere- and cube-shaped bodies

The data regarding the sinking velocity of the sphere and the cube plotted in the same way as in Fig. 10.29 (the range of the data being $20 \lesssim Re \lesssim 150$) showed that the drag coefficient defined by

$$C_D = \alpha \frac{g(s_p - s_\ell)D}{w^2 s_\ell} \qquad (10.118)$$

is an order of magnitude smaller than in the case of the pipe (Sumer et al., 1999). Here, $\alpha = 4/3$ in the case of the sphere and $\alpha = 1$ in the case of the cube.

10.6. SINKING OF ARMOUR BLOCKS IN LIQUEFIED SOIL

Practically no significant difference has been observed between the C_D data for the sphere and that for the cube. The following empirical expression for the drag coefficient has been given:

$$C_D = \frac{7 \times 10^6}{\sqrt{Re_p}} \text{ for } Re_p > O(10) \qquad (10.119)$$

Again, as in the case of the pipeline, caution must be exercised when the above equation is extrapolated to large Re_p numbers.

Remarks on practical applications

See the remarks made under Section 10.5.1 in conjunction with the sinking/floatation of pipelines.

10.6.2 The case of momentary liquefaction

Sakai, Gotoh and Yamamoto (1994) report an extensive series of laboratory experiments where the block subsidence due partly to the momentary liquefaction and partly to the oscillatory flow action has been investigated.

Maeno and Nago (1988) present the results of an experimental study where the effect of a progressive wave is simulated by an oscillating water table. A concrete, rectangular-prism-shaped block sitting initially on the surface of the soil gradually sank, as the oscillating movement of the water table continued.

Gratiot and Mory (2000), from their study on momentary liquefaction referred to earlier in Section 10.4.3, draw conclusions in relation to mine burials.

Although the problem of sinking/subsidence of armour blocks, sea mines etc. in soils subject to momentary liquefaction has been recognized widely, no study is yet available investigating this problem in a systematic manner under laboratory/field conditions.

The following practice may be recommended to make assessments about burial/sinking depth of armour blocks:

1. Adopt Eq. 10.104 to calculate the pore pressure, p;

2. Insert p from the previous step together with the bed pressure p_1 from Eq. 10.97 into Eq. 10.99, to get the effective shear stress σ';

3. Check if $\sigma' \leq 0$, the criterion for the momentary liquefaction;

4. Find z depths for which $\sigma' \leq 0$. This will give the depth of the soil layer in which the momentary liquefaction occurs;

5. This depth can, to a first approximation, be taken as the depth of sinking of the armour blocks.

10.7 Appendix I. Relationships among soil properties

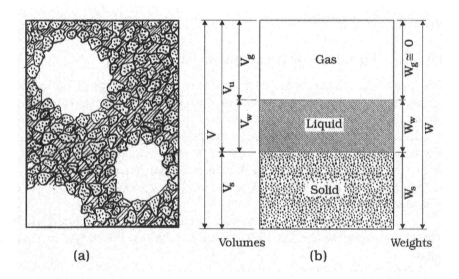

Figure 10.32: (a) Close-up picture of unsaturated soil (adapted from Sills et al., 1991). (b) Definition sketch.

(From Lambe and Whitman, 1969. See Fig. 10.32 for the definition sketch)

(1) *Volume*
Porosity:

$$n = \frac{V_v}{V} \qquad (10.120)$$

10.7. APPENDIX I. RELATIONSHIPS AMONG SOIL PROPERTIES

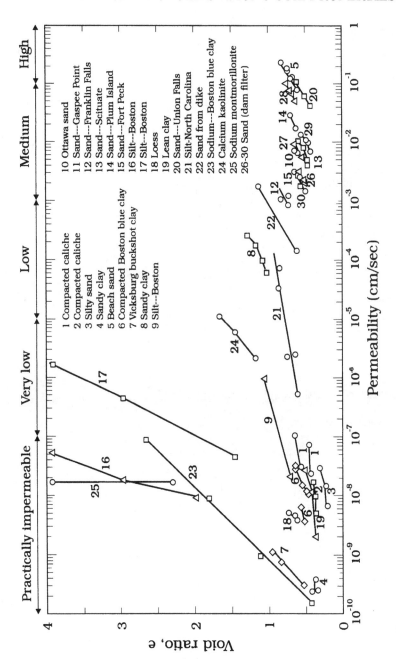

Figure 10.33: Permeability for various kinds of soil. Lambe and Whitman (1969).

Void ratio:
$$e = \frac{V_v}{V_s} \tag{10.121}$$

Degree of saturation:
$$S_r = \frac{V_w}{V_v} \tag{10.122}$$

$$n = \frac{e}{1+e}; \quad e = \frac{n}{1-n} \tag{10.123}$$

(2) *Weight*
Water content:
$$w = \frac{W_w}{W_s} \tag{10.124}$$

(3) *Specific gravity*
Mass:
$$s_t = \frac{\gamma_t}{\gamma_0} \tag{10.125}$$

Water:
$$s_w = \frac{\gamma}{\gamma_0} \tag{10.126}$$

Solids:
$$s = \frac{\gamma_s}{\gamma_0} \tag{10.127}$$

in which γ_0 = Unit weight of water at 4^0 C $\approx \gamma$.

(4) *Unit weight*
Total:
$$\gamma_t = \frac{W}{V} = \frac{s + S_r e}{1+e}\gamma =$$
$$= \frac{1+w}{1+e}s\gamma \tag{10.128}$$

Solids:
$$\gamma_s = \frac{W_s}{V_s} \tag{10.129}$$

Water:
$$\gamma = \frac{W_w}{V_w} \tag{10.130}$$

Dry:
$$\gamma_d = \frac{W_s}{V} = \frac{s}{1+e}\gamma = \tag{10.131}$$

$$= \frac{s\gamma}{1 + ws/S_r} = \frac{\gamma_t}{1+w}$$

Submerged (unsaturated soil):

$$\gamma' = \gamma_t - \gamma = \tag{10.132}$$

$$= \frac{s - 1 - e(1 - S_r)}{1 + e}\gamma$$

Submerged (saturated soil):

$$\gamma' = \gamma_t - \gamma = \tag{10.133}$$

$$= \frac{s-1}{1+e}\gamma$$

10.8 Appendix II. Ranges of soil properties

(1) *Soil components* (Lambe and Whitman, 1969, p. 36)
Table 10.2. Soil components

Soil component	Grain size (mm)
Boulder	>300
Cobble	300-150
Coarse gravel	76-20
Fine gravel	20-5
Coarse sand	5-2
Medium sand	2-0.4
Fine sand	0.4-0.074
Silt[1]	<0.074
Clay[2]	<0.074

[1] Particles smaller than 0.074 mm identified by behaviour; slightly or non-plastic regardless of moisture, and exhibit little or no strength when air-dried.

[2] Particles smaller than 0.074 mm identified by behaviour; it can be made to exhibit plastic properties within a certain range of moisture and exhibits considerable strength when air dried.

(2) *Void ratio, e and porosity, n* (Lambe and Whitman, 1969, p. 31)

Table 10.3. Void ratio and porosity for granular soils

Description	e_{max}	e_{min}	n_{max}	n_{min}
Uniform spheres	0.92	0.35	0.476	0.26
Standard Ottowa sand	0.80	0.50	0.44	0.33
Clean uniform sand	1.0	0.40	0.50	0.29
Uniform inorganic silt	1.1	0.40	0.52	0.29
Silty sand	0.90	0.30	0.47	0.23
Fine to coarse sand	0.95	0.20	0.49	0.17
Micaceous sand	1.2	0.40	0.55	0.29
Silty sand and gravel	0.85	0.14	0.46	0.12

(3) *Relative density, D_r* (Lambe and Whitman, 1969, p. 31)
$D_r = 0 - 0.15$ for very loose, $0.15 - 0.35$ for loose, $0.35 - 0.65$ for medium, $0.65 - 0.85$ for dense, and $0.85 - 1.00$ for very dense granular soils.

(4) *Specific gravity of soil grains, s* (Lambe and Whitman, 1969, p. 30)
$s = \gamma_s/\gamma_0 \approx \gamma_s/\gamma = 2.65$ for quartz.

(5) *Coefficient of lateral earth pressure, k_0* (Lambe and Whitman, 1969, p. 100)
k_0 typically has a value between 0.4 and 0.5 for a sand deposit which is formed by an accumulation of sediment from above. Lambe and Whitman (1969, p. 100) note, however, that k_0 may well reach a value of 3, if a soil deposit has been heavily loaded in the past.

(6) *Shear modulus, G* (Yamamoto et al., 1978)
The shear modulus varies from $G = 4.8 \times 10^2$ kN/m^2 (silt and clay) to 4.8×10^5 kN/m^2 (dense sand).

(7) *Poisson's ratio, ν* (Lambe and Whitman, 1969, p. 160)
For the early stages of a first loading of a sand (when particle rearrangements are important) ν typically has values of about 0.1 to 0.2. During cyclic loading, however, ν becomes more of a constant with values from 0.3 to 0.4.

(8) *Coefficient of permeability, k* (Lambe and Whitman, 1969, p. 286)
Fig. 10.33 presents laboratory permeability test data on a variety of soils.

10.9 Appendix III. Hsu & Jeng coefficients

Hsu and Jeng (1994) have developed an analytical solution to the Biot equations (Eqs. 10.30, 10.32 and Eq. 10.36) for an unsaturated, anisotropic soil of finite depth. For the case of a saturated, isotropic soil exposed to a progressive wave, this solution is given in Section 10.2.2 (Eqs. 10.59-10.71). The solution includes a number of coefficients. These coefficients are given below.

$$C_i = \frac{D_i}{D_0} \text{ for } i = 1, ..., 6 \tag{10.134}$$

in which the D coefficients are given by

$$D_j = C_{j0} + C_{j1}e^{-2\lambda d} + C_{j2}e^{-(\lambda+\delta)d} + C_{j3}e^{-4\lambda d} + C_{j4}e^{-2\delta d} \tag{10.135}$$

$$+C_{j5}e^{-2(\lambda+\delta)d} + C_{j6}e^{-(3\lambda+\delta)d} + C_{j7}e^{-(4\lambda+2\delta)d} \text{ for } j = 0, 1, ..., 6$$

The coefficients C_{ji} are

$$C_{00} = (\delta - \lambda)^2(\delta - \delta\nu + \lambda\nu)B_1$$

$$C_{01} = -2\delta\left[\left(\lambda^2\nu - \delta^2 + \delta^2\nu\right)^2 + \lambda^4(1-2\nu)^2 + 2\lambda^2 d^2(1-\nu)^2(\delta^2 - \lambda^2)^2\right]$$
$$+4\lambda^2 d(\delta^4 - \lambda^4)$$

$$C_{02} = -8\delta\lambda^2(1-2\nu)\left[\lambda d(\delta^2 - \lambda^2)(1-\nu) - \delta^2(1-\nu) + \lambda^2\nu\right]$$

$$C_{03} = (\delta + \lambda)^2(\delta - \delta\nu - \lambda\nu)B_2$$

$$C_{04} = C_{03}$$

$$C_{05} = C_{01} - 8\lambda^2 d(\delta^4 - \lambda^4)(1-\nu)(1-2\nu)$$

$$C_{06} = C_{02} + 16\delta\lambda^3 d(\delta^2 - \lambda^2)(1-\nu)(1-2\nu)$$

$$C_{07} = C_{00}$$

$$C_{11} = 2\lambda^2 d(\delta + \lambda)(\delta - \delta\nu - \lambda\nu)B_3$$

$$C_{12} = 4\delta\lambda^3 d(1-2\nu)(\delta^2 - \delta^2\nu - \lambda^2\nu)$$

$$C_{15} = 2\lambda^2 d(\delta - \lambda)(\delta - \delta\nu + \lambda\nu)B_4$$

$$C_{20} = (\delta - \lambda)^2(\delta - \delta\nu + \lambda\nu)B_1$$

$$C_{21} = C_{03} + (\delta + \lambda)^2(\delta - \delta\nu - \lambda\nu)B_5$$

$$C_{22} = 4\delta\lambda^2(1-2\nu)\left[2\delta^2(1-\nu) - 2\lambda^2\nu - \lambda d(1-\nu)(\delta^2 - \lambda^2)\right]$$
$$C_{24} = (\delta + \lambda)^2(\delta - \delta\nu - \lambda\nu)B_2$$
$$C_{25} = C_{00} + (\delta - \lambda)^2(\delta - \delta\nu + \lambda\nu)B_6$$
$$C_{31} = 2\lambda^2 d(\delta - \lambda)(\delta - \delta\nu + \lambda\nu)B_3$$
$$C_{33} = C_{14}$$
$$C_{35} = 2\lambda^2 d(\delta + \lambda)(\delta - \delta\nu - \lambda\nu)B_4$$
$$C_{36} = -C_{12}$$
$$C_{37} = C_{10}$$
$$C_{41} = C_{25} - 2C_{00}$$
$$C_{42} = -C_{26}$$
$$C_{43} = -C_{24}$$
$$C_{45} = C_{21} - 2C_{03}$$
$$C_{46} = 4\delta\lambda^2(1-2\nu)\left[2\lambda^2\nu - 2\delta^2(1-\nu) - \lambda d(1-\nu)(\delta^2 - \lambda^2)\right]$$
$$C_{47} = -C_{20}$$
$$C_{51} = -4\lambda^2 d(1-2\nu)B_3$$
$$C_{52} = -2\lambda^2 d(1-2\nu)(\delta + \lambda)(\delta - \delta\nu - \lambda\nu)$$
$$C_{56} = -2\lambda^2 d(1-2\nu)(\delta - \lambda)(\delta - \delta\nu + \lambda\nu)$$
$$C_{62} = -C_{56}$$
$$C_{64} = C_{53}$$
$$C_{65} = -4\lambda^2 d(1-2\nu)B_4$$
$$C_{66} = -C_{52}$$
$$C_{67} = C_{50}$$
$$C_{10}, C_{13}, C_{14}, C_{16}, C_{17} = 0$$
$$C_{23}, C_{26}, C_{27} = 0$$
$$C_{30}, C_{32}, C_{34} = 0$$
$$C_{40}, C_{44} = 0$$
$$C_{50}, C_{53}, C_{54}, C_{55}, C_{57} = 0$$

$$C_{60}, C_{61}, C_{63} = 0$$

in which the B coefficients are given by

$$B_1 = \lambda^2 \nu - (1-\nu)(\delta^2 + \delta\lambda + \lambda^2)$$
$$B_2 = -\delta^2 + \delta\lambda - \lambda^2 + \delta^2\nu - \delta\lambda\nu + 2\lambda^2\nu$$
$$B_3 = (\delta^3 d - \lambda^2 - \delta\lambda^2 d)(1-\nu) + \lambda^2\nu$$
$$B_4 = (\delta^3 d + \lambda^2 - \delta\lambda^2 d)(1-\nu) - \lambda^2\nu$$
$$B_5 = 2\delta\lambda d(\delta - \lambda)(1-\nu)$$
$$B_6 = 2\delta\lambda d(\delta + \lambda)(1-\nu)$$

10.10 References

1. Alba, P.D., Seed, H.B. and Chan, C.K. (1976): Sand liquefaction in large-scale simple shear tests. J. Geotechnical Engineering Division, ASCE, vol. 102, No. GT9, 909-927.

2. ASCE Pipeline Floatation Research Council (1966): ASCE preliminary research on pipeline floatation. J. Pipeline Division, ASCE, vol. 92, No. PL1, 27-71.

3. Barends, F.B.J. and Calle, E.O.F. (1985): A method to evaluate the geotechnical stability of off-shore structures founded on a loosely packed seabed sand in a wave loading environment. Behaviour of Offshore Structures, Proc. 4th International Conf., Delft, The Netherlands, Elsevier Science Publishers B.V. Amsterdam, 643-652.

4. Biot, M.A. (1941): General theory of three-dimensional consolidation. J. Appl. Physics, vol. 12, 155-164.

5. Chaney, R.C. and Fang H.Y. (1991): Liquefaction in the coastal environment: An analysis of case histories. Marine Geotechnology, vol. 10, 343-370.

6. Cheng, H.-D. and Liu, P. L.-F. (1986): Seepage force on a pipeline buried in a poroelastic seabed under wave loading. Applied Ocean Research, vol. 8, No. 1, 22-32.

7. Cheng, L., Sumer, B.M. and Fredsøe, J. (2001): Solutions of pore pressure buildup due to progressive waves. International Journal of Numerical and Analytical Methods in Geomechanics, vol. 25, issue 9, 885-907.

8. Christian, J.T., Taylor, P.K., Yen, J.K.C. and Erali D.R. (1974): Large diameter underwater pipeline for nuclear plant designed against soil liquefaction. Offshore Technology Conference, May 6-8, 1974, Houston, Texas, OTC 2094, 597-606.

9. Clukey, E.C., Kulhawy, F.H. and Liu, P.L.-F. (1983): Laboratory and field investigation of sediment interaction. Cornell University, Itacha, New York, Geotechnical Engineering Report 83-9, and Joseph H. DeFrees Hydraulics Laboratory Report 83-1, October 1983.

10. Clukey, E.C., Kulhawy, F.H. and Liu, P.L.-F. (1985): Response of silts to wave loads: Experimental study. In: Strength Testing of marine sediments, Laboratory and In-Situ Measurements, ASTM STP 883, R.C. Chaney and K.R. Demars, Eds., American Society for testing and materials, Philadelphia, 381-396.

11. Dalrymple, R.A. and Liu, P.L.-F. (1982): Gravity waves over a poroelastic seabed. Proc. Ocean Structural Dynamics Symposium' 82, 181-195.

12. Damgaard, J.S. and Palmer, A.C. (2001): Pipeline stability on a mobile and liquefied seabed: A discussion of magnitudes and engineering implications. Proc. OMAE 01, 20th International Conf. on Offshore Mechanics and Arctic Engineering, June 3-8, 2001, Rio de Janeiro, Brazil.

13. Dawson, T.H. (1978): Wave propagation over a deformable sea floor. Ocean Engng., vol. 5, 227-234.

14. Dawson, T.H., Shuyada, J.N. and Coleman, J.M. (1981): Correlation of field measurements with elastic theory of seafloor response to surface waves. Proc. 13th Offshore Technology Conference, Houston, TX, vol. 1, 201-210.

15. de Groot, M.B. and Meijers, P. (1992): Liquefaction of trench fill around a pipeline in the seabed. BOSS 92: Behaviour of Offshore Structures, London, 1333-1344.

10.10. REFERENCES

16. de Groot, M.B., Lindenberg, J. and Meijers, P. (1991): Liquefaction of sand used for soil improvement in breakwater foundations. Proc. Int. Symp. Geo-coast' 91, Yokohama, 555-560.

17. de Wit, .J. (1995): Liquefaction of cohesive sediments caused by waves. Communications on Hydraulic Engineering, Report No. 95-2, Delft University of Technology, Faculty of Civil Engineering, June 1995.

18. de Wit, J. and Kranenburg, C. (1992): Liquefaction and erosion of China Clay due to waves and current. Proc. 23rd International Conference on Coastal Engineering, 4-9 October, 1992, Venice, Italy, 2937-2947.

19. de Wit, J., Kranenburg, C. and Battjes, J.A. (1994): Liquefaction and erosion of mud due to waves and current. Abstract Book of the 24th International Conference on Coastal Engineering, ICCE' 94, 23-28. October 1994, Kobe. Japan, 278-279.

20. de Wit, J. and Kranenburg, C. (1997): The wave-induced liquefaction of cohesive sediment beds. Estuarine, Coastal and Shelf Science, vol. 45, 261-271.

21. Dean, R.G. and Dalrymple, R.A. (1984): Water Wave Mechanics for Engineers and Scientists. Prentice-Hall, Inc., New Jersey.

22. Dunlap, W., Bryant, W.R., Williams, G.N. and Suheyda, J.N. (1979): Storm wave effects on deltaic sediments—Results of SEASWAB I and II. Port and Ocean Engineering Under Arctic Conditions (POAC 79), Norwegian Institute of Technology, vol. 2, 899-920.

23. Foda, M.A. (1995): Sea Floor Dynamics. Advances in Coastal and Ocean Engineering, vol. 1, Ed. P.L.-F. Liu, World Scientific, 77-123.

24. Foda, M.A. and Tzang, S.-Y. (1994): Resonant fluidization of silty soil by water waves. J. Geophysical Res., vol. 99, No. C10, 20,463-20,475.

25. Gade, H.G. (1958): Effects of non-rigid impermeable bottom on plane surface waves in shallow water. J. Marine Res., vol. 16, 61-82.

26. Goda, Y. (1994): A plea for engineering-minded research efforts in harbor and coastal engineering. International Conference on hydro-Technical Engineering for Port and Harbor Construction, Hydro-Port'94, October 19-21, 1994, Yokosuka, Japan, 1-21.

27. Gratiot, N. and Mory, M. (2000): Wave induced sea bed liquefaction with application to mine burial. Proc. Tenth International Offshore and Polar Conference, ISOPE-2000, May 28-June 2, 2000, Seattle, USA, vol. 2.

28. Happel, J. and Brenner, H. (1973): Low Reynolds Number Hydrodynamics. Second Edition, Noordhoff, Leyden, The Netherlands.

29. Herbich, J.B., Schiller, R.E., Dunlap, W.A. and Watanabe, R.K. (1984): Seafloor Scour, Design Guidelines for Ocean-Founded Structures, Marcel Dekker, Inc.

30. Hsiao, S.V. and Shemdin, O.H. (1980): Interaction of ocean waves with a soft bottom. J. Physics of Oceans, vol. 10, 605-610.

31. Hsu, J.R.S. and Jeng, D.S. (1994): Wave-Induced soil response in an unsaturated anisotropic seabed of infinite thickness. International Journal for Numerical and Analytical Methods in Geomechanics, vol. 18, 785-807.

32. Hunt, J.N. (1959): On the damping of gravity waves propagated over a permeable surface. J. Marine Research, vol. 16, 61-82.

33. Isaacson, M. (1979): Wave-induced forces in the diffraction regime. In: Mechanics of Wave-induced Forces on Cylinders, (Ed. T.L. Shaw). Pitman Advanced Publishing Program, 68-89.

34. Jeng, D.S. (1997): Wave-induced seabed instability in front of a breakwater. Ocean Engng., vol. 24, No. 10, 887-917.

35. Lambe T.W. and Whitman, R.V. (1969): Soil Mechanics. John Wiley and Sons, Inc.

36. Law, A.W.K. (1993): Wave-induced effective stress in seabed and its momentary liquefaction. Discussion to the paper by Sakai et al. (1992), J. Waterway, Port, Coastal and Ocean Engineering, ASCE, 119, No.6, 694-695.

10.10. REFERENCES

37. MacPherson, H. (1980): The attenuation of waves over a non-rigid bed. J. Fluid Mech., vol. 97, 721-742.

38. Madsen, O.S. (1978): Wave-induced pore pressures and effective stresses in a porous bed. Geotechnique, vol. 28, 377-393.

39. Maeno, S., Magda, W. and Nago, H. (1999): Floatation of buried pipeline under cyclic loading of water pressure. Proceedings of the 9th International Offshore and Polar Engineering Conference and Exhibition, Brest, France, May 30- June 4, 1999, vol. II, 217-225.

40. Maeno S. and Nago, H. (1988): Settlement of a concrete block into a sand bed under water pressure variation. In: Modelling Soil-Structure Interactions, (Eds. Kolkman et al.). Balkema, Rotterdam, 67-76.

41. Magda, W. (1997): Wave-induced uplift force on a submarine pipeline buried in a compressible seabed. Ocean Engineering, vol. 24, No. 6, 551-576.

42. Magda, W., Maeno S. and Nago, H. (1998): Wave-induced pore-pressure response on a submarine pipeline buried in seabed sediments - Experiment and numerical verification. The Journal of the Faculty of Environmental Science and Technology, Okayama University, vol. 3, No. 1, January 1998.

43. Mallard, W.W. and Dalrymple, R.A. (1977): Water waves propagating over a deformable bottom. Proc. 9th Offshore Technology Conference, Houston, TX, vol. 3, 141-146.

44. McDougal, W.G. and Sollitt, C.K. (1984): Geotextile stabilization of seabed: theory. Engineering Structures, vol. 6, 211-216.

45. McDougal, W.G., Tsai, Y.T., Liu, P.L-F. and Clukey, E.C. (1989): Wave-induced pore water pressure accumulation in marine soils. J. Offshore Mechanics and Arctic Engineering, ASME, vol. 111, 1-11.

46. Mei, C.C. and Foda, M.A. (1981): Wave-induced responses in a fluid filled poroelastic solid with a free surface A boundary layer theory. Geophysics, J. of the Royal Astr. Society, vol. 66, 597-631.

47. Miyamoto, T., Yoshinaga, S., Soga, F., Shimizu, K., Kawamata, K. and Sato, M. (1989): Seismic prospecting method applied to the detection of offshore breakwater units settling in the seabed. Coastal Engineering in Japan, vol. 32, No. 1, 103-112.

48. Moshagen, H. and Tørum, A. (1975): Wave induced pressures in permeable seabeds. J. Waterways, Harbors and Coastal Engineering Division, ASCE, vol. 101, No. WW1, 49-57.

49. Murray, J.D. (1965): Viscous damping of gravity waves over a permeable bed. J. Geophysical Res., vol. 70, 2325-2331.

50. Peacock, W.H. and Seed, H.B. (1968): Sand liquefaction under cyclic loading simple shear conditions. J. Soil Mechanics and Foundations Engineering Division, ASCE, vol. 94, No. SM3, 689-708.

51. Putnam, J.A. (1949): Loss of wave energy due to percolation in a permeable sea bottom. Trans. American Geophysical Union, vol. 30, 349-356.

52. Rahman, M.S., Seed, H.B. and Booker, J.R. (1977): Pore pressure development under offshore gravity structures. J. Geotechnical Eng. Division, ASCE, vol. 103, No. GT12, 1419-1436.

53. Reid, R.D. and Kajiura, K. (1957): On the damping of gravity waves over a permeable seabed. Trans. American Geophysical Union, vol. 38, 662-666.

54. Sakai, T., Hatanaka, K. and Mase, H. (1992): Wave-induced effective stress in seabed and its momentary liquefaction. J. Waterway, Port, Coastal and Ocean Engineering, ASCE, vol. 118, No. 2, 202-206. See also Discussions and Closure in vol. 119, No. 6, 692-697.

55. Sakai, T., Gotoh, H. and Yamamoto, T. (1994): Block subsidence under pressure and flow. Proc. 24th Conference on Coastal Engineering (ICCE 94), 23-28. October, 1994, Kobe, Japan, 1541-1552.

56. Sassa, S. and Sekiguchi, H. (1999): Wave-induced liquefaction of beds of sand in a centrifuge. Geotechnique, vol. 49, No. 5, 621-638.

57. Schlichting H. (1979): Boundary-Layer Theory. 7th ed., McGraw-Hill.

58. Seed, H.B. and Booker, J.R. (1976): Stabilization of Potentially Liquefiable Sand Deposits Using Gravel Drain Systems. Report No. EERC 76-10, Earthquake Engineering Research Center, University of California, Berkeley, Calif., Apr. 1976.

59. Seed, H.B. and Rahman, M.S. (1978): Wave-induced pore pressure in relation to ocean floor stability of cohesionless soil. Marine Geotechnology, 3, No. 2, 123-150.

60. Sekiguchi, H., Kita, K. and Okamoto, O. (1995): Response of poroelastoplastic beds to standing waves. Soils and Foundations, vol. 35, No. 3, 31-42, Japanese Geotechnical Society.

61. Siddhartan, R. and Norris, G.M. (1993): Analysis of offshore pipeline floatation during storms in liquefiable soils. Proc. Third International Offshore and Polar Engineering Conference, Singapore, 6-11. June, 1993. vol. II, 106-113.

62. Sills, G., Wheeler, S.J., Thomas, S.D. and Gardner, T.N. (1991): Behaviour of offshore soils containing gas bubbles. Geotechnique, vol. 41, No. 2, 227-241.

63. Silvis, F. (1990): Wave induced liquefaction of seabed below pipeline. The 4th Young Geotechnical Engineers' Conference, Delft, The Netherlands, 18-22- June 1990.

64. Sleath, J.F.A. (1970): Wave-Induced pressures in beds of sand. J. Hydraulic Div., ASCE, vol. 96, 367-378.

65. Sneddon, I (1957): Elements of Partial Differential Equations. McGraw-Hill, New York.

66. Spierenburg, S.E.J. (1987): Seabed Response to Water Waves. Ph. D. dissertation, Delft University of Technology, The Netherlands.

67. Sumer, B.M. and Cheng, N.-S. (1999): A random-walk model for pore pressure accumulation in marine soils. Proc. 9th International Offshore and Polar Engineering Conference, ISOPE-99, Brest, France, 30. May-4. June, 1999, vol. 1, 521-526.

68. Sumer, B.M. and Fredsøe, J. (1997): Hydrodynamics Around Cylindrical Structures, World Scientific, xiii + 530 p.

69. Sumer, B.M., Fredsøe, J., Christensen, S. and Lind, M. T. (1999): Sinking/Floatation of pipelines and other objects in liquefied soil under waves. Coastal Engineering, vol. 38, 53-90.

70. Tanaka, Y. (1996): Liquefaction of reclaimed lands along Osaka Bay by Great Hanshin Earthquake (1995. 1. 17). Proc. 6th international Offshore and Polar Engineering Conference, Los Angeles, USA, 26-31. May. 1996, vol. 1, 20-28.

71. Terzaghi, K. (1948): Theoretical Soil Mechanics. London: Chapman and Hall, John Wiley and Sons, Inc., NY.

72. Tzang, S.Y. (1998): Unfluidized soil responses of a silty seabed to monochromatic waves. Ocean Engng., vol. 35, 283-301.

73. Tzang, S.Y., Hunt, J.R. and Foda, M.A. (1992): Resuspension of seabed sediments by water waves. Abstract Book of the 23rd International Conference on Coastal Engineering, ICCE' 92, 4-9 October, Venice, Italy, 1992, 69-70.

74. van Kessel, T. and Kranenburg, C. (1998): Wave-induced liquefaction and flow of subaqueous mud layers. Coastal Eng., vol. 34, 109-127.

75. van Kessel, T., Kranenburg, C. and Battjes, J.A. (1996): Transport of fluid mud generated by waves on inclined beds. Proc. 25th International Conference on Coastal Engineering, ICCE' 96, 1996, Orlando, U.S.A., vol. 3, 3337-3348.

76. Verruijt, A. (1969): Elastic storage of aquifers. In: Flow through Porous Media (ed. R.J.M. De Wiest), Chap. 8, Academic Press.

77. Yamamoto, T. (1977): Wave-induced instability in seabed. Proc. ASCE Special Conference, Coastal Sediments '77, Charleston, SC, 898-913.

78. Yamamoto, T. (1978): Seabed instability from waves. Proc. 10th Offshore Technology Conference, Houston, TX, vol. 3, 1819-1828.

79. Yamamoto, T. and Suzuki, (1980): Stability analysis of seafloor foundations. Proc. Coastal Engineering Conference, 1799-1818.

10.10. REFERENCES

80. Yamamoto, T. (1981a): Wave-induced pore pressures and effective stresses in homogenous seabed foundations. Ocean Engng., vol. 8, 1-16.

81. Yamamoto (1981b): Ocean waves spectrum transformations due to seabed interactions. Proc. 13th Offshore Technology Conference, Houston, TX, vol. 1, 249-258.

82. Yamamoto, T., Koning, Sellmeijer, H. and van Hijum, E. (1978): On the response of a poro-elastic bed to water waves. J. Fluid Mech., vol. 87, part 1, 193-206.

83. Zen, K. and Yamazaki, H. (1993). Wave-induced liquefaction in a permeable seabed. Report of Port and Harbour Research Institute, Japan, vol. 31, 155-192.

Appendix A

Small amplitude, linear waves

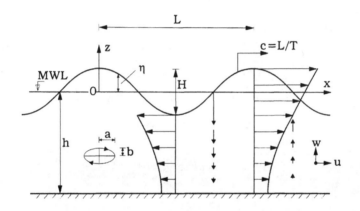

Figure A.1: Definition sketch.

Basic equation:
$$\nabla^2 \phi = \phi_{xx} + \phi_{zz} = 0 \tag{A.1}$$

Bed boundary condition:
$$w = \phi_z = 0 \text{ at } z = -h \tag{A.2}$$

Kinematic, free-surface boundary condition:
$$\left(\frac{\partial \phi}{\partial z}\right)_{z=0} = \frac{\partial \eta}{\partial t} \tag{A.3}$$

Dynamic, free-surface boundary condition:
$$\left(\frac{\partial \phi}{\partial t} + g\eta\right)_{z=0} = C(t) \tag{A.4}$$

Water surface elevation:
$$\eta = \frac{H}{2}\cos(\omega t - kx) \tag{A.5}$$

Potential function:
$$\phi = -\frac{Hc}{2}\frac{\cosh(k(z+h))}{\sinh(kh)}\sin(\omega t - kx) \tag{A.6}$$

Wave celerity:
$$c = L/T = \omega/k \tag{A.7}$$

$$k = 2\pi/L = \text{ wave number}$$

$$\omega = 2\pi/T = 2\pi f = \text{ angular wave frequncy}$$

$$f(=1/T) \text{ being the wave frequency}$$

Dispersion relation:
$$\omega^2 = gk\tanh(kh) \tag{A.8}$$

or
$$(2\pi f)^2 = gk\tanh(kh) \tag{A.9}$$

g being the acceleration due to gravity

Horizontal particle velocity:
$$u = \phi_x = \frac{\pi H}{T}\frac{\cosh(k(z+h))}{\sinh(kh)}\cos(\omega t - kx) \tag{A.10}$$

or
$$u = \phi_x = \frac{gkH}{4\pi f}\frac{\cosh(k(z+h))}{\cosh(kh)}\cos(\omega t - kx) \tag{A.11}$$

Vertical particle velocity:
$$w = \phi_z = -\frac{\pi H}{T}\frac{\sinh(k(z+h))}{\sinh(kh)}\sin(\omega t - kx) \tag{A.12}$$

or
$$w = \phi_z = \frac{gkH}{4\pi f}\frac{\sinh(k(z+h))}{\sinh(kh)}\sin(\omega t - kx) \tag{A.13}$$

Horizontal amplitude of particle motion:
$$a = \frac{H}{2}\frac{\cosh(k(z+h))}{\sinh(kh)} \tag{A.14}$$

Vertical amplitude of particle motion:
$$b = \frac{H}{2}\frac{\sinh(k(z+h))}{\sinh(kh)} \tag{A.15}$$

Pressure:
$$\frac{p}{\rho} = -gz - \phi_t \tag{A.16}$$

$-gz$ = hydrostatic pressure, and $-\phi_t$ = excess pressure

ρ being the density of water

Excess pressure:
$$\frac{p^+}{\rho} = -\phi_t = g\frac{H}{2}\frac{\cosh(k(z+h))}{\cosh(kh)}\cos(\omega t - kx) \tag{A.17}$$

Wave energy per unit area:
$$E = \frac{1}{L}(2\int_0^L (\rho g \eta dx)\frac{\eta}{2}) = \frac{1}{16}\rho g H^2 \tag{A.18}$$

Author Index

Abramowitz, M., 293
Aguirre-Pe, J., 63, 66, 69
Ajiwibowo, H., 410
Alach, D., 174
Alba, P.D., 468
Angus, N.M., 12, 275
Antonia, R.A., 44
Arneborg, L., 353, 365, 366, 368, 369
Arnskov, , M., 70
ASCE Pipeline Floatation Research Council 1966, 487, 494

Bagnold, R.A., 337
Baker, C.J., 151, 152, 153, 154, 155, 158, 167, 168
Baker, R.A., 176, 179
Baquerizo, A., 346
Barbas, S.T., 94
Barends, F.B.J., 473
Barkdoll, B., 175
Barnett, M.R., 408
Bartels, A., 331, 332
Batchelor, G.K., 337
Battjes, J.A., 473
Bayram, A., 263, 264
Bearman, P.W., 17, 28, 170
Belcher, S.E., 217
Bergh, H., 424
Bernetti, R., 49, 52, 80, 95, 117, 123, 124, 137
Bijker, E.A., 32,
Bijker, E.W., 39, 61, 62, 65, 124, 205, 334
Bijker, R., 95, 108, 110, 111, 137
Biot, M.A., 450, 451, 454, 455
Bisset, D.K., 44

Blaauw, H.G., 427
Blevins, R.D., 109
Booker, J.R., 473
Brenner, H., 496
Breusers, H.N.C., 6, 7, 174, 175, 183, 213, 243, 245, 246, 383, 419
Briaud, J.-L., 149, 184, 185, 186, 187, 188
Briley, W.R., 219
Bringaker, K.G., 33
Brown, R.J., 94, 108
Browne, L.W.B., 44
Bruschi, R., 49, 95, 117, 138
Bruyn, C.A., 205
Bryant, W.R., 445
Bryndum, M., 122, 137
Brørs, B., 125, 126, 130, 131
Bundgaard, K., 264, 358
Burger, A.M., 108

Calle, E.O.F., 473
Carpenter, K., 409, 410
Carreiras, J., 202, 203, 204, 262
Carter, L.W., 94
Carter, T.G., 334
Cederwall, K., 424
Cevik, E., 46, 53, 54
Chabert, J., 180
Chan, C.K., 468
Chaney, R.C., 448
Chang, F.M., 377
Chang, J.L.C., 219
Chao, J.L., 38, 66, 117, 118, 123, 124, 125
Chao, S.Y., 157
Chapman, B., 387
Chen, Z., 137, 187

Cheng, H.-D., 454
Cheng, L., 117, 124, 125, 135, 462, 463, 476
Cheng, N.-S., 34, 475
Chesnutt, C.B., 408
Chiew, Y.M., 16, 21, 29, 30, 62, 63, 65, 110, 119, 174, 175, 178, 210, 215, 218
Chou, J.H., 157
Chow, W.Y., 248
Christensen, S., 469
Christian, J.T., 445, 487
Christiansen, N., 156, 179, 205, 206, 248
Chua, L.C.H., 34, 187
Cimbali, W., 95
CIRIA, 401
Clukey, E.C., 462, 473
Coastal Engineering Manual, 5, 364, 402, 412
Cokgor, S., 217
Coleman, S.E., 6, 7, 12, 13, 149, 175, 178, 183, 212, 213, 214
Crossley, C.W., 94

Dahlberg, R., 278, 318, 319, 322, 323
Dalrymple, R.A., 333, 457, 463
Dalton, C., 151
Damgaard, J.S., 487
Dargahi, B., 151, 152, 222, 225
Dawson, T.H., 463
Dean, R., 333, 457
de Best, A., 334
de Groot, M.B., 473, 487, 494
de Wit, J., 473
Deigaard, R., 10, 48, 53, 74, 121, 201, 211, 314, 344, 370, 404, 405, 415
Deng, G.B., 219
DHI, 438
DHI/Snamprogetti, 182, 279, 280, 281, 282, 283, 284, 285
DIF, 101
Dongol, D.M.S., 175
Drago, M., 138
Dunlap, W.A., 39, 301, 353, 445

Eadie, R.W., IV, 173, 195, 196, 248, 301
Engeldinger, P., 180
Engelund, F.A., 367
Ennemark, F., 214
Erali, D.R., 445
Escaramelia, M., 419
Ettema, R., 174, 176, 178, 182

Fang, H.Y., 448
Foda, M.A., 454, 463, 473, 485
Fog, N.G., 277, 278
Fowler, J.E., 334, 341, 350, 356, 399, 402, 403, 406, 407, 408, 412
Fredsøe, J., 10, 11, 16, 21, 24, 26, 27, 28, 31, 32, 33, 34, 35, 39, 41, 43, 44, 45, 46, 48, 49, 50, 51, 52, 53, 54, 55, 56, 57, 59, 66, 67, 68, 69, 70, 71, 72, 73, 74, 75, 76, 80, 88, 89, 90, 91, 93, 95, 96, 97, 98, 100, 101, 102, 103, 104, 106, 107, 112, 117, 119, 121, 122, 124, 125, 136, 150, 152, 156, 162, 169, 170, 174, 179, 187, 194, 195, 196, 197, 198, 201, 202, 205, 206, 211, 214, 217, 219, 220, 224, 248, 249, 250, 251, 252, 253, 255, 256, 257, 258, 259, 260, 261, 264, 290, 291, 292, 293, 294, 295, 296, 297, 298, 299, 300, 301, 302, 303, 304, 305, 306, 307, 308, 309, 310, 311, 312, 314, 324, 325, 329, 330, 344, 347, 348, 349, 350, 351, 353, 354, 355, 356, 357, 358, 359, 362, 363, 365, 366, 367, 369, 370, 371, 372, 373, 374, 375, 376, 377, 378, 379, 380, 381, 382, 383, 384, 385, 386, 387, 389, 404, 405, 415, 462, 469, 498
Fuchs, R.A., 291, 292, 293, 311
Fuehrer, M., 429, 437, 438, 439

Gade, H.G., 463
Gardner, T.N., 449
Gimenez-Curto, L.A., 369
Gironella, X., 360
Gislason, K., 370, 371
Gjörsvik, O., 33
Goda, Y., 445

Goodier, 449
Gormsen, C., 247
Gotoh, H., 503
Graf, W.H., 152
Graham, J.M.R., 170
Grapon, A., 370
Gratiot, N., 487, 503
Gravesen, H., 80, 214, 432
Guler, I., 331
Gunbak, A.R., 46, 88, 110, 331, 369
Gökce, T., 46, 88, 110, 331

Hales, L.Z., 364
Hamill, G.A., 423, 424, 427, 428, 429, 430, 431, 432, 433, 434
Hannah, C.R., 243, 245, 246, 272
Hansen, E.A., 31, 32, 44, 45, 46, 49, 54, 55, 56, 61, 80, 82, 85, 86, 87, 88, 117, 119, 120, 121, 122, 123, 137, 353, 365
Hansen, S.B., 205
Happel, J., 496
Hebsgaard, M., 214, 215
Henderson, F.M., 64
Hennessy, P.V., 38, 66, 117, 118, 123, 124, 125
Herbich, J.B., 6, 39, 114, 116, 117, 173, 195, 196, 248, 301, 353, 445, 487
Hjorth, P., 11, 114, 151, 152, 159, 161, 174
Hoffmans, G.J.C.M., 7, 175
Horikawa, K., 418
Hsiao, S.V., 463
Hsu, J.R.C., 331, 343, 344, 345, 346, 460, 461, 462, 463, 476, 509
Hughes, D.A.B., 424
Hughes, S.A., 112, 329, 334, 341, 350, 356, 358, 381, 382, 383, 389, 390, 391
Hulsbergen, C.H., 108, 110, 111
Hunt, J.C.R., 217
Hunt, J.N., 463
Hunt, J.R., 473
Hwang, K.N., 416

Ibrahim, A.A., 39, 60, 64

Imberger, J., 174, 176
Irie, I., 334, 336
Isaacson, M., 109, 150, 170, 288, 289, 290, 293, 300, 485
Ishida, H., 220, 222
Iwata, K., 301

Jacobsen, J., 33
Jacobsen, V., 122
Jeng, D.S., 460, 461, 462, 463, 476, 487, 509
Jensen, B.L., 31, 34, 36, 38, 43, 44, 125, 132, 135, 136
Jensen, H.R., 43, 125, 136
Jensen, R., 31, 34,
Jiao, G., 138
Jo, C.H., 59
Johnston, H.T., 423
Juhl, J., 353, 365
Jui, J., 412
Jønsson, P.H., 277, 278

Kaa, E.J., 427
Kadib, A.L., 418
Kajiura, K., 463
Kamphuis, J.W., 390, 391, 412, 413
Karahan, E., 10
Katsui, H., 287, 301, 305
Katz, J., 134
Kaul, U.K., 219
Kawamata, K., 445
Kennedy, J.F., 118
Kim, C.J., 301
Kim, H., 416
Kim, T.H., 416
Kita, K., 473
Kjeldsen, S.P., 33, 39, 40, 70, 71
Kjellesvig, H.M., 219, 220
Klomp, W.H.G., 16, 115, 137
Kloos, M., 331
Klopman, G., 355
Ko, S.C., 353
Kobayashi, T., 189, 191, 219, 220, 248, 301

Komar, P.D., 415
Koning, 457
Kovacs, A., 226
Kozakiewicz, A., 59
Kranenburg, C., 473, 482
Kraus, N.C., 401, 402, 410, 411
Kristiansen, Ø., 39, 40, 57
Kroezen, M., 108, 109
Kulhawy, F.H., 473
Kwak, D., 219

Lagasse, P.F., 149
Lambe, T.W., 480, 481, 504, 505, 507, 508
Larroudé, Ph., 202, 262
Larsen, T., 247
Larson, M., 263, 264, 411
Lauchlan, C.S., 218
Laursen, E.M., 180, 182, 183
Laursen, T., 217, 378
Law, A.W.K., 486
Lee, T.H., 416
Leeuwenstein, W., 32, 39, 80, 89, 125, 132
Leopardi, G., 95
Li, F., 117, 124, 125, 135
LICENGINEERING A/S, 276, 277
Lillycrop, W.J., 329, 382, 389, 390
Lind, M.T., 469
Lindenberg, J., 108, 473
Littlejohn, P.S.G., 39
Liu, H.K., 377
Liu, L.-F.P., 334, 454, 462, 463, 473
Longuet-Higgins, M.S., 344
Losada, I.J., 350
Losada, M.A., 346, 347, 350, 369
Lovera, F., 118
Lowe, J.P., 408, 409
Lucassen, R.J., 33, 39, 40, 44, 49

MacCamy, R.C., 291, 292, 293, 311
Machemehl, J.L., 59
Mackinnon, P.A., 429
MacPherson, H., 463

Madsen, O.S., 463
Maeno, S., 499, 503
Magda, W., 499
Mallard, W.W., 463
Mantz, P.A., 336
Mao, Y., 11, 16, 21, 32, 33, 34, 39, 40, 44, 54, 55, 57, 61, 70, 71, 88, 117, 124, 130, 131
MATHSOFT 1997, 293
Mayer, S., 370
McConnell, K., 419
McDonald, H., 219
McDougal, W.G., 401, 402, 410, 411, 412, 413, 415, 462, 463, 469, 471, 473
McGarvey, J.A., 429, 430, 431
Mei, C.C., 334, 454, 463, 485
Meijers, P., 473, 487, 494
Melaaen, M.C., 219, 220
Melville, B.W., 6, 7, 12, 13, 149, 174, 175, 176, 177, 178, 179, 180, 181, 182, 183, 184, 210, 212, 213, 214, 218, 220, 221
Meneveau, C., 134
Menter, F.R., 224
Meyer-Peter, E., 81
Michelsen, J.A., 224
Miyaike, Y., 301
Miyamoto, T., 445
Moncada-M, A.T., 63, 66, 69
Moore, R.L., 12, 275
Moreno, L., 360
Mory, M., 202, 262, 263, 487, 503
Moshagen, H., 463
Mostafa, E.A., 182
Murray, J.D., 463
Müller, R., 81
Müller, W. von, 31, 117
Mærsk Olie & Gas, 276, 277

Nadaoka, K., 334, 338
Nago, H., 499, 503
Nakagawa, H., 337
Nalluri, C., 39
Naylor, P., 170
Nezu, I., 337
Nicollet, G., 174

Niedoroda, A.W., 151
Nishimura, H., 417, 418, 419
Norris, G.M., 487

Obasaju, E.D., 170
O'Connor, B.A., 416
Oda, K., 189, 191, 248, 301
O'Donoghue, T., 350
Offshore, August Issue 1988, p.48, 116
Okamoto, O., 473
Olsen, N.R.B., 219, 220
Orgill, G., 94, 95
Oumeraci, H., 112, 329, 360

Palmer, A.C., 487
Panchang, V.G., 220, 221
Pankchik, B., 432, 437, 440, 441
Park, W., 416
Parker, G., 226
Parola, A.C., 213
Peacock, W.H., 464, 465, 466, 467, 468, 472
Pedersen, C., 217
Peerbolte, E.B., 32
Pena, C., 360
Phelp, D., 331
PIANC (1997), 440
Piquet, J., 219
Pluim-van der Velden, E.T.J.M., 61, 62, 65
Pohl, H., 429
Posey, C.J., 263
Powell, K.A., 401, 402, 408, 409, 410
Putnam, J.A., 463

Quarrain, R.M.M., 424

Rahman, M.S., 114, 473
Rajaratnam, N., 419, 427
Rakha, K.A., 412, 413
Rance, P.C., 287, 301, 305, 312, 316, 317, 318, 319, 320, 321, 322

Raudkivi, A.J., 6, 7, 174, 175, 183, 213, 220, 221, 243, 245, 246, 383, 419
Reid, R.D., 463
Richardson, E.V., 149
Richardson, J.E., 220, 221
Robakiewicz, W., 429
Rodi, W., 128
Rogers, S.E., 219
Roll, P., 92, 94
Romisch, K., 429, 437, 438, 439
Roulund, A., 152, 175, 220, 222, 223, 224, 225, 226, 227, 228
Ryan, D., 424, 435, 436, 437

Sabol, S.A., 149
Saito, E., 301
Sakai, T., 485, 486, 487, 499, 503
Sanchez-Arcilla, A., 360, 361
Sarpkaya, T., 109, 162, 170, 293
Sassa, S., 499
Sato, M., 445
Sato, S., 301
Schepis, J., 174
Schiller, R.E., Jr., 39, 301, 353, 408, 445
Schlichting, H., 298, 366, 497
Schwichtenberg, B.R., 358
Schwind, R., 151
Seabra-Santos, F.J., 202, 262
Seed, H.B., 113, 464, 465, 466, 467, 468, 472, 473
Sekiguchi, H., 473, 499
Sellmeijer, H., 457
Shemdin, O.H., 463
Shen, H.W., 174
Shibayama, T., 301
Shields, A., 336
Shimizu, K., 445
Shore Protection Manual, 390, 399
Sichmann, T., 16
Siddhartan, R., 487
Sidek, F.J., 60, 64
Sierra, J.P., 360
Sills, G., 449, 504
Silva, R., 350
Silvester, R., 331, 343, 344, 345, 346

Silvis, F., 487
Silvis, S., 95
Skinner, M.M., 377
Sleath, J.F.A., 463
Smagorinsky, J., 134
Smed, P.F., 137
Smith, J.M., 411
Sneddon, I., 475
Soga, F., 445
Sollitt, C.K., 411, 463
Song, C.C.S., 220, 222
Sotberg, T., 138
Spangenberg, S., 214
Spierenburg, S.E.J., 473
Stansby, P.K., 103
Starr, P., 103
Staub, C., 80, 95, 137
Stegun, I.A., 293
Stewart, D.P., 424
Sturtevant, 415
Suheyda, J.N., 445
Sumer, B.M., 11, 16, 19, 20, 21, 22, 23, 24, 25, 26, 27, 28, 29, 31, 33, 34, 35, 36, 37, 41, 43, 44, 45, 46, 49, 50, 51, 52, 54, 55, 56, 57, 58, 59, 60, 66, 67, 68, 69, 70, 71, 80, 88, 96, 97, 98, 100, 101, 102, 103, 104, 105, 106, 107, 112, 121, 124, 125, 130, 132, 135, 136, 137, 150, 152, 156, 161, 162, 163, 164, 165, 166, 167, 168, 169, 170, 171, 172, 173, 174, 176, 179, 183, 187, 189, 190, 191, 192, 193, 194, 195, 196, 197, 198, 202, 205, 206, 207, 208, 209, 210, 216, 217, 219, 220, 221, 223, 224, 248, 249, 250, 251, 252, 253, 255, 256, 257, 258, 259, 260, 261, 264, 265, 266, 267, 268, 269, 270, 271, 272, 290, 291, 292, 293, 294, 295, 296, 297, 298, 299, 300, 301, 302, 303, 304, 305, 306, 307, 308, 309, 310, 311, 312, 314, 324, 325, 329, 330, 336, 337, 347, 348, 349, 350, 351, 353, 354, 355, 356, 357, 358, 359, 360, 362, 363, 365, 369, 370, 371, 372, 373, 374, 375, 376, 377, 378, 379, 380, 381, 382, 383, 384, 385, 386, 387, 389, 405, 462, 469, 473, 475, 478, 479, 480, 481, 482, 488, 489, 490, 491, 492, 493, 494, 495, 497, 498, 500, 501, 502
Sutherland, A.J., 175, 176, 177, 178, 179, 180, 181, 183, 184
Sutherland, J., 112, 350, 387, 388, 389
Suzuki, 463
Sybert, 263
Sørensen, L.S., 370
Sørensen, N.N., 224

Tanaka, Y., 481, 498
Taylor, P.K., 445
Terzaghi, K., 17, 20, 100, 101, 450
Thomas, S.D., 449
Thompson, P.L., 149
Thomsen, J., 432
Tison, L.J., 180
Toch, A., 180
Tonda, P.L., 115
Toue, T., 287, 301, 305, 415
Truelsen, C., 16
Tsai, Y.T., 462
Tseng, M.-H., 220, 222
Tsuchia, Y., 345
Tsujimoto, T., 213
Tzang, S.Y., 473, 474
Tørum, A., 39, 57, 112, 360, 463

Umeda, S., 220, 222
US Army Corps of Engineers, 360

Valentini, V., 49, 95, 117
van Beek, F.A., 125, 132
van der Meer, J.W., 115, 355
van Dijk, R.N., 278
van Hijum, E., 457
van Kessel, T., 473, 482
Vellinga, P., 108
Venkatadri, C., 180
Venturi, M., 49, 95, 117, 138
Verges, D., 360, 361
Verheij, H.J., 7, 175, 429
Verruijt, A., 455

Vincenzi, M., 95

Wang, H., 415
Wang, R.-K., 195, 196, 248, 301
Watanabe, A., 418
Watanabe, R.K., 39, 301, 353, 445
Westerhortmann, J.H., 59, 63
Wheeler, S.J., 449
Whitehouse, R.J.S., 6, 112, 175, 312, 316, 360, 387, 401, 408, 409
Whitman, R.V., 480, 481, 504, 505, 507, 508
Wichers, J.E.W., 334
Wilcox, D.C., 224
Williams, G.N., 445
Williamson, C.H.K., 43, 170, 173, 253, 254, 255, 256
Wind, H.G., 32, 125, 132
Worman, A., 217

Xie, S.L., 334, 335, 337, 338, 339, 340, 352, 359, 362
Yu, D.P., 217

Yalin, M.S., 10
Yamamoto, T., 457, 458, 462, 463, 485, 503, 508
Yamazaki, H., 487
Yassin, A.A., 182
Yen, C.-L., 220, 222
Yen, J.K.C., 445
Yoshinaga, S., 445
Yu, H.-S., 301
Yuhi, M., 220, 221, 222, 223
Yuksel, Y., 46, 53, 54
Yulistiyanto, B, 152

Zaleski-Zamenhof, L.C., 323
Zdravkovich, M.M., 17, 28, 240, 241, 242, 255
Zen, K., 487

Subject Index

Amplification factor, 7
Amplification in bed shear stress, 7

Basic concepts, 5
Bed shear velocity, 10
   Undisturbed bed shear velocity, 10
Biot equations, 450
Breakwater, 329
Breakwater scour
  At the head of breakwater, 371
    Rubble-mound breakwater, 377
      Scour depth, 382, 384
      Scour protection, 385
    Vertical-wall breakwater, 371
      Scour depth, 375
  At the trunk section, 333
    Mathematical modelling, 365
    Rubble-mound breakwater, 347
      Irregular waves, 355
      Scour depth, 351
      Time scale of scour, 356
    Scour protection, 361
    Steady streaming, 334
    Submerged breakwaters, 360
    Vertical-wall breakwater, 333
      Influence of angle of attack, 342
      Irregular waves, 341
      Scour depth, 338

Clear-water scour, 9
Consolidation, coefficient, 470
Cohesive sediment, 184
Complex configurations, 275

Darcy's law, 454
Discrete-Vortex models, 135

Effective stress, 453
Equilibrium scour depth, 8
Erosion, 5

Friction velocity, see bed-shear velocity

Global scour, 11
Global and local scour at pile groups, 264
Group of piles, 239 (see pile group)

Hook's law, 450

Initiation of motion at the bed, 10
Initiation of suspension from the bed, 337

Jetties, 390

Keulegan-Carpenter number, 42
k-ε simulation, 126

Large-eddy simulation (LES), 132
Large piles, 287
Large-pile scour
  Diffraction effect, 288
  Phase-resolved flow, 291
  Scour, 301
    Combined waves and current, 316
    Influence of KC and D/L
    Mechanism, 302
  Steady streaming, 294
Liquefaction, 445
  Assessment of liquefaction potential, 476, 486
  Equation governing the buildup of pore pressure, 469
  Floatation of pipelines, 487
  Momentary liquefaction, 448, 483, 499, 503
    Assessment of liquefaction potential, 486
    Saturated soil, 484
    Sinking of armour blocks, 500
    Sinking of pipelines, 488
    Unsaturated soil, 485
  Pysics of liquefaction, 446
  Residual liquefaction, 446, 464, 488, 500
    Assessment of liquefaction potential, 476
    Behaviour of pressure, large times, 480
    Floatation velocity, 496
    History of wave exposure, 482
    Irregular waves, 481
    Peacock and Seed experiment, 464
    Sinking of armour blocks, 500
    Sinking of objects, 487
    Sinking of pipelines, 488
    Sinking velocity, 496
  Sinking of armour blocks, 500
  Sinking of pipelines,
    Momentary liquefaction, 499
    Residual liquefaction, 488
  Stresses in soil, progressive wave, 455
    Finite soil depth, 460
    Large soil depth, 455
Live-bed scour, 9

Local scour, 11

Mathematical modelling
  Breakwaters, 365
  Piles, 218
  Pipelines, 117
    Potential-flow models, 117
    Advanced models, 125
    Integrated models, 137
Mode of sand transport, 335
  No-suspension mode of transport, 336
  Suspension mode of transport, 336
Momentary liquefaction, 448, 483, 499, 503

Obliquely incident waves, 342
Offshore structures, 276
Onset of scour below pipelines, 16
  Criterion, 23
  Current case, 19
  Effect of change in flow regime, 26
  Mechanism, 16
  Wave case, 21
Overburden pressure, 472

Pile-group scour
  Global and local scour at pile groups, 264
  In steady current, 239
    Three-pile group, 247
    Two-pile group, 240
  In waves, 248
    Effect of KC number, 260
    Four-pile group, 258
    Three-pile group, 257
    Two-pile group, 249
Pile scour,
  Contraction of streamlines, 174
  Flow around a slender pile, 150
  Horseshoe vortex in steady current, 151
    Bed shear stress, 159
    Effect of boundary-layer thickness, 153
    Effect of pile geometry, 156

Subject index

Effect of Re number, 154
Transition to turbulence, 158
Horseshoe vortex in waves, 160
  Bed shear stress, 167
  Existence of horseshoe vortex, 162
  Lifespan of a horseshoe vortex, 164
  Separation position, 166
Lee-wake vortex flow, 169
Pile group, 239
Scour, 174
Scour in steady current, 174
  Cohesive sediment, 184
  Scour depth, 175
Scour in waves, 188
  In combined waves and current, 195
  In breaking waves, 203
  Irregular waves, 194
  Scour depth, 190
Scour protection, 212
  Failure, 214
Time scale, 206
Pipeline scour,
  Backfilling in free-span areas, 77
  Change in wave climate, 75
  Combined waves and current, 49
  Effect of angle of attack, 59
  Effect of armouring, 60
  Effect of cohesive sediment, 61
  Effect of multiple pipelines, 59
  Effect of pipe position in vertical, 55
  Effect of pipe roughness, 54
  Effect of sagging, 88
  Effect of Shields Parameter, 54
  Effect of vibrations, 57
  Effect of water depth, 62
  Free-span length, 94
  Irregular waves, 49
  Lee-wake erosion, 33
  Onset of scour, 16
  Scour depth, 38
  Scour depth in steady current, 41
  Scour depth in waves and tidal flows, 45
  Self-burial, free-span areas, 77
  Self-burial, span shoulders, 96
  Shoaling conditions, 53
  Sinking at span shoulders, in currents, 97

Sinking at span shoulders, in waves, 103
Sinking at span shoulders, time scale, 106
Sinking in liquefied soil, 487, 499
Steady current, 38
Stimulated self-burial, 108
Three-dimensional scour, 76
Time scale, 69
Tunnel erosion, 30
Two-dimensional scour, 32
Waves and tidal flows, 42
Width of scour hole, 66
Piping, 17
Poro-elastic soil, 450, 453

Quay walls, 424

Residual liquefaction, 446, 464, 488, 500

Scale effects, 111
Scour at/around
  Breakwater, 329
  Cone-shaped object, 205, 324
  Jetties, 390
  Large pile, 287
  Pile, 149
  Pile-supported offshore structures, 276
  Pipeline, 15
  Seawall, 399
  Stone, 181
Scour protection
  Breakwater, 361, 385
  Failure, 214
  Pile, 212
  Pipeline, 113
  Seawall, 419
  Ship propeller, 436
Seawall scour
  Normally incident breaking waves, 402
  Normally incident nonbreaking waves, 415
  Scour protection, 419
  Wave overtopping, 417

Seepage flow, 16
Shields parameter, 10
   Critical value of Shields parameter
      Initiation of motion at bed, 10
      Initiation of suspension from bed, 337
Ship-propeller scour, 423
   Confined propeller, 431
   Quay wall, 424
   Scour protection, 436
   Unconfined propeller, 426
Sinking of objects, liquefied soil, 487, 499
Soil properties, 504
Spoiler, 110
Steady streaming, 294, 334

Time scale of scour, definition, 8
Time scale of scour
   Breakwater, 356
   Pile, 206
   Pipeline, 69, 85, 106

Vertical-wall breakwater, 333